Stable Isotope Forensics

Stable Isotope Forensics

An Introduction to the Forensic Application of Stable Isotope Analysis

Wolfram Meier-Augenstein

*Centre for Anatomy & Human Identification
University of Dundee, UK
SCRI, Invergowrie, Dundee, UK*

WILEY-BLACKWELL

A John Wiley & Sons, Ltd., Publication

This edition first published 2010, © 2010 by John Wiley & Sons Ltd.

Wiley-Blackwell is an imprint of John Wiley & Sons, formed by the merger of Wiley's global Scientific, Technical and Medical business with Blackwell Publishing.

Registered office:
John Wiley & Sons Ltd, The Atrium, Southern Gate, Chichester, West Sussex, PO19 8SQ, UK

Other Editorial Offices:
9600 Garsington Road, Oxford, OX4 2DQ, UK
111 River Street, Hoboken, NJ 07030-5774, USA

For details of our global editorial offices, for customer services and for information about how to apply for permission to reuse the copyright material in this book please see our website at www.wiley.com/wiley-blackwell

The right of the author to be identified as the author of this work has been asserted in accordance with the Copyright, Designs and Patents Act 1988.

All rights reserved. No part of this publication may be reproduced, stored in a retrieval system, or transmitted, in any form or by any means, electronic, mechanical, photocopying, recording or otherwise, except as permitted by the UK Copyright, Designs and Patents Act 1988, without the prior permission of the publisher.

Wiley also publishes its books in a variety of electronic formats. Some content that appears in print may not be available in electronic books.

Designations used by companies to distinguish their products are often claimed as trademarks. All brand names and product names used in this book are trade names, service marks, trademarks or registered trademarks of their respective owners. The publisher is not associated with any product or vendor mentioned in this book. This publication is designed to provide accurate and authoritative information in regard to the subject matter covered. It is sold on the understanding that the publisher is not engaged in rendering professional services. If professional advice or other expert assistance is required, the services of a competent professional should be sought.

Library of Congress Cataloguing-in-Publication Data

Meier-Augenstein, Wolfram.
   Stable isotope forensics : an introduction to the forensic application of stable isotope analysis / Wolfram Meier-Augenstein.
      p. ; cm.
   Includes bibliographical references and index.
   ISBN 978-0-470-51705-5
   1. Chemistry, Forensic.   2. Stable isotopes.   I. Title.
   [DNLM: 1. Forensic Sciences–methods.   2. Isotopes–analysis. W 700 M511s 2010]
   RA1057.M45 2010
   363.25–dc22
                                                                                                 2009041399

ISBN: 978-0-470-51705-5

A catalogue record for this book is available from the British Library.

Typeset in 10.5/12.5pt Minion by Aptara Inc., New Delhi, India.

First Impression   2010

Contents

Series Foreword: Developments in Forensic Science *Niamh Nic Daéid*   ix

Foreword: *Sean Doyle*   xi

List of Figures   xiii

List of Tables   xix

Introduction: Stable Isotope 'Fingerprinting' or Chemical 'DNA': A New Dawn for Forensic Chemistry?   xxi

I   **How It Works**   1
- I.1   What are Stable Isotopes?   3
- I.2   Natural Abundance Variation of Stable Isotopes   5
- I.3   Chemically Identical and Yet Not the Same   8
- I.4   Isotope Effects, Mass Discrimination and Isotopic Fractionation   10
  - I.4.1   Physical Chemistry Background   10
  - I.4.2   Fractionation Factor α and Enrichment Factor ε 11
  - I.4.3   Isotopic Fractionation in Rayleigh Processes   13
  - I.4.4   Isotopic Fractionation Summary   14
- I.5   Stable Isotopic Distribution and Isotopic Fractionation of Light Elements in Nature   16
  - I.5.1   Hydrogen   16
  - I.5.2   Oxygen   19
    - I.5.2.1 ^{18}O in Bone Bio-apatite and Source Water   20
    - I.5.2.2   Bone Remodelling   23
    - I.5.2.3   Bone Diagenesis   24
      - I.5.2.3.1   Diagenetic Changes of Bio-apatite   24
  - I.5.3   Carbon   25
  - I.5.4   Nitrogen   27
    - I.5.4.1   Food Chain and Trophic Level Shift   29
    - I.5.4.2   Diagenetic Changes of Structural Proteins   32
  - I.5.5   Sulfur   33
- I.6   Stable Isotope Forensics in Everyday Life   36
  - I.6.1   'Food Forensics'   37
    - I.6.1.1   Authenticity and Provenance of Single-Seed Vegetable Oils   38
    - I.6.1.2   Authenticity and Provenance of Beverages   39
    - I.6.1.3   Authenticity and Provenance of other Premium Foods   41

	I.6.2 Counterfeit Pharmaceuticals	42
	I.6.3 Environmental Forensics	43
	I.6.4 Wildlife Forensics	46
	I.6.5 Anti-Doping Control	47
I.7	Summary of Part I	49
I.8	Set problems	50
References		51

II Instrumentation and Analytical Techniques — 65

II.1	Mass Spectrometry versus Isotope Ratio Mass Spectrometry	67
II.2	Instrumentation and δ Notation	72
	II.2.1 Dual-Inlet Isotope Ratio Mass Spectrometry	74
	II.2.2 Continuous Flow Isotope Ratio Mass Spectrometry	74
	II.2.3 Bulk Material Stable Isotope Analysis	77
	II.2.4 Compound-Specific Stable Isotope Analysis	78
	II.2.4.1 CSIA and Compound Identification	79
	II.2.4.2 Position-Specific Isotope Analysis	81
	II.2.4.3 CSIA of Polar, Non-Volatile Organic Compounds	83
II.3	Isotopic Calibration and Quality Control in Continuous Flow Isotope Ratio Mass Spectrometry	85
	II.3.1 Two-Point or End-member Scale Correction	86
	II.3.1.1 Scale Correction of Measured δ^2H Values	87
	II.3.1.2 Scale Correction of Measured δ^{13}C Values	88
II.4	Statistical Analysis of Stable Isotope Data within a Forensic Context	91
	II.4.1 Chemometric Analysis	91
	II.4.2 Bayesian Analysis	94
II.5	Forensic Stable Isotope Analytical Procedures	100
	II.5.1 FIRMS Network	101
II.6	Generic Considerations for Stable Isotope Analysis	102
	II.6.1 Generic Considerations for Sample Preparation	102
	II.6.2 Generic Considerations for BSIA	104
	II.6.2.1 Isobaric Interference	104
	II.6.3 Particular Considerations for ^{2}H-BSIA	105
	II.6.3.1 Keeping Your Powder Dry	105
	II.6.3.2 Total δ^2H versus True δ^2H Values	106
	II.6.3.2.1 ^{2}H Isotope Analysis of Human Hair	108
	II.6.3.3 Ionization Quench Effect	113
	II.6.4 Generic Considerations for CSIA	116
	II.6.4.1 Isotopic Calibration during GC/C-IRMS	116
	II.6.4.2 Isotope Effects in GC/C-IRMS during Sample Injection	117
	II.6.4.3 Chromatographic Isotope Effect in GC/C-IRMS	118
II.7	Summary of Part II	121
II.8	Set Problems	122
II.A	How to Set Up a Laboratory for Continuous Flow Isotope Ratio Mass Spectrometry	123
	II.A.1 Pre-Installation Requirements	124
	II.A.2 Laboratory Location	124
	II.A.3 Temperature Control	125
	II.A.4 Power Supply	125
	II.A.5 Gas Supply	126

	II.A.6	Forensic Laboratory Considerations	129
	II.A.7	Finishing Touches	130
	References		136

III Stable Isotope Forensics: Case Studies and Current Research — 143

- III.1 Forensic Context — 145
- III.2 Distinguishing Drugs — 149
  - III.2.1 Natural and Semisynthetic Drugs — 149
    - III.2.1.1 Marijuana — 149
    - III.2.1.2 Morphine and Heroin — 150
    - III.2.1.3 Cocaine — 152
  - III.2.2 Synthetic Drugs — 154
    - III.2.2.1 Amphetamines — 154
    - III.2.2.2 MDMA: Synthesis and Isotopic Signature — 157
      - III.2.2.2.1 Three Different Synthetic Routes – Controlled Conditions — 157
      - III.2.2.2.2 One Synthetic Route – Variable Conditions — 164
    - III.2.2.3 Methamphetamine: Synthesis and Isotopic Signature — 164
  - III.2.3 Conclusions — 167
- III.3 Elucidating Explosives — 169
  - III.3.1 Bulk Isotope Analysis of Explosives and Precursors — 170
    - III.3.1.1 Ammonium Nitrate — 171
    - III.3.1.2 Hexamine, RDX and Semtex — 172
    - III.3.1.3 Hydrogen Peroxide and Peroxides — 176
  - III.3.2 Isotopic Product/Precursor Relationship — 179
  - III.3.3 Potential Pitfalls — 182
  - III.3.4 Conclusions — 183
- III.4 Matching Matchsticks — 184
  - III.4.1 ^{13}C-Bulk Isotope Analysis — 185
  - III.4.2 ^{18}O-Bulk Isotope Analysis — 186
  - III.4.3 ^{2}H-Bulk Isotope Analysis — 187
  - III.4.4 Matching Matches from Fire Scenes — 188
  - III.4.5 Conclusions — 189
- III.5 Provenancing People — 190
  - III.5.1 Stable Isotope Abundance Variation in Human Tissue — 191
  - III.5.2 The Skull from the Sea — 194
  - III.5.3 A Human Life Recorded in Hair — 197
  - III.5.4 Found in Newfoundland — 201
  - III.5.5 The Case of 'The Scissor Sisters' — 207
  - III.5.6 Conclusions — 211
- III.6 Stable Isotope Forensics of other Physical Evidence — 214
  - III.6.1 Microbial Isotope Forensics — 214
  - III.6.2 Paper, Plastic (Bags) and Parcel Tape — 215
    - III.6.2.1 Paper — 215
    - III.6.2.2 Plastic and Plastic Bags — 216
    - III.6.2.3 Parcel Tape — 218
  - III.6.3 Conclusions — 221
- III.7 Summary — 222

III.A 'Play True?': Stable Isotopes in Anti-doping Control or *Quis custodiet ipsos custodes?* 224
    III.A.1 Testosterone Metabolism and ^{13}C Isotopic Composition 226
    III.A.2 Analytical Methodology: Gas Chromatography and Peak Identification 230
  III.B Sample Preparation Procedures 236
    III.B.1 Preparing Silver Phosphate from Bio-apatite for ^{18}O Isotope Analysis 236
    III.B.2 Acid Digest of Carbonate from Bio-apatite for ^{13}C and ^{18}O Isotope Analysis 238
    III.B.3 Standard Protocol for Preparing Hair Samples for ^{2}H Isotope Analysis 240
  References 242

Government Agencies and Institutes with Dedicated Stable Isotope Laboratories — 253

Acknowledgements — 255

Recommended Reading — 257

Author's Biography — 261

Index — 263

Series Foreword
Developments in Forensic Science

The world of forensic science is changing at a very fast pace. This is in terms of the provision of forensic science services, the development of technologies and knowledge, and the interpretation of analytical and other data as it is applied within forensic practice. Practicing forensic scientists are constantly striving to deliver the very best for the judicial process, and as such need a reliable and robust knowledge base within their diverse disciplines. It is hoped that this book series will be a valuable resource for forensic science practitioners in the pursuit of such knowledge.

The Forensic Science Society is the professional body for forensic practitioners in the United Kingdom. The Society was founded in 1959 and gained professional body status in 2006. The Society is committed to the development of the forensic sciences in all of its many facets, and in particular to the delivery of highly professional and worthwhile publications within these disciplines through ventures such as this book series.

Dr Niamh Nic Daéid
Series Editor

Foreword

I am delighted to be able to write the foreword for this, the first textbook of stable isotope forensics.

The breadth of material covered is wide, ranging from fundamentals to policy issues, and therefore this text will be of benefit to practitioners, researchers and investigators, indeed to anyone who has an interest in this new forensic discipline.

The year 2001 saw the formation of the Forensic Isotope Ratio Mass Spectrometry (FIRMS) Network. Since then much has been achieved in terms of advancing the forensic application of stable isotope analysis, this textbook being the latest significant step.

These advances have been made in the face of considerable challenges resulting from the novelty and complexity of the technique. Isotope forensics has already proved a powerful tool in the investigation and prosecution of high-profile crimes, including terrorism. Stable isotope analysis enables questions regarding the source and history of illicit and other forensic materials to be addressed – questions which might otherwise remain unanswered.

Isotope forensics is now being widely adopted for profiling illicit materials and human provenancing. Stable isotope analysis has already been used successfully in two major terrorist trials in the United Kingdom, and in a variety of investigations and trials in the United Kingdom, Europe and the United States.

Dr Meier-Augenstein is to be commended for his vision in recognizing the forensic potential of stable isotopes, for his energy in developing and optimizing the methodology, and in promoting the technique to end-users. He is also well aware of the risk of contributing to a miscarriage of justice and recognizes that only an appropriate regulatory framework can significantly mitigate that risk.

The development of suitable databases of reference materials and appropriate tools for evaluation remain significant tasks; once complete the next decade should see isotope forensics taking a deserved place in mainstream forensic science and, to a greater extent, contributing to the efficient and effective delivery of justice.

Sean Doyle
Past Chair of the FIRMS Network
Principal Scientist, Forensic Explosives Laboratory, Defence Science
and Technology Laboratory
September 2009

List of Figures

Part I

Figure I.1 δ^2H and $\delta^{18}O$ values of whole wood and plant sugars (beet and cane sugar) in the framework of the global meteoric water line 18

Figure I.2 Correlation plot of source water δ^2H values versus δ^2H values of fresh soft fruit water 19

Figure I.3 Correlation graphs according to Daux *et al.*, Longinelli, and Luz and Kolodny for $\delta^{18}O_{phosphate}$ versus $\delta^{18}O_{water}$ and the resulting different solutions of $\delta^{18}O_{water}$ for the same $\delta^{18}O_{phosphate}$ 23

Figure I.4 Bivariate graph plotting $\delta^{15}N$ versus $\delta^{13}C$ values of scalp hair samples volunteered by residents in different countries reflecting their regionally different diet 26

Figure I.5 Schematic representation of typical $\delta^{15}N$ values in relation to trophic level 28

Figure I.6 ^{13}C isotopic composition of various food and animal tissue 29

Figure I.7 Bivariate graph plotting $\delta^{15}N$ versus $\delta^{13}C$ values from scalp hair of a vegan and an omnivore who has a relatively strong meat component in their diet 31

Figure I.8 Approximate $\delta^{13}C$ values for ^{13}C isotopic composition of various body pools and tissue 32

Figure I.9 Natural variation in ^{13}C isotopic composition of single-seed vegetable oils and selected fatty acids isolated from these oils 39

Figure I.10 Bivariate plot of δ^2H and $\delta^{13}C$ values of ethanol from a selection of European white wines, including one suspect sample of wine labelled as vintage Austrian wine 41

Figure I.11 Isotopic bivariate plot of $\delta^{13}C$ and $\delta^{15}N$ values of the API folic acid from three different manufacturers at three different locations 44

Figure I.12	Isotopic bivariate plot of $\delta^{13}C$ and $\delta^{18}O$ values of the API naproxen from six different manufacturers at four different locations	44

Part II

Figure II.1	Schematic (top) and picture (bottom) of a modern IRMS magnetic sector instrument with a multicollector analyser	70
Figure II.2	Schematic (top) and picture (bottom) of a typical EA-IRMS system	75
Figure II.3	Schematic (top) and picture (bottom) of a TC/EA-IRMS system	78
Figure II.4	Schematic (top) and picture (bottom) of a GC/C-IRMS system	80
Figure II.5	Schematic (top) and picture (bottom) of a GC(-MS)/C-IRMS hybrid system	82
Figure II.6	Trivariate plots of measured $\delta^{13}C$, $\delta^{2}H$ and $\delta^{15}N$ values of 10 Ecstasy tablets from eight different seizures in two different European countries	92
Figure II.7	HCA (furthest neighbour, Euclidean distance) using $\delta^{2}H$, $\delta^{13}C$, $\delta^{15}N$ and $\delta^{18}O$ values as well as MDMA content from 10 Ecstasy tablets from eight different seizures in two different European countries	93
Figure II.8	Plot of PCA score factors for the first two principal components of multivariate data from farmed and wild European sea bass	93
Figure II.9	Means of $\delta^{2}H$, $\delta^{13}C$ and $\delta^{18}O$ observations for each of 51 samples of white paints plotted over the smoothed bivariate density (darker equates to higher density) for each variate pair	97
Figure II.10	Costech Zero-Blank autosampler as used in our laboratory for BSIA by EA- or TC/EA-IRMS	107
Figure II.11	(Top) Untreated ammonium nitrate from six different sources. Note the already 'wet' appearance of the sample in the top right corner. (Bottom) The same samples after an 8-day exposure to ambient atmosphere	109
Figure II.12	Effect of argon concurrently present in the ion source on measured $\delta^{13}C$ values of same sized aliquots of CO_2 (accepted $\delta^{13}C_{VPDB}$: −32.56‰)	114
Figure II.13	Illustration of the potential interference on a H_2 peak caused by a partial overlap with a following N_2 peak	115
Figure II.14	Comparison of peak heights, peak shape and retention time of a H_2 peak in the absence (left) and presence (right) of N_2	115

LIST OF FIGURES

Figure II.15	Illustration of the time displacement caused by the 'inverse' chromatographic isotope effect between the $^{13}CO_2$ and $^{12}CO_2$ aspects of a compound CO_2 peak and the resulting S-shaped 45/44 ratio signal	119
Figure II.A.1	Pressure triggered change-over unit for helium supply	128
Figure II.A.2	Laboratory gas delivery manifold fed from an external gas supply	129

Part III

Figure III.1	Morphine and heroin	151
Figure III.2	Cocaine	154
Figure III.3	Six amphetamine powders from the 18 seizures isotopically profiled in Figure III.4	156
Figure III.4	Bivariate plot of $\delta^{15}N$ versus $\delta^{13}C$ values for 18 amphetamine seizures	157
Figure III.5	Schematic synthetic route for PMK from safrole	158
Figure III.6	Schematic synthetic routes for MDMA from PMK	158
Figure III.7	Bivariate plot of $\delta^{15}N$ versus $\delta^{13}C$ values of 18 MDMA batches from three different synthetic routes	159
Figure III.8	Bivariate plot of $\delta^{2}H$ versus $\delta^{13}C$ values of 18 MDMA batches from three different synthetic routes	159
Figure III.9	Three-dimensional plot of $\delta^{2}H$ versus $\delta^{15}N$ versus $\delta^{13}C$ values of MDMA hydrochloride samples synthesized from aliquots of the same precursor PMK but by three different routes of reductive amination	162
Figure III.10	HCA of 18 batches of MDMA; three variables, Euclidean distance, single linkage	163
Figure III.11	Schematic synthetic routes 'Emde' and 'Nagai' for methamphetamine from ephedrine or pseudoephedrine	166
Figure III.12	Bivariate plot of $\delta^{15}N$ versus $\delta^{18}O$ values for ammonium nitrate prills from various sources (country of origin shown where known)	171
Figure III.13	Ammonium nitrate prills from various sources	172
Figure III.14	Detailed bivariate plot of $\delta^{15}N$ versus $\delta^{13}C$ values for the explosive RDX from two different sources demonstrating homogeneity of the samples and reproducibility of isotope analysis	174

Figure III.15	Hexamine to RDX	174
Figure III.16	Three-dimensional plot of δ^2H versus δ^{15}N versus δ^{13}C values for the RDX precursor hexamine	175
Figure III.17	Dendrogram of a HCA (single linkage, Euclidean distance) for the trivariate stable isotope dataset of 14 hexamine samples	176
Figure III.18	Changing δ^2H values of a hydrogen peroxide solution with increasing dilution	178
Figure III.19	Changing δ^{18}O values of a hydrogen peroxide solution with increasing dilution	178
Figure III.20	Bivariate plot of δ^2H versus δ^{13}C values from matchsticks recovered at the crime scene and seized from the suspect's house as well as from the controls	188
Figure III.21	Global map for δ^{18}O values in precipitation	191
Figure III.22	You are what and where you eat and drink – a few stable isotopes in the human body	192
Figure III.23	Diagram of the area of skull submitted for isotope analysis	195
Figure III.24	Sample of scalp hair as submitted for sequential stable isotope analysis	198
Figure III.25	Time-resolved changes in ^{15}N isotopic composition of the victim's scalp hair	199
Figure III.26	Time-resolved changes in ^{2}H isotopic composition of the victim's scalp hair	200
Figure III.27	Geographic life history for the last 17 months prior to death as gleaned from ^{2}H isotope analysis of scalp hair segments from the body found at Minerals Road, Conception Bay South, Newfoundland	205
Figure III.28	Poster based on information derived from, amongst other sources, stable isotope analysis for a public appeal for information regarding the murder victim found at Minerals Road, Conception Bay South, Newfoundland	207
Figure III.29	Geographic life trajectory of the murder victim found in the Dublin Royal Canal based on ^{18}O isotope analysis of bone phosphate extracted from his femur	210
Figure III.30	(Top) Global map with highlighted areas with model predictions for δ^{18}O values in precipitation ranging from −10.1 to −7.6‰ illustrating the constraining power of stable isotope profiling in aid of human provenancing. (Bottom)	

	Zoomed-in version of the global map focusing on Central Europe and the United Kingdom	213
Figure III.31	Bivariate plots of δ^2H versus δ^{13}C values of (top) intact (untreated) brown parcel tape samples and (bottom) treated brown parcel tape samples (i.e. backing material only)	220
Figure III.A.1	Schematic of the metabolic pathway of testosterone	227

List of Tables

Part I

Table I.1	Stable isotopes of light elements and their typical natural abundance	6
Table I.2	Representative but not concise list of international reference materials for stable isotope ratio mass spectrometry (IRMS) administered and distributed by the IAEA (Vienna, Austria)	7
Table I.3	Isotopic abundance of ^{13}C and ^{2}H in sugar from different sources and geographic origin	9

Part II

Table II.1	Comparison of MS and IRMS system performance when applied to stable isotope analysis at near-natural abundance levels	68
Table II.2	Key dates in instrument research and development influencing design and evolution of commercially available CF-IRMS systems	76
Table II.3	Sample batch sequence composition in BSIA favouring high sample throughput under stable experimental conditions using ^{2}H isotope analysis of water as an example	86
Table II.4	Sample VSMOW – SLAP δ^{2}H scale correction	87
Table II.5	Organic ^{13}C reference materials available from the IAEA	89
Table II.6	Sample two-end-member VPDB δ^{13}C scale corrections showing the effect on appropriate and inappropriate choice of end-members	89
Table II.7	List of 51 white architectural paints from different sources	95
Table II.8	Percentage distributions for the likelihood ratios from each comparison	99
Table II.A.1	List of useful tools and equipment in an IRMS laboratory	130

Table II.A.2	List of secondary organic standards for stable isotope analysis (courtesy of Arndt Schimmelmann, University of Indiana)	132

Part III

Table III.1	Observed ranges for δ^2H, $\delta^{13}C$, $\delta^{15}N$ and $\delta^{18}O$ values of natural and hemisynthetic drugs	15
Table III.2	Summary δ^2H, $\delta^{13}C$ and $\delta^{15}N$ values of MDMA hydrochloride samples synthesized from aliquots of the same precursor PMK, but by three different synthetic routes of reductive amination	161
Table III.3	Reported δ^2H, $\delta^{13}C$ and $\delta^{15}N$ values of ephedrine hydrochloride and pseudo-ephedrine from various sources	165
Table III.4	Summary of fraction factors α and enrichment factors ε for individual hexamine/RDX precursor/product pairs	181
Table III.5	Results of stable isotope analysis of the tissue samples studied in the case of the unidentified body found at Minerals Road, Conception Bay South, Newfoundland	204
Table III.A.1	Athlete's versus reported $\Delta\delta^{13}C$ values for pathway-linked testosterone metabolites	229

Introduction

Stable Isotope 'Fingerprinting' or Chemical 'DNA': A New Dawn for Forensic Chemistry?

Starting with the conclusion first, I would say neither of the above two terms is appropriate, although I am convinced information locked into the stable isotopic composition of physical evidence may well represent a new dawn for forensic chemistry.

The title for this general introduction is a deliberate analogy to the term 'DNA Fingerprinting' coined by Professor Sir Alec J. Jeffreys. I seek to draw the reader's attention to the remarkable analogy between the organic, life-defining material DNA and the more basic (and, on their own, lifeless) chemical elements in their various isotopic forms when examined in the context of forensic sciences, and human provenancing in particular. At the same time, it has also been my intention to alert readers from the start to the dangers of expecting miracles of stable isotope forensics. DNA evidence is at its most powerful when it can be matched against a comparative sample or a database entry and the same is true to a degree for the information locked into the isotopic composition of a given material. One could argue that the random match probability of 1 : 1 billion for a DNA match based on 10 loci and the theoretical match probability of an accidental false-positive match of a multi-isotope signature were also seemingly matched with multivariate or multifactor probabilistic equations being the common denominator for both. If we consider a material such as hair keratin and we make the simplifying assumption this material may exist naturally in as many different isotopic states per element as there are whole numbers in the natural abundance range for each isotope given in δ units of per mil (‰) (Fry, 2006), we can calculate a hypothetical figure for the accidental match probability of such a multi-element isotope analysis that is comparable to that of a DNA fingerprint.

For example, the widest possible natural abundance range for carbon-13 (^{13}C) is 110‰ (Fry, 2006), so for the purpose of this example we could say keratin can assume 110 different integer ^{13}C values. Analysing hair keratin for its isotopic composition with regard to the light elements hydrogen (H), carbon (C), nitrogen (N), oxygen (O) and sulfur (S) could thus theoretically yield a combined specificity ranging from 1 : 638 million to 1 : 103.95 billion. In fact, one can calculate that the analysis of hair keratin for its isotopic composition with regard to hydrogen, carbon, nitrogen, oxygen and sulfur would theoretically yield a combined specificity of 1 : 1 billion, thus suggesting a 'stable isotope fingerprint' based on these four letters of the chemical alphabet may have the same accidental match probability as a DNA fingerprint that ultimately is based on the four letters of the DNA alphabet, A (adenine), C (cytosine), G (guanine) and T (thymine) (see Box). However, it should be stressed that it has as yet not been fully explored if this hypothetical level of random match probability and, hence, level of discrimination is actually achievable given that actually assumed natural abundance ranges of organic materials are usually much narrower than the widest possible range. We will learn more about that in the course of this book. Thus, forensic scientists and statisticians such as Jurian Hoogewerff and Jim Curran suggest more conservative estimates, and put the potentially realized random match probability of stable isotope fingerprints at levels between 1 : 10 000 and 1 : 1 million, depending on the nature and history of the material under investigation. However, even at these lower levels, stable isotope profiling is a potentially powerful tool.

Analogies between DNA and stable isotopes of light elements

Biological DNA versus Chemical 'DNA'

Alphabet of Biological DNA comprises the letters	Alphabet of Chemical 'DNA' comprises the letters
A	^2H
C	^{13}C
G	^{15}N
T	^{18}O
[U]	[^{34}S]

Random match probability of Biological DNA is approximately 1 : 1 billion (1×10^9) for a DNA profile based on 10 loci.
Random match probability of a five-element stable isotope profile can theoretically range from 1 : 693 million (6.93×10^8) to as high as 1 : 1.04×10^{11}.
Note this is for illustrative purposes only and does not denote any equivalence between DNA bases and chemical elements.

While one can make a good case that isotopic abundances of ^2H, ^{13}C, ^{15}N and ^{34}S are independent variables, and figures representing their abundance range can hence be combined in a probabilistic equation, the same is not entirely the case for ^2H and ^{18}O, which when originating both from water behave like dependent variables. More relevant to this issue is the question if and to what degree isotopic abundance varies for any given material or compound. While across all materials and compounds known to man ^{13}C isotopic abundance may indeed stretch across a range of 110 units, its range in a particular material such as coca leaves may only extend to 7 units (Ehleringer et al., 2000).

Another reason why the analogy between DNA fingerprinting and stable isotope profiling should only be used in conjunction with qualifying statements is the fact that both a DNA fingerprint and a physical fingerprint are immutable – they do not change over time. Drawing on an example from environmental forensics, calling a gas chromatography or gas chromatography-mass spectrometry profile from a sample of crude oil spillage a fingerprint of that oil is a misnomer since ageing processes such as evaporation will lead to changes in the oil's composition with regard to the relative abundance of its individual constituents. Incidentally, due to isotopic fractionation during evaporation the isotopic composition of any residual compound will have changed as compared to its isotopic composition at the point of origin. A more apt analogy would therefore be the use of the term stable isotope signature. Just as a person's signature can change over time or under the burden of stress, so can the stable isotopic composition of the residual sample have changed by the time it ends up in our laboratories. Furthermore, in the same way a forensic expert relies on more than one physicochemical characteristic as well as drawing on experience and contextual information to arrive at an interpretation regarding similarity or dissimilarity, the stable isotope scientist combines measured data with experience, expertise and contextual information to come to a conclusion as to what the stable isotope signature does or does not reveal.

Despite these caveats it is easy to see why the prospect of potentially having such powerful a tool at one's disposal for combating crime and terrorism has caused a lot of excitement in both the end-user and scientific communities. However, if the history of applying DNA fingerprinting in a forensic context has taught us anything then it is this – great potential is no substitute for good forensic science and good forensic science cannot be rushed or packaged to meet externally driven agendas. At first there was no great interest in this new forensic technique; however, after a few spectacular successes demand for what seemed to be the silver bullet to connect suspect perpetrators to victims or crime scenes increased faster than research, still concerned with answering underlying fundamental questions, could keep up with – and history has all but repeated itself recently on the subject of low template DNA. Good forensic science cannot be rushed, but is the outcome of good forensic science research and, in turn, becomes the foundation of good forensic practice. While the former requires proper funding, the latter requires proper regulation, and both requirements must be addressed and met.

Not surprisingly, therefore, even at the time of writing this book we still have a mountain to climb if we are to turn stable isotope forensics into a properly validated forensic analytical tool or technique that is fit-for-purpose. Even though this technique has been successfully applied in a number of high-profile criminal cases where salient questions could be answered by comparative analysis, this should not blind us to the fact that a considerable amount of time, effort, money and careful consideration still has to be spent to develop and finely hone this technique into the sharp investigative tool it promises to be.

Similar to DNA, data have to be generated and databases have to be compiled for a statistically meaningful underpinning of this technique and the interpretation of its analytical results. Equally important, if not more so, all the steps from sample collection, storage and preparation through to the analytical measurement and final data reduction have to be carefully examined either to avoid process artefacts or, if unavoidable, to quantify such artefacts and develop fit-for-purpose correction protocols to avoid stable isotope forensics suffering the same fate as low template DNA.

One way of ensuring appropriate and well-advised use of this technique in a forensic context is to advise and instruct upcoming generations of forensic scientists in this technique as early as possible. Fortunately, in spite of the aforementioned drawbacks, this is possible for two main reasons; (i) Thanks to end-user interest, there is a sufficient amount of actual case work and associated background research, and their results provide part of the foundations on which this book is built. (ii) Contrary to the misconception of many an analytical chemist, there is a huge body of knowledge and insight gained in scientific areas ranging from archaeology, biochemistry, environmental chemistry, geochemistry, palaeoecology to zoology, to name but a few, that is based on stable isotope chemistry and stable isotope analytical techniques.

In this book, the theory, instrumentation, potential and pitfalls of stable isotope analytical techniques are discussed in such a way as to provide an appreciation of this analytical technique. To this end some of the physical chemistry background relating to such aspects as mass discrimination, isotopic fractionation and mass balance is only touched upon, while some of the practical consequences of the aforementioned on the analytical process, the kind of information obtainable or the level of uncertainty associated with stable isotope data from a particular type of sample are discussed in finer detail. There are a number of excellent books and review articles dealing with the fundamental principles of stable isotope techniques, both from the instrumentation side and a physical chemistry point of view, which the interested reader is strongly encouraged to use for further study. These books and review articles are listed separately in the 'Recommended Reading' section at the back of this book.

In the main, what follows will focus on stable isotopes of light elements of which all organic material is comprised, and why and how stable isotope composition of an organic material can yield an added dimension of information with regard to 'Who, Where and When?'.

References

Ehleringer, J.R., Casale, J.F., Lott, M.J. and Ford, V.L. (2000) Tracing the geographical origin of cocaine. *Nature*, **408**, 311–312.

Fry, B. (2006) *Stable Isotope Ecology*, Springer, New York.

Part I
How it Works

Chapter I.1
What are Stable Isotopes?

Of the 92 natural chemical elements, almost all occur in more than one isotopic form – the vast majority of these being stable isotopes, which do not decay, unlike radioisotopes, which are not stable and, hence, undergo radioactive decay. In this context, 'almost all' means with the exception of 21 elements, including fluorine and phosphorous, which are mono-isotopic. The word isotope was coined by Professor Frederick Soddy at the University of Glasgow, and borrows its origin from the two Greek words *isos* (ισοζ) meaning 'equal in quantity or quality' and *topos* (τoπoζ) meaning 'place or position', with isotope thus meaning 'in an equal position' (of the periodic table of chemical elements). Frederick Soddy was later awarded the Nobel Prize in Chemistry in 1921 for his work on the origin and nature of isotopes. By coining this term he referred to the fact that isotopes of a given chemical element occupy the same position in the periodic table of elements since they share the same number of protons and electrons, but have a different number of neutrons. Therefore, as is so often mistakenly thought, the word isotope does not denote radioactivity. As mentioned above, radioactive isotopes have their own name – radioisotopes. Non-radioactive or stable isotopes of a given chemical element share the same chemical character and only differ in atomic mass (or mass number A), which is the sum of protons and neutrons in the nucleus.

Moving from the smallest entity upwards, atoms are comprised of positively charged protons and neutral neutrons, which make up an atom's nucleus, and negatively charged electrons, which make up an atom's shell ('electron cloud'). Due to charge balance constraints, the number of protons is matched by the number of electrons. A chemical element and its position in the periodic table of elements is determined by the number of protons in its nucleus. The number of protons determines the number of electrons in the electron cloud, and the configuration of this electron cloud in turn determines chemical characteristics such as electronegativity and the number of covalent chemical bonds a given element can form. Owing to this link, the number of protons in the atomic nucleus of a given chemical element is always the same and is denoted by the atomic number Z, while the number of neutrons (in its nucleus) may vary. Since the number of neutrons (N) has no effect on the number of electrons in the electron cloud surrounding an atom the overall chemical properties of an element are not affected. In other words, a chemical element like carbon will always behave like carbon irrespective if the number of neutrons in its nucleus is N or $N + 1$. However, differences in mass-dependent properties can cause compounds containing different amounts of carbon with N or $N + 1$ neutrons or at different positions to behave subtly differently, both chemically and physically.

Mass number $A (= Z + N)$ and atomic number Z are denoted as whole numbers in superscript and subscript, respectively, to the left of the element symbol. So carbon-12 comprised of six protons and six neutrons would be written as $_6^{12}C$, while carbon-13 that is comprised of six protons and seven neutrons would be written as $_6^{13}C$. In general practice different isotopes of the same chemical element are denoted by mass number and chemical symbol only (e.g. ^2H or ^{13}C).

For example, the simplest of chemical elements, hydrogen (H) in its most abundant isotopic form has a nucleus comprised of a single proton and therefore has the atomic mass of 1 (in atomic mass units (amu)) and this is indicated by adding a superscript prefix to the element letter (i.e. ^1H). The less abundant, by one neutron heavier hydrogen isotope is therefore denoted as ^2H, although one will also find the symbol D being used since this stable hydrogen isotope has been given the name deuterium. The discovery of this isotope won Harold C. Urey the Nobel Prize in Chemistry in 1934 and Urey is today regarded as one, if not *the* father of modern stable isotope chemistry.

Staying with hydrogen as an example, one could say ^1H and its sibling deuterium, ^2H (or D), are identical twins but are of different weight and of different abundance. Deuterium (^2H) is the heavier twin whose weight differs from that of hydrogen (^1H) by 1 amu. Deuterium is also the less abundant of the two hydrogen isotopes. The same is true for the carbon twins. Here, sibling ^{13}C is the heavier twin, weighing 1 amu more than its sibling ^{12}C, and as for the two hydrogen isotopes, the heavier ^{13}C is the less abundant of the two carbon isotopes. Where the normal weight versus overweight twin analogy has its limitations is the matter of abundance or occurrence, but only for as long as we stay with the example of two complete twins. We will revisit the twin example in the following chapter after a brief excurse on the natural abundance of stable isotope and natural abundance level variations.

Chapter I.2
Natural Abundance Variation of Stable Isotopes

The isotope abundances of all elements were fixed when the Earth was formed and, on a global scale, have not changed since. Figures usually quoted in chemistry textbooks for isotope abundance refer to these global values, such that when considering the entire carbon mass of the Earth system the natural abundance of ^{12}C and the one neutron heavier ^{13}C is 98.892 and 1.108 atom%, respectively (Table I.1). However, what tends to be overlooked by most and, hence, not be taught to students in chemistry classes is the fact that compartmental isotope abundance of light elements is not fixed, but is in a continuous state of flux due to mass discriminatory effects of biological, biochemical, chemical and physical processes. For instance, when looking at individual carbon pools one finds some with a higher abundance of ^{13}C, such as marine carbonate sediments, whereas others are more depleted in ^{13}C, such as hydrocarbons found in crude oil.

Expressed in 'atomic percent' (i.e. the percentage of one kind of atom relative to the total number of atoms in units of atom%) and staying with the example of ^{13}C, these differences are very small, with the range covered amounting to approximately 0.11 atom%. To express these minute variations, the δ notation in units of per mil (‰; one part per 1000) has been adopted to report changes in isotopic abundance as a per mil deviation compared with a designated isotopic standard (Equation I.1).

Various isotope standards are used for reporting isotopic compositions (Commission on Isotopic Abundances and Atomic Weights (CIAAW), www.ciaaw.org; National Institute of Standards and Technology (NIST), https://www-s.nist.gov/srmors/tables/view_table.cfm?table=104-10.htm; and International Atomic Energy Agency (IAEA), http://curem.iaea.org/catalogue/SI/index.html). By virtue of Equation I.1, the δ values of each of the standards are by definition 0‰. Carbon stable isotope ratios were originally reported relative to the PDB (Pee Dee Belemnite) standard. Since this reference material became exhausted, VPDB carbonate (Vienna Pee Dee Belemnite) has become the new international anchor for the ^{13}C scale. The oxygen stable isotope ratios of carbonates are also commonly expressed relative to VPDB. Stable oxygen and

Table I.1 Stable isotopes of light elements and their typical natural abundance.

Chemical element	Major abundant isotope	First minor abundant isotope	Second minor abundant isotope	Isotope ratio first minor/major for scale calibration material
Hydrogen	^1H 99.985 atom%	^2H 0.015 atom%		0.00015576 VSMOW
Carbon	^{12}C 98.89 atom%	^{13}C 1.11 atom%		0.0112372 VPDB
Nitrogen	^{14}N 99.63 atom%	^{15}N 0.37 atom%		0.0036765 Air
Oxygen	^{16}O 99.76 atom%	^{18}O 0.20 atom%	^{17}O 0.04 atom%	0.0020052 VSMOW
Sulfur	^{32}S 95.02 atom%	^{34}S 4.22 atom%	^{33}S 0.76 atom%	0.0450045 VCDT

hydrogen isotopic values are reported relative to VSMOW (Vienna Standard Mean Ocean Water), while sulfur and nitrogen isotope values are reported relative to VCDT (Vienna Canyon Diablo Troilite) and Air (atmospheric air), respectively.

Use of VSMOW and VPDB as standard reference points means that measurements have been normalized according to IAEA guidelines for expression of δ values relative to traceable reference materials on internationally agreed per mil scales (Coplen, 1994, 1996; Coplen et al., 2006a, 2006b) (Table I.2):

$$\delta_s = \left(\frac{[R_S - R_{STD}]}{R_{STD}} \right) \times 1000 \; (‰) \tag{I.1}$$

where R_S is the measured isotope ratio of the heavier isotope over the lighter (e.g. $^{13}C/^{12}C$ or $^2H/^1H$) for the sample and R_{STD} is the measured isotope ratio for the standard (e.g. VPDB or VSMOW). To give a convenient rule-of-thumb approximation, in the δ notation, a difference in ^{13}C abundance of 0.011 atom% corresponds to a change in $\delta^{13}C$ value of 10‰. In other words, a change in ^{13}C abundance from 1.0893 to 1.0783 atom% corresponds to a change in $\delta^{13}C$ value from −20 to −30‰ on the VPDB scale, respectively. Depending on how accurately and precise ^{13}C composition at natural abundance level can be measured by modern analytical instruments, for organic materials measured differences of 0.3‰ can be statistically significant.

Let us now revisit the twin analogy once more to picture what natural abundance means in praxis. Obviously, the abundance ratio of any given pair of twins is 1 : 1 or 50 : 50 (i.e. when meeting any one twin in a crowd where both are known to be present, one has an even chance of speaking either to twin A or twin B). However, if we consider a hypothetical case where both twins were victims of a major explosion, the probability of any given body part belonging to either twin now becomes a function of the number of pieces each body has been divided into. The same in a way is true for chemical elements and their 'overweight' twins. If one would take apart a lump of sugar to its molecular level, one would find that depending on circumstances (in this case which plant had

Table I.2 Representative but not concise list of international reference materials for stable isotope ratio mass spectrometry (IRMS) administered and distributed by the IAEA (Vienna, Austria).

International reference material	Code	$\delta^{13}C_{VPDB}$ (‰)[a]	$\delta^{15}N_{AIR}$ (‰)	$\delta^{2}H_{VSMOW}$ (‰)	$\delta^{18}O_{VSMOW}$ (‰)
TS-limestone	NBS-19	+1.95			−2.20
Lithium carbonate	LSVEC	−46.6			−26.6
Oil	NBS-22	−30.031[a]		−118.5	
Sucrose	IAEA-CH-6	−10.449[a]			
Polyethylene foil	IAEA-CH-7	−32.151[a]		−100.3	
Wood	IAEA-C4	−24.0			
Wood	IAEA-C5	−25.5			
Wood	IAEA-C9	−23.9			
Sucrose	IAEA-C6	−10.8			
Oxalic acid	IAEA-C7	−14.5			
Oxalic acid	IAEA-C8	−18.3			
Caffeine	IAEA-600	−27.771[a]	(+1.0)[b]		
L-Glutamic acid	USGS 40	−26.389[a]	−4.5		
L-Glutamic acid	USGS 41	+37.626[a]	+47.6		
Cellulose	IAEA-CH-3	−24.724[a]			
Ammonium sulfate	IAEA-N1		+0.4		
Ammonium sulfate	IAEA-N2		+20.3		
Potassium nitrate	IAEA-NO-3		+4.7		
Water	VSMOW			0	0
Water	GISP			−189.5	−24.8
Water	SLAP			−428	−55.5
Benzoic acid	IAEA-601				+23.3
Benzoic acid	IAEA-602				+71.4

[a] Note the $\delta^{13}C$ values given in this table include the latest values published by the IAEA as of 30 November 2006.
[b] This $\delta^{15}N$ value is based on data from one laboratory only.
NBS, National Bureau of Standards; LSVEC, lithium isotope reference material originally prepared by H. Svec, Iowa State University, USA; USGS, US Geological Survey; GISP, Greenland Ice Sheet Precipitation; SLAP, Standard Light Antarctic Precipitation; see text for other abbreviations.

produced the sugar) one would have a 98.9617 or 98.9015% chance of finding ^{12}C if the sugar would be beet sugar or cane sugar, respectively. Similarly, one would have a 1.0833 or 1.0985% chance of finding ^{13}C in carbon from beet sugar or cane sugar, respectively. Thus, generally speaking, one always has a better chance of encountering ^{12}C than ^{13}C, meaning ^{12}C has a higher abundance than its heavier isotope ^{13}C. However, on a case-by-case basis one finds that chemically identical substances such as sugar can exhibit different isotopic compositions where a variation in ^{12}C abundance is accompanied by a proportionate yet opposite variation in ^{13}C. In this case, beet sugar contains more ^{12}C and less ^{13}C than cane sugar; conversely, cane sugar contains more ^{13}C and less ^{12}C than beet sugar (Hobbie and Werner, 2004; Meier-Augenstein, 1999; Rossmann et al., 1997). The chemical and physicochemical reasons behind these differences will be discussed in Chapter I.4.

Chapter I.3
Chemically Identical and Yet Not the Same

Analytical methods traditionally applied in forensic science laboratories establish a degree of commonality between one substance and another by identifying their constituent elements, functional groups, and by elucidating their chemical structures. Thus, for two samples of sugar all of the aforementioned data will correspond and it can be concluded that they are chemically indistinguishable – they are indeed both sugar. However, it can be argued that although two substances in question are chemically indistinguishable they may not be the same (e.g. they may have come from different sources or be of different origin). Attention is drawn to the following – whenever we speak of *source* and *origin* of a natural product such as sugar, by *source* we mean from which particular plant the sugar was sourced (i.e. ultimately made), whereas by *origin* we mean its geographic origin (i.e. where the plant was grown and harvested). In other words, by differentiating between source and origin the distinction is being made where two substances do not share the same provenance then they are not truly identical even if chemically they are indistinguishable. This assertion can be contested by stable isotope analysis either to protect people from being convicted of a crime they have not committed such as drug trafficking or, staying with the example of drugs, to convict people who may be prepared to admit to the lesser offence of possession for personal use while in fact they are drug dealers or drug traffickers.

How is this possible? For reasons we will touch upon in Section I.4, two chemically indistinguishable compounds will be isotopically distinguishable if they do not share the same origin or are derived from a different source. In the case of sugar, traditionally the two main sources of sugar are sugar cane and sugar beet. With the help of stable isotope profiling it is perfectly straightforward to determine if a sugar sample is either cane sugar or beet sugar. In addition, with concomitant use of $d^{13}C$ and d^2H values, it is even possible to say where approximately in the world the sugar cane or sugar beet was grown and cultivated (Table I.3).

Stable Isotope Forensics: An Introduction to the Forensic Application of Stable Isotope Analysis Wolfram Meier-Augenstein
© 2010 John Wiley & Sons, Ltd

Table I.3 Isotopic abundance of ^{13}C and ^{2}H in sugar from different sources and geographic origin.

Sample	$\delta^{13}C_{VPDB}$ (‰)	$\delta^{2}H_{VSMOW}$ (‰)
Sugar (sugar beet; Poland)	−25.42	−71.0
Sugar (sugar beet, Sweden)	−26.84	−93.4
Sugar (sugar cane, Brazil)	−11.76	−21.4
Sugar (sugar cane; South Africa)	−11.10	−6.7

Chapter I.4
Isotope Effects, Mass Discrimination and Isotopic Fractionation

I.4.1 Physical Chemistry Background

If for a given compound a non-quantitative chemical reaction or a physicochemical process such as vapourization has taken place, this will be subject to mass discrimination (or associated with an isotope effect), which will cause a change in isotope abundance and, hence, result in isotopic fractionation. In principle, two different types of isotope effects can cause isotopic fractionation: kinetic isotope effects (kinetic as in chemical reaction kinetics) and thermodynamic isotope effects. In general, mass discrimination is caused by differences in the vibration energy levels of bonds involving heavier isotopes as compared to bonds involving lighter isotopes.

Differences in the zero-point energy of chemical bonds containing one heavy isotope and one light isotope relative to bonds containing two light isotopes are reflected by differences in the rates of cleavage of these bonds because differences in zero-point energy results in differences in bond energy. For example, for hydrogen gas the bond strengths of $^1H-^1H$, $^1H-^2H$ and $^2H-^2H$ are 436.0, 439.4 and 443.5 kJ/mol, respectively. Thus, $^2H-^2H$ bonds are broken at a slower rate than $^1H-^2H$ bonds, which in turn are broken at a slower rate than $^1H-^1H$ bonds. It is usually observed that the product of a chemical reaction involving bond cleavage will be isotopically lighter in the element(s) forming that bond compared to the corresponding isotopic composition of the initial precursor or source substrate.

I think it would be useful at this point to revisit some basic principles of physics and a mainstay of analytical chemistry instrumentation – infrared spectroscopy – to illustrate the relation between the reduced mass of a chemical two-atom system, bond length and bond strength (also known as bond energy), which are ultimately responsible for the mass discrimination that leads to the wide range of isotopic composition of natural and synthetic compounds.

The rotational (or vibrational) kinetic energy E_{vib} of a rigid body can be expressed in terms of its moment of inertia I and its angular velocity ω:

$$E_{vib} = \frac{1}{2} I \omega^2 \qquad (I.2)$$

The (scalar) moment of inertia of a point mass m rotating about a known axis r is defined by:

$$I = mr^2 \qquad (I.3)$$

For a system comprising two masses (or two atoms) m_1 and m_2 joined by, say, a spring (or chemical bond) of length r and if this system rotates around an axis intersecting a point on that spring (or bond), the mass term m in Equation I.3 is replaced by the reduced mass μ of this system, which is given by:

$$\mu = \frac{m_1 m_2}{m_1 + m_2} \qquad (I.4)$$

The vibrational or rotational energy of a molecule can be measured by its infrared absorbance. In the world of quantum physics where rotating or vibrating systems assume discrete energy levels, the associated discrete packets of energy differences ΔE can be expressed by the rotational constant B, the difference between two infrared absorption bands:

$$\Delta E_{\Delta j=1} = 2B = \frac{h}{4\pi^2 c I} \qquad (I.5)$$

where h is the Planck constant and c is the speed of light.

The infrared spectra of gaseous hydrochloric acid (HCl) are a fine example of how differences in isotopic make-up, and therefore differences in μ and r, and, hence, in I, result in differences in ΔE between neighbouring infrared absorption bands (e.g. for $^1H-^{35}Cl$ and $^2H-^{35}Cl$). Since we are able to measure B and we can calculate μ, it is possible to calculate r, the bond length for the different HCl isotopologues. Measured values for B for the aforementioned HCl species are given as 10.44 and 5.39 cm^{-1}, respectively. For the interested reader this exercise is included in the Set Problems in Chapter I.8.

I.4.2 Fractionation Factor α and Enrichment Factor ε

This difference in bond length and, hence, bond strength between bonds involving different isotopes of the same chemical element that already results in measurable differences in spectroscopic characteristics also leads to different reaction rates for a bond when different isotopes of the same element are involved (Melander and Saunders, 1980). The most significant isotope effect is the kinetic or primary isotope effect, whereby a bond containing the chemical elements under consideration is broken or formed in the rate-determining step of the reaction (Rieley, 1994), such as the reaction between two amino acids leading to the formation of the peptide bond R-CO–NH-R' involving the carboxyl carbon of amino acid R and the amino nitrogen of amino acid R'.

The second type of isotope effect is associated with differences in physicochemical properties such as infrared absorption, molar volume, vapour pressure, boiling point and melting point. Of course, these properties are all linked to the same parameters as those mentioned for the kinetic isotope effect – bond strength, reduced mass and, hence, vibration energy levels. However, to set it apart from the kinetic isotope effect, this effect is referred to as the thermodynamic isotope effect (Meier-Augenstein, 1999) because it manifests itself in processes where chemical bonds are neither broken nor formed. Typical examples of such processes in which the results of thermodynamic isotope effects can be observed are infrared spectroscopy and any kind of two-phase partitioning (e.g. liquid–liquid extraction) or phase transition (e.g. liquid to gas, i.e. distillation or vapourization). The thermodynamic isotope effect, or physicochemical isotope effect, is the reason for the higher infrared absorption of $^{13}CO_2$ as compared to $^{12}CO_2$, for the vapourization of ocean surface water resulting in clouds (i.e. water vapour) being depleted in both ^2H and ^{18}O compared to ocean surface water, and for the isotopic fractionation observed during chromatographic separations.

Another way of describing any isotope effect is to say that the reaction rate constant or equilibrium constant 'k' of a given reaction or transformation:

$$\text{Precursor (or Source)} \xrightarrow{k} \text{Product}$$

is in fact comprised of two subtly different reaction rate constants k_L and k_H for the light (L) and heavy (H) isotope-containing molecules or 'isotopologues' (Sharp, 2007), respectively, that make up the precursor or source compound. The ratio of these reaction rate constants yields the fraction factor α:

$$\alpha = \frac{k_H}{k_L} \qquad (I.6)$$

Since a molecule with a light isotope at the bond involved in the reaction usually reacts slightly faster (because breaking this light isotope bond requires slightly less energy) than a molecule with a heavy isotope in the same position (because breaking this heavy isotope bond requires slightly more energy) the ratio α of k_H/k_L is normally less than 1. For example, a reaction rate constant k_L that is 2% faster than the corresponding k_H translates into a fraction factor α of 0.98, thus already indicating that the product will be isotopically lighter compared to the precursor. In other words, with respect to the heavier isotope under consideration, the δ value of the product will be lower than the δ value of the precursor or source.

Enrichment factors ε or fractionation values derived from α values by $\varepsilon = \alpha - 1$ show this difference immediately in units of the δ notation:

$$\varepsilon = (\alpha - 1) \times 1000 \; (‰) \qquad (I.7)$$

Using the above example of an α value of 0.98 one can easily calculate an ε value of −20‰ for this hypothetical reaction, meaning the δ value of the product will be 20‰ lower than the δ value of the precursor or source: $\delta_{Product} = \delta_{Source} - 20‰$. In most cases the straightforward difference between $\delta_{Product}$ and δ_{Source} yields an enrichment factor very similar to that calculated from α values. However, since it is quite simple to calculate α values from measured δ values (Equation I.8), determination of enrichment factors from α values is recommended:

$$\alpha = \frac{\delta_{Product} + 1000}{\delta_{Source} + 1000} \tag{I.8}$$

It should be noted that in some scientific disciplines such as biology and ecology the fraction factor $\alpha_{L/H}$ is defined by the ratio of the reaction rates of the lighter isotopologue over the heavier:

$$\alpha_{L/H} = \frac{k_L}{k_H} \tag{I.9}$$

Adopting this convention leads to positive fractionation values Δ for reactions or transformations in which L reacts faster than H. Since both conventions essentially describe the same phenomenon using the same principles, converting α into $\alpha_{L/H}$ and ε into Δ is quite simple: $\alpha = 1/\alpha_{L/H}$ and $\varepsilon = -\Delta$.

In Chapter I.6 we will have a look at some of the aforementioned scenarios in more detail to develop an appreciation for how and why we can exploit the different ways in which isotopic fractionation plays out in the context of forensic science.

I.4.3 Isotopic Fractionation in Rayleigh Processes

One of the most important isotopic fractionation processes is the change in isotopic composition of a reservoir because of the removal of an increasing fraction of its contents. A typical example of such a removal of a compound (= sink) from a reservoir without an additional input that we can find in any chemical laboratory is when a solvent or any other suitably volatile synthetic or natural compound is purified by distillation. Another example would be the evaporation of water from a lake in an arid region during the dry season (i.e. no additional water input from a river or through rainfall). This particular type of Rayleigh process is therefore also referred to as Rayleigh distillation.

At a given time t a given reservoir comprises a total number of molecules N and a ratio of the rare (e.g. ^2H or ^{13}C) to the abundant (e.g. ^1H or ^{12}C) molecular concentration R (i.e. the isotope ratio). In other words, $N/(1+R)$ is the number of the most abundant isotopic molecules and $RN/(1+R)$ is the number of rare isotopic molecules. If we remove dN molecules with an accompanying fractionation factor α, the mass balance for the rare isotope can be written as:

$$\frac{R}{1+R}N = \frac{R+dR}{1+R+dR}(N+dN) - \frac{\alpha R}{1+\alpha R}dN \quad (I.10)$$

To simplify Equation I.10 somewhat we make the following approximations: (i) the total number of molecules is equal to the number of the most abundant isotopic molecules and (ii) all denominators in Equation I.10 are taken to be equal to $1+R$. The mass balance for the rare isotope thus simplifies to:

$$RN = (R+dR)(N+dN) - \alpha R\, dN \quad (I.11)$$

By neglecting the product of differentials and separating the variables, the mass balance for the rare isotope becomes:

$$\frac{dR}{R} = \frac{(\alpha-1)dN}{N} \quad (I.12)$$

Integration of Equation I.12 while applying the boundary conditions for time $t=0$ of $R=R_0$ when $N=N_0$ yields:

$$\frac{R}{R_0} = \left(\frac{N}{N_0}\right)^{\alpha-1} \quad (I.13)$$

We can write Equation I.13 using δ values with respect to a standardized reference:

$$\delta = (1+\delta_0)\left(\frac{N}{N_0}\right)^{\varepsilon} - 1 \quad (I.14)$$

Here, N/N_0 represents the remaining fraction of the original reservoir while R_0 and δ_0 refer to the original isotopic composition. We also need to remember that ε ($=\alpha-1$) is a very small number (e.g. $\varepsilon = 0.00938$ if the fractionation factor $\alpha = 1.00938$ as in the case of the ^{18}O equilibrium fractionation between water in the liquid phase and water in the vapour phase at a temperature T of 25 °C).

I.4.4 Isotopic Fractionation Summary

Appreciation of the phenomenon that is isotopic fractionation is crucial for developing an understanding of what we can detect and measure, the potential magnitude of the isotope effect we may be able to exploit, and, last but not least, how it may impact on results due to the potential of fractionation during sample preparation. For this reason, at this point we summarize the key facts about isotopic fractionation in Box I.1.

Box I.1 Isotopic fractionation

Equilibrium isotopic fractionation

- Equilibrium fractionation describes isotopic exchange reactions that occur between two different phases of a compound at a rate that maintains equilibrium, as with the transformation of water vapour to liquid precipitation.

Isotopic fractionation during physicochemical processes

- Mass differences give rise to fractionation during physicochemical processes (diffusion, evaporation, two-phase partitioning, molecular spectroscopy).

- Fractionation during physicochemical processes is again the result of differences ultimately influenced by the reduced mass of a system such as different velocities of isotopic molecules of the same compound in the gas phase (kinetic energy) or differences in absorption of energy (e.g. infrared radiation (vibrational energy)).

- Associated isotope effects are called secondary isotope effects or thermodynamic isotope effects since chemical bonds are neither formed nor broken during these processes.

- Note: Some textbooks refer to mass discrimination during evaporation as a kinetic effect, although from a chemistry point of view this is incorrect.

Isotopic fractionation during (bio-)chemical processes

- Mass differences give rise to fractionation during chemical processes, whereby a bond containing the atom or its isotope is broken or formed in the rate-determining step of the reaction.

- Fractionation during chemical processes is again the result of differences ultimately influenced by the reduced mass of a system, primarily bond strength (energy) and bond length.

- Associated isotope effects are called primary isotope effects or kinetic isotope effects since the degree of fractionation is determined by the reaction kinetics of the particular chemical reaction.

- Note: If a partner in a chemical reaction is fully consumed (i.e. reacts completely (quantitatively)), no mass discrimination/isotopic fractionation is observed in respect of this compound.

Chapter I.5
Stable Isotopic Distribution and Isotopic Fractionation of Light Elements in Nature

I.5.1 Hydrogen

The vast majority of hydrogen in nature is found in the hydrosphere, which is often called the 'water sphere' as it includes all the Earth's water that is found in streams, lakes, oceans, ice, soil, groundwater and in the air. The total volume of water on Earth amounts to $1\,385\,984 \times 10^3$ km^3 and 96.54% of this volume is ocean water, making seawater the main reservoir of hydrogen in nature. Not surprisingly, VSMOW is the standard reference point for hydrogen (and oxygen) stable isotope analysis.

As we have seen, hydrogen occurs in nature in two stable isotope varieties, ^{1}H and ^{2}H (or D). The global or average natural abundances on Earth for ^{1}H and ^{2}H are 99.9844 and 0.01557 atom%, respectively (Sharp, 2007). From these natural abundance values the average terrestrial ^{2}H/^{1}H isotope ratio can be easily calculated as 0.00015572. The ^{2}H/^{1}H isotope ratio for the primary standard of the ^{2}H scale, VSMOW, is given as 0.00015576, so one can see why VSMOW was chosen to anchor the ^{2}H scale. (Incidentally, the IAEA who administer and distribute calibration materials and reference materials for stable isotope analysis strongly suggest only calibration materials should be referred to as primary standards.)

The 'hydrological cycle' traces the movement of water and energy between the various water stores and Earth's spheres (i.e. the lithosphere, atmosphere, biosphere and hydrosphere). Ocean water can be looked at as the starting point of the hydrologic cycle since clouds formed from evaporating ocean water transport water from the oceans to the continents. Isotopically speaking, the hydrologic cycle is one big Rayleigh process or combination of Rayleigh processes.

Isotopic fractionation associated with Rayleigh processes (see Section I.4.3) such as evaporation, condensation and precipitation of meteoric water ultimately results

I.5: STABLE ISOTOPIC DISTRIBUTION AND FRACTIONATION OF LIGHT ELEMENTS IN NATURE

in drinking water having a different isotopic composition depending on geo-location. Depending upon latitude, altitude, temperature and distance to the open seas, observed δ^2H values of meteoric water (precipitation) and, hence, fresh water can range from +20 to −230‰ across the world with the 'heavier' or less negative δ^2H values being typical of coastal/near equatorial regions and the 'lighter' or more negative δ^2H values being typical of inland/high altitude/high latitude regions. Geographical information system (GIS) maps and contour maps of meteoric ^2H and ^{18}O isotope abundance are in the public domain, and can be accessed via the Internet from resources such as www.waterisotopes.org, which is maintained by Gabriel Bowen at Purdue University, and the hydrogeology section of the IAEA (http://isohis.iaea.org/userupate/waterloo/index.html), while information on ^{18}O isotope abundance in global seawater can be found on the web site of NASA's Goddard Institute for Space Studies (http://www.giss.nasa.gov/data/o18data).

To understand this phenomenon let us apply some of what we have learned in the preceding chapters. The isotopic composition of water vapour (i.e. a cloud) over seawater with an isotopic composition of $\delta^2H = \delta^{18}O = 0$‰ versus VSMOW is somewhat lighter than a strictly theoretical calculation from isotopic equilibrium with the water would predict since strictly speaking evaporation of seawater is a non-equilibrium process. However, once a cloud with a given vapour composition has formed, the vapour and its resultant precipitation remain in isotopic equilibrium because the formation of precipitation occurs from saturated vapour. Consequently, δ^2H and $\delta^{18}O$ values of precipitation both change in unison as dictated by the ratio of their respective fractionation values $^2\varepsilon$ and $^{18}\varepsilon$ (Mook, 2000).

Let us assume the ^2H isotopic composition of water vapour in a cloud formed above the ocean near the equator has a δ^2H_v value of −84.44‰, the temperature is a balmy 25 °C and the first precipitation out of this cloud occurs. The fractionation factor α for ^2H between water vapour and liquid water in equilibrium at this temperature is 1.0793 (Mook, 2000; Majoube, 1971). Rearranging Equation I.8 we can calculate the δ^2H_l value of this precipitation as:

$$\delta^2H_l = 1.0793(-84.44 + 10^3) - 10^3 = -11.84\text{‰}$$

For ^{18}O, the fractionation factor α between water vapour and liquid water in equilibrium at 25 °C is 1.00938. The typical ^{18}O isotopic composition of equatorial atmospheric vapour corresponds to a $\delta^{18}O_v$ value of −12.03‰. Inserting these values into Equation I.8 and solving for $\delta^{18}O$ of the water falling as precipitation yields a $\delta^{18}O_l$ value of −2.76‰:

$$\delta^{18}O_l = 1.00938(-12.03 + 10^3) - 10^3 = -2.76\text{‰}$$

Assuming for reasons of simplicity that evaporation and condensation in nature occur in isotopic equilibrium, the relation between δ^2H and $\delta^{18}O$ values of precipitation would be determined solely by their equilibrium fractionations $^2\varepsilon$ and $^{18}\varepsilon$. While values for $^2\varepsilon$ and $^{18}\varepsilon$ change with temperature, the ratio of $^2\varepsilon/^{18}\varepsilon$ is virtually constant over a temperature range spanning 35 °C. The $^2\varepsilon/^{18}\varepsilon$ ratio for $^2\varepsilon$ and $^{18}\varepsilon$ values at 25 °C yields a value of 7.9. For $^2\varepsilon$ and $^{18}\varepsilon$ values covering the temperature range from -5 to $+30\,°C$ the average $^2\varepsilon/^{18}\varepsilon$ ratio yields an average value of 8.2 (Mook, 2000).

While studying isotopic composition of precipitation from different parts of the world, Craig (1961) and Dansgaard (1964) found a relation between δ^2H and $\delta^{18}O$ values of precipitation (Craig, 1961). The correlation equation describing this relation is referred to as the Global Meteoric Water Line (GMWL):

$$\delta^2H = 8\delta^{18}O + 10 \qquad (I.15)$$

In the context of what we have learned so far, the slope of 8 for the global meteoric water line (Equation I.15) is obviously not a random number, but can be explained by the ratio of the equilibrium fractionations $^2\varepsilon$ and $^{18}\varepsilon$ for the formation of precipitation.

For plants, water is the only available hydrogen precursor pool for the biosynthesis of carbohydrates, lipids and proteins. While biosynthesis of lipids seems to be associated with a considerable isotopic fractionation of about $-200‰$ compared to the δ^2H value of source water (Fry, 2006), δ^2H values of carbohydrates such as sugars and cellulose seem to be largely unaffected, showing a very strong correlation with δ^2H values of source water (Figure I.1). With knowledge of the underlying fractionation processes associated with evapotranspiration in plant leaves (the capillary flow of soil water through the xylem does not incur isotopic fractionation) and biosynthetic pathways, it is possible to use δ^2H of plant material as an indicator of source water (i.e. local precipitation) and, hence, provenance.

Another and more direct source of information regarding ^2H isotopic composition of source water and, hence, geographic provenance of highly popular fruit such as

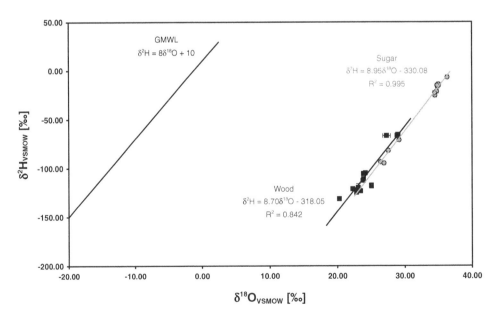

Figure I.1 δ^2H and $\delta^{18}O$ values of whole wood and plant sugars (beet and cane sugar) in the framework of the global meteoric water line.

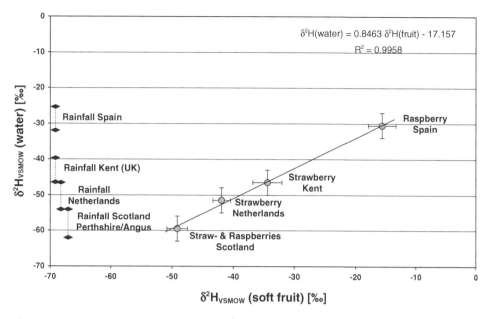

Figure I.2 Correlation plot of source water δ^2H values versus δ^2H values of fresh soft fruit water.

blackberries, raspberries, red or black currents and strawberries is the water contained in or stored by the fruit itself. Almost all fruit, but soft fruit in particular, contains between 85 and 91% water, so fruit water δ^2H values of freshly pressed or pulped fruit should be very closely correlated with δ^2H values of source water. Pilot data from a preliminary study carried out in our laboratory seem to support this hypothesis (Figure I.2). Of course, only further studies will tell if the correlation really is as strong as these pilot data seem to suggest and if analysing for δ^2H values of fruit water may provide a 'forensic' tool to detect mislabelling of, for example, cheap imports as highly priced local or regional produce (*cf.* Section I.6.1).

In contrast to plants, for most animals, particularly for land-living mammals and humans, the body's hydrogen pool is fuelled by three precursor pools: drinking water (including water in beverages in the case of human beings), water stored in food and hydrogen chemically bound in food, although in most cases this can be simplified to water and hydrogen chemically bound in food. This is an important point to remember since it has a bearing on our ability to ascertain geographic origin and geographic movement from ^{2}H isotope analysis of non-remodelling body tissue such as hair and nail.

I.5.2 Oxygen

Oxygen as O_2 comprises approximately 21% by volume of the Earth's atmosphere and in the form of water it is at 88.8% by mass the major component of the world's oceans. In fact, by mass oxygen is the Earth's most abundant light chemical element.

Oxygen exists in the form of three stable isotopes in nature, ^{16}O, ^{17}O and ^{18}O, with relative abundances of 99.7621, 0.0379 and 0.2000 atom%, respectively (Sharp, 2007). Given its low abundance, other than in a few specialist subject areas, analysis of ^{17}O isotope abundance receives much less attention than ^{18}O in studies exploiting oxygen isotopes as natural tracers, although the link between ^{17}O and ^{18}O abundance plays an important role when correcting the contribution of CO_2 isotopologue $^{12}C^{17}O^{16}O$ to mass 45 when analysing ^{13}C isotope abundance based on the abundance of CO_2 isotopologue $^{13}C^{16}O_2$ (Craig, 1961). As with hydrogen, the hydrological cycle and, hence, water control most of the processes and fluxes involving oxygen, its transfer, and incorporation into organic and inorganic materials. However, atmospheric oxygen that as O_2 comprises approximately 21% of the Earth's atmosphere also plays an important role as a precursor pool for many biochemical reactions. Due to its volume and gaseous nature, atmospheric O_2 is regarded as a constant pool with a constant $\delta^{18}O$ value of 23.5‰ versus VSMOW.

Biochemical reactions or transformations using atmospheric O_2 as a precursor pool are hydroxylation reactions such as converting proline units in collagen fibres into hydroxyproline or the metabolic oxidation/hydroxylation catalysed by monoaminoxidase and cytochrome P450 enzymes. In plants, O_2 is the precursor pool used by mono-oxygenases in conjunction with $NADH/H^+$ to introduce hydroxyl groups into aromatic carboxylic acids to yield hydroxyl aromatic carboxylic acids such as cumaric acid, which is made from cinnamic acid. The majority of these aromatic acids will eventually be used to form lignin.

Pool mixing and isotopic fractionation during biochemical processes are also the reason why cellulose and other plant sugars typically exhibit $\delta^{18}O$ values that are 25–30‰ enriched compared to source water (see Figure I.1). One must also not forget that oxygen bound in CO_2 is yet another precursor pool for oxygen that is accessible to and made use of by plants.

I.5.2.1 ^{18}O in Bone Bio-apatite and Source Water

For land-based mammals and humans, the three main precursor pools feeding into the mixing cauldron that is body water/blood are water, atmospheric oxygen and oxygen chemically bound in food. Unless one is able to model the various fluxes to and from the mixing pool of body water, and can thus find approximations for the various fractionation factors between these pools, it will be rather difficult to glean meaningful information on ^{18}O isotopic composition of source water from tissue such as hair or nail. On the other hand, recent work carried out by Jim Ehleringer's group on North American hair deriving a model to account for contributions of atmospheric oxygen and dietary oxygen is very promising indeed (Ehleringer et al., 2008a).

However, there is one compartment for which the relationship between blood (i.e. body water) and tissue $\delta^{18}O$ values is well understood – bone (or the phosphate fraction of bone bio-apatite, to be precise) (Longinelli, 1984; Luz, Kolodny and Horowitz, 1984). The good news is that for all mammals the ^{18}O isotopic fractionation between body water (i.e. blood water) and bone phosphate is nearly constant, with a fractionation factor α of approximately 1.021 between bone phosphate ^{18}O and blood water ^{18}O in equilibrium at body temperature. The not so good news is that the relationship between $\delta^{18}O_{blood}$ and $\delta^{18}O_{water}$ differs from species to species (Longinelli, 1984), which for different species leads to different relationships between $\delta^{18}O_{phosphate}$ and $\delta^{18}O_{water}$. However, a wide knowledge base exists that allows us to attribute $\delta^{18}O$ values of source water to measured $\delta^{18}O$ values of bone phosphate:

$$\delta^{18}O_{phosphate} = 0.79 \delta^{18}O_{water} + 21.61 \text{ (mouse)} \qquad (I.16)$$

$$\delta^{18}O_{phosphate} = 0.86\ \delta^{18}O_{water} + 22.71 \text{ (pig)} \qquad (I.17)$$

$$\delta^{18}O_{phosphate} = 1.01\ \delta^{18}O_{water} + 24.90 \text{ (cattle)} \qquad (I.18)$$

$$\delta^{18}O_{phosphate} = 1.13\ \delta^{18}O_{water} + 25.55 \text{ (deer)} \qquad (I.19)$$

$$\delta^{18}O_{phosphate} = 1.48\ \delta^{18}O_{water} + 27.21 \text{ (sheep)} \qquad (I.20)$$

A major important lesson to learn from the above correlation equations (Dangela and Longinelli, 1993) is that animal studies in one species (e.g. mice) are not a good proxy for another species (e.g. deer) or indeed humans. Another point to remember is the fact that $\delta^{18}O$ values of phosphate (PO_4^{2-}) are reported on the VSMOW scale. However, in an exception to the general rule that all $\delta^{18}O$ values should be reported against VSMOW, $\delta^{18}O$ values of carbonates (CO_3^{2-}) are reported against VPDB, with the VPDB scale for ^{18}O being defined by $\delta^{18}O_{NBS-19}$ versus VPDB = −2.20‰ (Mook, 2000).

The conversion of a $\delta^{18}O_{VPDB}$ value of a given carbonate sample X into a $\delta^{18}O_{VSMOW}$ value is given by (Friedman and O'Neil, 1977):

$$\delta^{18}O_{VSMOW}(X) = 1.03086\ \delta^{18}O_{VPDB}(X) + 30.86 \qquad (I.21)$$

Of interest from a forensic and investigative point of view is of course the relationship between $\delta^{18}O$ values of human bone or tooth phosphate and source water since this can provide information on provenance or geographic origin of, for example, a mutilated or badly decomposed human body, especially when looking at permanent teeth which once fully formed and erupted do not remodel unlike skeletal bones. Such information will be useful in focusing an investigation and thus police resources by directing searches in DNA, fingerprint or missing persons' databases to those national databases most likely to produce a match. Unfortunately, research into the relationship between $\delta^{18}O$ values of human bone phosphate and source water in modern humans is hampered (at least in the United Kingdom) by restrictions imposed by The Human Tissue Act 2004 on research involving human tissue irrespective of the research nature or aims. Again, referring solely to the situation in the United Kingdom, The Scottish Anatomy Act 2006 may offer a *modus vivendi* permitting such research to be carried out. Discussions if this possibility indeed exists are taking place as this book is being written.

In response to the existing need for information in support of victim identification, current endeavours to aid police forces in their investigations of violent crimes (Meier-Augenstein and Fraser, 2008) still make use of work carried out in the early 1980s

(Longinelli, 1984; Luz, Kolodny and Horowitz, 1984) even though more recent work correlating $\delta^{18}O$ values of human bone phosphate and source water exists (Daux *et al.*, 2008). The correlation between $\delta^{18}O$ values of human bone phosphate and $\delta^{18}O$ values of source water based on the least-squares best fit of measured $\delta^{18}O$ values (Longinelli, 1984) was obtained using bone belonging to people who died between the end of the nineteenth century and 1950, and therefore still has some relevance to work carried out today:

$$\delta^{18}O_{phosphate} = 0.64\ \delta^{18}O_{water} + 22.37\ (\text{human}) \qquad (I.22)$$

In my own studies and actual criminal case work I have found Longinelli's correlation linking $\delta^{18}O$ values of human bone phosphate to $\delta^{18}O$ values of source water to yield far more reliable or meaningful results than the correlation reported by Luz *et al.* I attribute this finding to two factors. (i) Samples for Longinelli's study were drawn from a collective of people who had lived in Central Europe (i.e. in a moderate or temperate climate). (ii) The sample pool on which Luz *et al.* based their correlation equation was very much focused on people who had lived in the East and Northeast Africa (i.e. in a very warm if not to say arid climate). A recent study re-evaluated work carried out thus far on the link between the ^{18}O record comprised in bio-apatite and $\delta^{18}O$ values of source water. It also proposed a unified correlation equation (Equation I.23) that is not too dissimilar to the equation originally proposed by Longinelli (Daux *et al.*, 2008):

$$\delta^{18}O_{phosphate} = 0.65\ \delta^{18}O_{water} + 21.89\ (\text{human}) \qquad (I.23)$$

Figure I.3 gives a graphical illustration of how the correlations proposed by the three groups compare and, using data from an actual case, what influence these differences can have in practical terms when faced with the task of coming to a conclusion on geographic provenance of a victim based solely on $\delta^{18}O$ values of bone phosphate.

The slight difference in slope and offset between Equations I.23 and I.22 is the result of Daux *et al.* accounting for the contribution of water from both uncooked and cooked food. Since $\delta^{18}O$ values of food cooked with water tend to be higher than $\delta^{18}O$ values of actual source water, the ^{18}O isotopic signature of total ingested water for present-day people tends to be 1.05–1.20‰ higher than that of source water. As to whether the relation between $\delta^{18}O$ values of human bone phosphate and source water as given by Equation I.23 may have to be revised (again) in the future, this can only be answered by continuing systematic studies using bone and tooth samples from present-day people as well as their remains. Thus, if you, dear reader, are a student or a postgraduate engaged in forensic sciences or related areas of science, there is still many a question requiring answers and, hence, offering a lot of opportunities for further studies and research. Let me take this opportunity to give you two examples of bone-related stable isotope questions in the following two chapters that are of interest to stable isotope forensics.

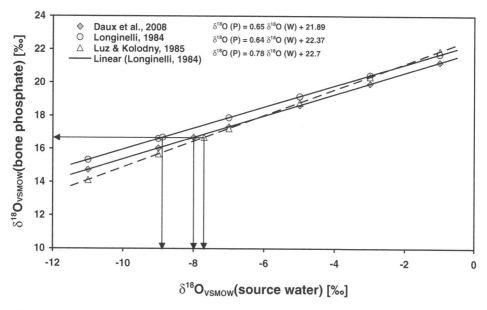

Figure I.3 Correlation graphs according to Daux *et al.*, Longinelli, and Luz and Kolodny for $\delta^{18}O_{phosphate}$ versus $\delta^{18}O_{water}$ and the resulting different solutions of $\delta^{18}O_{water}$ for the same $\delta^{18}O_{phosphate}$.

I.5.2.2 Bone Remodelling

When teeth are not available for forensic analysis, such as in cases of disarticulated skeletal or human remains or, and this is sadly happening with increasing frequency, in cases of deliberately dismembered bodies, mutilated to prevent identification of the victim and, by proxy, the perpetrator, the only tissue with a 'memory' measurable in years rather than weeks or months is human bone. Typical figures quoted in some texts or found on the Internet give bone remodelling rates of 10% per year, which is an averaged figure for the entire human skeleton. However, in reality bone remodelling rates differ from bone to bone. Rib bone is said to remodel completely every 5 years, whereas load-bearing bones such as the femur appear to remodel completely only every 20–25 years (Frost, 1990; Hedges *et al.*, 2007; Meier-Augenstein and Fraser, 2008; Pearson and Lieberman, 2004). Having robust remodelling rates for all bones or, more realistically, at least a selection of bones providing a good representation of the entire human skeleton and, hence, the time record contained therein would be very desirable for the construction of entire life history trajectories or geographic life histories. This natural archive or repository would provide invaluable forensic information not only for murder enquiries, but also for disaster victim identification, especially when victims come from a multinational, multiethnic background as is potentially the case in terrorist bombings of popular tourist destinations or multiethnic settlements such as major cities.

I.5.2.3 Bone Diagenesis

Another area of current and future research needs concerns the question if and to what degree bone diagenesis occurs in circumstances of clandestine body deposition or burial conditions such as a water-logged grave resulting in diagenetic changes of the ^{13}C and/or ^{18}O isotopic composition of bone mineral and in diagenetic changes of the ^{13}C and/or ^{15}N isotopic composition of bone collagen. Causes and mechanisms of diagenetic changes in bone, dentine and tooth enamel can vary considerably, with the major environmental influences being mechanical, chemical (water and pH) and microbiological weathering, and even bacterial remodelling of bone (Bell, Cox and Sealy, 2001; Hedges, 2002; Lee-Thorp, 2008).

I.5.2.3.1 Diagenetic Changes of Bio-apatite

Work by Koch *et al.* showed that δ^{18}O values of carbonate-bound oxygen in bio-apatite (of tusks) do not suffer from diagenetic changes in the case of modern and fairly recent samples (Koch, Fisher and Dettman, 1989). Of course the question remains what exactly is meant by 'recent' since chemical changes in bone can occur in only a few years? The C–O bond is thought to be vulnerable to exchange with water via the carbonate/water/CO$_2$ equilibrium, making ^{18}O isotopic composition of bone carbonate susceptible not only to diagenetic changes, but also to artefacts caused by sample treatment (Koch, Tuross and Fogel, 1997). While this places a question mark on the extent of diagenetic changes to δ^{18}O values of bone carbonate, we are on firmer ground when dealing with carbonate laid down in teeth (Sharp, Atudorei and Furrer, 2000). For example, a study of archaeological sheep teeth showed excellent agreement between δ^{18}O values for source water calculated on the basis of directly measured phosphate δ^{18}O values and phosphate δ^{18}O values calculated using Equation I.21 (Henton, Meier-Augenstein and Kemp, 2009). As part of the same study, carbonate and phosphate fractions from the same teeth were analysed for their ^{18}O isotopic composition and the linear regression solution ($r^2 = 0.99$; Equation I.24) for the resulting δ^{18}O$_{phosphate}$ versus δ^{18}O$_{carbonate}$ plot was in very good agreement with a published correlation equation linking bone/tooth carbonate δ^{18}O values to bone/tooth phosphate δ^{18}O values in the absence of carbonate diagenesis (Equation I.25):

$$\delta^{18}O_{phosphate} = 1.066\ \delta^{18}O_{carbonate} - 11.34 \qquad (I.24)$$

$$\delta^{18}O_{phosphate} = 0.98\ \delta^{18}O_{carbonate} - 8.5\ (Iacumin\ et\ al., 1996) \qquad (I.25)$$

Be it in teeth or in bone, once laid down as bio-apatite, phosphate appears to preserve its original equilibrium δ^{18}O value with body water and seems to be completely impervious to diagenetic changes (Lee-Thorp and Sponheimer, 2003; Sharp, 2007; Shemesh, Kolodny and Luz, 1983). However, this approach is based on the suggestion that both P—O and P=O bonds are so strong as to render PO$_4^{2-}$ all but immune to diagenetic changes and we know now of several post-mortem pathways that can also affect apatite phosphate ions and, hence, their ^{18}O isotopic composition. These pathways include recrystallization, which is almost inevitable for bone, and microbial attack that can swiftly break P—O bonds. The most pervasive change to bone microstructure is caused by bacterial activity that, depending on the severity of microbial attack, can even completely alter the profile of bone density values (Bell, Cox and Sealy, 2001).

That being said, for post-mortem intervals typically encountered in a criminal investigation, bio-apatite in bone and teeth offers us a window into a person's geographic past, even including changes in diet and location based on intra-bone changes in stable isotopic composition (Balasse, Bocherens and Mariotti, 1999; Meier-Augenstein and Fraser, 2008).

I.5.3 Carbon

Going from the most simplest organic compounds comprising only one carbon atom (C_1 bodies) such as methane or plant-derived methyl halides such as methyl chloride (Keppler *et al.*, 2004, 2006, 2007, 2008; Keppler and Rockmann, 2007) to sediments or marine carbonate, the variation in natural abundance of ^{13}C covers approximately 0.11 atom% or 110‰ (Fry, 2006). This wide range reflects the varying degree of mass discrimination associated with the different photosynthetic pathways used by plants for carbon assimilation and fixation. As we have seen previously (*cf*. Chapter I.3) in terms of ^{13}C isotopic abundance, beet sugar is not the same as cane sugar. In sugar beet, CO_2 fixation results in the formation of a C_3 body, 3-phosphoglycerate (3-PGA), an organic compound comprising three carbon atoms. This photosynthetic pathway of CO_2 fixation mediated by the enzyme ribulose-1,5-bisphosphate carboxylase/oxygenase (RuBisCO) is known as the Calvin–Benson cycle. Plants using the 3-PGA pathway for CO_2 fixation such as sugar beet are commonly called C_3-plants. However, in an adaptive response to hot climatic conditions some plants make use of a different pathway to increase the rate of glucose production. Here, CO_2 fixation by RuBisCO is compartmentalized and fuelled with CO_2 from a preceding carbon fixation step using the enzyme phosphoenolpyruvate carboxylase. This first key product of this process yields a C_4-dicarboxylic acid, oxaloacetate (hence the term C_4-plants). The C_4-dicarboxylic acid pathway is also known as the Hatch–Slack cycle. The products of these two different pathways are characterized by their different ^{13}C isotopic composition. Glucose in leaves of C_3-plants has a δ^{13}C value of about -28‰, whereas leaf glucose derived from C_4-plants exhibits a more positive δ^{13}C value of about -13‰. Since both plant types utilize atmospheric CO_2 with a δ^{13}C value of -8‰ for glucose production this is equivalent to a net isotopic fraction of only -5‰ for the Hatch–Slack cycle, while the corresponding net isotopic fractionation for the Calvin–Benson cycle is about -20‰ (Fry, 2006). The most important C_4-plants in terms of impact on dietary intake of ^{13}C by domestic animals and (directly or indirectly) by humans are sugar cane, maize (sweet corn), sorghum and millet. Owing to the extent to which C_4-plants pervade their staple diet, δ^{13}C values of tissue and human hair from North Americans and South Africans, for example, are more positive as compared to corresponding δ^{13}C values for

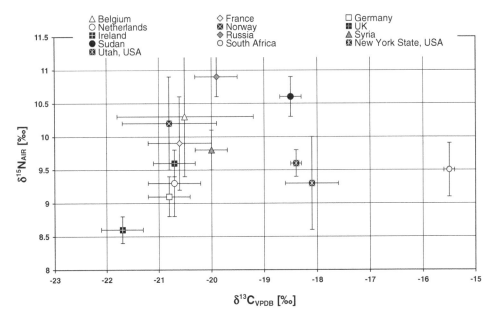

Figure I.4 Bivariate graph plotting $\delta^{15}N$ versus $\delta^{13}C$ values of scalp hair samples volunteered by residents in different countries reflecting their regionally different diet. Error bars are ±1σ of groups comprising four to 10 individuals per region. Based on data generated by or for my then PhD student Dr Isla Fraser as part of her PhD thesis.

Central Europeans, for example, with $\delta^{13}C$ values ranging from −18 to −15.5‰ (see Figure I.4).

The other major carbon fixation process in nature is the dissolution of CO_2 in seawater. Remembering our chemistry lessons we know the dissolution of CO_2 in water is pH dependent. Seawater has a pH of 8 and under these alkaline conditions over 99% of the CO_2 dissolved in seawater is present in the form of HCO_3^-. As a result the oceans contain about 50 times more CO_2 than the atmosphere, thus making them a major CO_2 sink. The net isotopic fractionation between atmospheric CO_2 ($\delta^{13}C = -8$‰) and total dissolved CO_2 in seawater ($\delta^{13}C = +1$‰) is +9‰. Fixation of seawater CO_2 ultimately results in marine particulate organic matter (POM), on which virtually all life in the oceans is based one way or another and the typical $\delta^{13}C$ value of marine POM is about −22‰, thus displaying an apparent net fractionation of −23‰ between dissolved seawater CO_2 and marine POM. In contrast, dissolved CO_2 in freshwater lakes with pH values of about 6 is a mixture of H_2CO_3 and HCO_3^-. Total dissolved CO_2 in freshwater lakes has a $\delta^{13}C$ value of approximately −15‰, which corresponds to a net isotopic fractionation of −7‰ against atmospheric CO_2. Freshwater POM exhibits a $\delta^{13}C$ value of about −35‰, which corresponds to a net isotopic fractionation of −20‰ between dissolved freshwater CO_2 and freshwater POM. This pronounced difference in ^{13}C isotopic composition between freshwater POM and marine POM is the reason why muscle tissue of marine plankton feeders such as mussels, prawns and whiting shows $\delta^{13}C$ values ranging from −21 to −19‰ (*cf.* Figure I.6 below). This leads us nicely on to the subject of isotopic fractionation associated with trophic ecology (i.e. food chains and food webs). However, before we have a closer look at this phenomenon with the name trophic level shift (fractionation) in Section I.5.4.1, let us continue conclude our studies of the major light elements by finding out a bit more about nitrogen.

I.5.4 Nitrogen

Unlike for oxygen, the main biosphere reservoir for nitrogen in nature is the Earth's atmosphere with nitrogen (as N_2) being its major constituent by volume, namely 78%. Given that atmospheric N_2 is evenly distributed, the $^{15}N/^{14}N$ ratio of nitrogen in air is constant at 0.0036765 and air nitrogen (AIR) is therefore used as our standard reference point for reporting $\delta^{15}N$ values. However, atmospheric nitrogen constitutes only 2% of all nitrogen on Earth. The vast proportion of nitrogen is bound in rocks, accounting for 97.98% of all nitrogen on Earth. Only 0.001% of all nitrogen is bound in organic matter. Again, unlike oxygen, the nitrogen contained in the Earth's atmosphere is not directly bio-available to most organisms, with the exception of a few nitrogen-fixating soil bacteria. The nitrogen cycle through the biosphere is driven by five major processes: nitrogen fixation, nitrogen uptake (through growing organisms), nitrogen mineralization (decay), nitrification and denitrification. Micro-organisms, particularly bacteria, are the major players in all of the aforementioned nitrogen transformation processes. For this reason, ^{15}N abundance in the biosphere typically spans a relatively narrow range from -20 to $+30$‰ for nitrogen bound as NH_4^+ or in animal waste, respectively (Sharp, 2007).

One reason for this relatively narrow range is that isotope fractionation associated with nitrogen fixation and mineralization (ammonification) of organic nitrogen to soil NH_4^+ is generally quite small. Due to natural variations associated with the pathways for these processes the ε values for isotope fractionation reported for nitrogen fixation range from -3 to $+3.7$‰, but is typically taken to be $+1$‰, while breakdown of organic matter to ammonium is given as 0 ± 1‰ (Fry, 2006; Sharp, 2007). The combination of nitrogen fixation, ammonification and nitrification (i.e. the oxidation of NH_4^+ to NO_2^- and ultimately NO_3^-) results in plants exhibiting $\delta^{15}N$ values that can range from -10 to $+10$‰, although typically $\delta^{15}N$ values of -6 to $+6$‰ are more generally observed while $\delta^{15}N$ values reported for soil can range from -10 to $+20$‰ (Hoefs, 2009) with $\delta^{15}N$ values of soil reflecting the ^{15}N isotopic composition of plant litter and overlaying vegetation (Handley and Scrimgeour, 1997). It should be noted that even though some nitrification and ammonification processes can be associated with large isotope effects, the corresponding reactions often progress in a quantitative fashion due to slow rates of nitrogen supply and limited amounts of nitrogen substrate, which means all available nitrogen is converted and no overall net fractionation is observed.

However, there are two factors influencing $\delta^{15}N$ values of materials we may encounter in the course of a criminal investigation requiring forensic stable isotope analysis – trophic ecology and anthropogenic activity. We have encountered the term trophic ecology before in Section I.5.3 and we will have a closer look at the associated

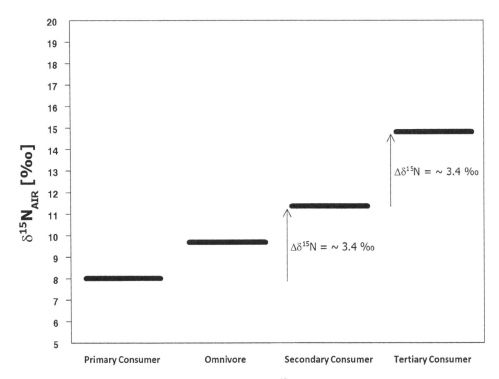

Figure I.5 Schematic representation of typical δ^{15}N values in relation to trophic level.

phenomenon of tropic level shift fractionation in Section I.5.4.1. For the moment it suffices to know that primary consumers such as cattle feeding on a primary producer such as grass will increase their body ^{15}N isotopic composition by 3–4‰ as compared to the ^{15}N signature of grass by excreting urine that is ^{15}N-depleted and comprises isotopically light (i.e. ^{14}N-enriched) ammonia (NH_3) and urea (OCN_2H_4) (Senbayram *et al.*, 2008). Moving up in the food chain from herbivore to omnivore to carnivore (i.e. from one trophic level to the next), δ^{15}N values of proteinogenous tissue will shift upwards by 3–4‰ when moving up each trophic level (Figure I.5).

Anthropogenic influences can be divided in two types of activity. One would be human agricultural activity and its influence on δ^{15}N values of cultivated soils through soil disturbance, but chiefly through the application of fertilizers (both synthetic and organic). Addition of organic (animal waste) fertilizer increases soil δ^{15}N values while the addition of synthetic nitrogenous fertilizer lowers soil δ^{15}N values since synthetic inorganic fertilizers tend to have δ^{15}N values ranging from −5 to +5‰. Synthetic fertilizers in turn are the result of the second type of anthropogenic activity, namely manufactured materials. Depending on synthetic route and process, isotopic fractionation of −14‰ can be observed between precursor and product δ^{15}N values (Lock and Meier-Augenstein, 2008) with δ^{15}N values for materials of forensic interest such as 3,4-methylenedioxy-*N*-methylamphetamine (MDMA) or nylon fibres ranging from −18 to +35‰.

I.5.4.1 Food Chain and Trophic Level Shift

To illustrate the meaning of the term trophic level let us imagine a simplified food web comprising potatoes, herbivorous pigs and pork-loving carnivorous humans, where pigs feed exclusively on potatoes and pork-loving humans consume nothing but pork chops and ham. Moving up one trophic level from the primary producer plant to the primary consumer herbivorous pig we observe an increase in $\delta^{13}C$ value from -27 (freeze dried potato) to $-25.5‰$ (lean pork chop) – a shift of $+1.5‰$ (Morrison et al., 2000). Moving up another trophic level to the secondary consumer carnivorous human we find for hair (or bone collagen) a $\delta^{13}C$ value of $-21.3‰$ – a shift of $+4.2‰$. However, if we would have analysed a mixed tissue sample from our pork-loving friend (who of course has mashed potatoes or French fries with his pork chop too), namely a sample comprised of skin, muscle and fat, we would most likely have observed a $\delta^{13}C$ value of about $-24.8‰$ (i.e. a shift of $+1.2‰$ compared to a mixed pool dietary ^{13}C signal of $-26‰$) (Figure I.6).

As already mentioned in Section I.5.4, a particularly remarkable feature of the nitrogen cycle is the positive isotopic fractionation of about $+3$ to $+4‰$ associated with moving up from one trophic level to the next (Minagawa and Wada, 1984). Since

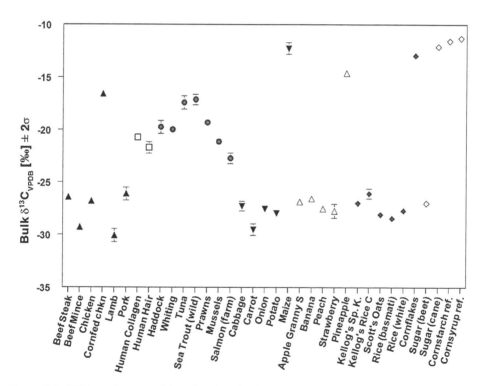

Figure I.6 ^{13}C isotopic composition of various food and animal tissue. Note that human collagen and human hair $\delta^{13}C$ values shown here are typical for people with a C_3-plant-dominated diet. Figure is based on data published in Morrison, D.J. et al., 2000 as well as author's own.

marine ecosystems have far longer food chains than terrestrial ecosystems, marine foods exhibit significantly higher $\delta^{15}N$ values (Schoeninger and DeNiro, 1984; Schoeninger, DeNiro and Tauber, 1983), resulting in the extremely high $\delta^{15}N$ values of more than 15‰ observed in people such as the Inuits with an almost exclusively animal source of dietary protein be it in the form of fish or in the form of seal meat, who could be categorized as tertiary consumers. In general, one can say that in humans $\delta^{15}N$ values of structural proteins such as bone collagen or hair keratin reflect the relative level of meat protein in our diet. Vegan hair exhibits lower $\delta^{15}N$ values than hair from ovo-lacto vegetarians or omnivores. In fact, hair $\delta^{15}N$ values from ovo-lacto vegetarians are virtually identical to those from omnivores. Hair from people for whom meat (and/or fish) forms a very large proportion of their diet has higher $\delta^{15}N$ values than omnivores and ovo-lacto vegetarians (O'Connell et al., 2001; O'Connell and Hedges, 1999; Petzke et al., 2005; Petzke, Boeing and Metges, 2005).

Freshwater animals also show positive tissue $\delta^{15}N$ values, although their tissue $\delta^{13}C$ values tend to be lower than those found in marine animals for reasons mentioned in Section I.5.3. Bivariate isotope datasets or graphs plotting $\delta^{15}N$ versus $\delta^{13}C$ values can therefore be used to distinguish marine-sourced foods from terrestrially sourced foods or to study trophic relationships in birds as compared to mammals (Kelly, 2000). Employing tissue $\delta^{13}C$ and $\delta^{15}N$ values in combination is also useful to differentiate trophic level fractionation-induced shifts in $\delta^{15}N$ values from shifts caused by physiological/metabolic changes in the human body, such as nutritional stress (starvation, crash dieting), infection or pregnancy (Fuller et al., 2004, 2005). For mixed tissue, purely trophic level shift-induced changes in $\delta^{15}N$ of 3–4‰ should be matched by a simultaneous change in $\delta^{13}C$ values of 1–2‰ when moving up from one trophic level to the next. As always there are exceptions to these rules of thumb, as in the case of constrained diets seen, for example, in corn-fed chickens where eventually all body pools and compartments approach the same level of ^{13}C isotopic composition as the diet (cf. Figure I.6). Changes in $\delta^{15}N$ values of synthesized protein such as hair keratin of about 1‰ over and above the 'normal' or expected trophic level shift change are indicative of physiological factors such as nutritional stress and are hypothesized to be the result of muscle protein catabolism to make up for the lack of dietary protein as a source of amino acids and, hence, nitrogen.

Let us now consider an example to see what kind of forensic information we can extract from the phenomenon that is trophic level shift. Picture this – a body of a man is found in a major city in Scotland and enquiries suggest he is one of a pair of identical twins but it cannot be established conclusively whether the body is James or John McDoe. However, during their enquiries the investigating team learn that James McDoe was a vegan while his twin brother John liked his meat. A quick stable isotope analysis of a sample of scalp hair taken from the victim's body reveals a $\delta^{15}N$ value of 7.5‰ while the corresponding $\delta^{13}C$ value is determined as $-21.96‰$. Both isotope abundance values, but in particular the $\delta^{15}N$ value, place the victim into the category of primary consumers, thus consistent with the victim being a vegan and therefore quite likely to be James McDoe (Figure I.7).

While the above examples illustrate the meaning of the term trophic level and what forensically useful information we can glean from human tissue isotopic composition as

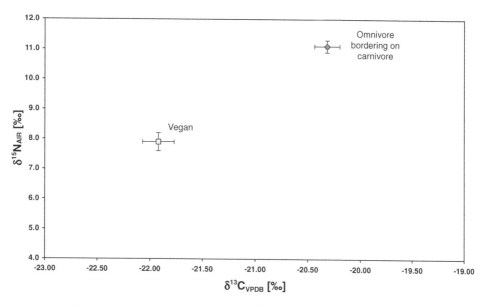

Figure I.7 Bivariate graph plotting $\delta^{15}N$ versus $\delta^{13}C$ values from scalp hair of a vegan and an omnivore who has a relatively strong meat component in their diet.

a result, it leaves us with the question why associated shifts in tissue isotopic composition with regard to ^{13}C can be so different.

The bicarbonate pool in human blood is driven by CO_2 produced by our body as the end product of metabolism. The $\delta^{13}C$ value of CO_2 exhaled in breath therefore reflects the average ^{13}C isotopic composition of diet due to the quasi-steady state between CO_2 generated during metabolism and its transfer into the blood stream. However, there is a fractionation of 4–5‰ between average dietary ^{13}C isotopic composition and, therefore, blood bicarbonate (released as CO_2 in exhaled breath) and ^{13}C isotopic composition of structural proteins such as keratin or bone collagen, while fractionation between blood bicarbonate and bone carbonate can range from 10 to 15‰ (Ambrose and Krigbaum, 2003).

In mixed tissue samples of consumer (i.e. a mix of protein and fat), a small trophic level shift enrichment or positive isotopic fractionation would be observed for $\delta^{13}C$ values of about 1–2‰ when compared to source food. However, $\delta^{13}C$ values of collagen (or keratin) preferentially reflect the protein component of the diet (Jim, Ambrose and Evershed, 2004) and therefore shifts of 4–5‰ are observed when comparing $\delta^{13}C$ values of structural proteins in humans to the average ^{13}C signature of total dietary intake (see Figures I.7 and I.8). It is not unusual to see an intra-individual difference of up to 1.4‰ between $\delta^{13}C$ values of keratin and bone collagen, with the $\delta^{13}C$ value for bone collagen being the higher of the two (O'Connell et al., 2001).

In conclusion, ^{13}C and ^{15}N isotope abundance in consumer tissues are known to correlate with diet isotope composition with $\delta^{15}N$ values becoming more positive with increasing trophic level. A strong relationship also exists between ^{2}H isotope abundance and trophic level with $\delta^{2}H$ values also becoming more positive when moving up

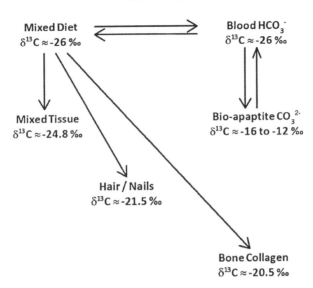

Figure I.8 Approximate δ^{13}C values for ^{13}C isotopic composition of various body pools and tissue. Note that all human tissue δ^{13}C values shown here are based on an omnivorous C_3-plant-dominated diet.

in trophic level from herbivores to carnivores (Birchall *et al.*, 2005). Thus, one has to be careful when interpreting stable isotope data from, for example, human tissue, taking into account the different influence and magnitude of trophic level shift fractionation between food protein source and consumer proteinogenic tissue as compared to apparent isotopic fractionation between the pooled isotopic composition of total dietary intake and individual (i.e. different) consumer tissue. On the other hand, knowledge of these different processes (Hedges and Reynard, 2007; Reynard and Hedges, 2008) can sometimes yield very powerful forensic information that otherwise would have been difficult, if not impossible, to obtain.

I.5.4.2 Diagenetic Changes of Structural Proteins

The best-studied protein prone to deterioration and diagenetic changes, and yet one of the longest surviving proteins after death, is collagen, particularly bone collagen. As far as palaeoecologists are concerned, the survival rate of collagen is negligible once death has occurred. However, for archaeologists, forensic archaeologists and forensic scientists, collagen is a wonderful repository from which information on dietary habits, nutritional status, geographic provenance and even approximate post-mortem interval can be mined. Survival rates for bone collagen or tooth dentine of up to 100 000 years have been repeatedly reported (Jones *et al.*, 2001), and contextual interpretation of the information contained therein has yielded insights into infant feeding and even the age of weaning in medieval societies (Fuller, Richards and Mays, 2001, 2003; Mays, Richards and Fuller, 2002; Richards, Mays and Fuller, 2002).

Keratin, the major constituent and structural protein comprising hair and nails is another useful source of information with regard to recent diet, nutritional status and geographic history, with the time frame covered depending on the length of hair or nail available for study (Ambler et al., 1999; Macko et al., 1999b; Nardoto et al., 2006). Since hair and nail are fully exposed to the environment, survival rates obviously very much depend on conditions and their usefulness as a source of forensic information on present-day people is limited to the living, recently deceased or well-preserved bodies. There are of course the fields of forensic archaeology and forensic anthropology where hair from well-preserved bodies either through freezing, as in the case of the Austrian Ice Man (Macko et al., 1999b), or mummification can yield fascinating insights ranging from information on life style and social status of an individual to religious practices of ancient societies (Wilson et al., 2007).

Strictly speaking, conclusions drawn from stable isotopic make-up of proteins such as collagen and keratin are only valid in the absence of any diagenetic changes such as chemical or microbiological degradation since this will invariably alter stable isotopic composition (Balzer et al., 1997; DeNiro, 1985; Schoeninger and DeNiro, 1982). One simple test to assess preservation of protein integrity is to measure its C/N ratio, which typically should be of the order of 1:3.5 through microchemical elemental analysis. However, this is more a rule-of-thumb test given that a protein's C/N ratio will depend on its amino acid profile and C/N ratios should therefore not be regarded as an exclusive criterion to assess state of protein preservation (Harbeck et al., 2006). A more reliable method to assess protein preservation is based on the fact that with the exception of glycine, all amino acids in the human body only appear in the L-enantiomeric form. Maintaining exclusive L-configuration is an energy-consuming process that ceases with death so an L-amino acid will eventually turn into a racemic mixture of its D and L form. There is even evidence to suggest that, at least for skin elastin, this process already starts during life as part of the natural ageing process (Ritz-Timme, Laumeier and Collins, 2003). Determining the level of racemization of key amino acid residues in collagen such as aspartic acid can provide reliable information on protein degradation in relation to the post-mortem interval (Dobberstein et al., 2008).

I.5.5 Sulfur

In comparison to the application involving stable isotopes of bio-relevant light elements such as carbon, hydrogen, nitrogen and oxygen, applications involving sulfur (i.e. its stable isotopes ^{32}S and ^{34}S) are rather under-represented in the scientific literature. This, however, is not necessarily a reflection on its usefulness, but can be more attributed to the fact that sulfur stable isotope analysis requires (i) a relatively large sample amount due to its low relative abundance in organic compounds, (ii) great analytical skill and (iii) ideally an instrumental set-up dedicated to sulfur isotope analysis. One could argue the latter two points are more relevant to continuous flow isotope ratio mass spectrometry (CF-IRMS; see Part II for full details of the various analytical techniques) due to the 'viscous' and hygroscopic nature of SO_2, the gas into which samples have to be converted for sulfur isotope analysis, given that recent advances in multicollector inductive coupled (MC-ICP)-MS have made it possible to measure ^{34}S isotope ratios at natural abundance level either by direct laser ablation (LA) or by direct introduction of a sample liquid into the plasma via a nebulizer (Santamaria-Fernandez and Hearn, 2008; Santamaria-Fernandez, Hearn and Wolff, 2008).

Of all bio-relevant light elements, the major sulfur stable isotope ^{34}S has the highest natural abundance at 4.215 atom%. The corresponding isotope ratio of 0.0430023 is quite close to that of primordial sulfur, which is represented by the ^{34}S/^{32}S ratio of 0.0450045 of the international primary isotope standard VCDT meteorite – a meteoritic troilite (FeS) found in a meteor crater near Flagstaff, Arizona. In comparison, dissolved sulfate (SO_4^{2-}), which is the chief component of the sulfur reservoir in Earth's oceans, has a ^{34}S isotopic composition of about 21‰ on the VCDT scale, on which by definition $\delta^{34}S_{VCDT} = 0.0$‰. In a reflection of global-scale fluctuations in sulfate reduction activities over geological time scales δ^{34}S values of dissolved marine sulfate have ranged from about 10–33‰. Continental uplift and preservation of marine sediments containing sulfur in different oxidation states, which can range from −II to +VI, have produced on land something like insular sulfur environments, each associated with different δ^{34}S values for sulfur in bedrock. Despite these large ranges in terrestrial δ^{34}S values in general, δ^{34}S values for continental vegetation seem to average near 2–6‰ over large areas and are quite distinct from the more ^{34}S-rich values of 17–21‰ found in marine plankton and seaweeds.

As discussed for carbon, hydrogen, nitrogen and oxygen, sulfur isotope signatures of animal or human tissue can be regarded as a reflection of the sulfur isotopic composition of the consumed diet (Macko *et al.*, 1999a; Richards *et al.*, 2003; Richards, Fuller and Hedges, 2001). In turn, the ^{34}S isotopic composition of diet is a reflection of ^{34}S isotope abundance in a given environment, incorporated into plant and animal tissue from sulfur sources such as bedrock weathering, atmospheric deposition and microbiological activity (Richards *et al.*, 2003). From a human provenancing point of view, tissue with a high collagen or keratin content has the highest potential to yield meaningful information given the relatively high sulfur content of either protein compared to other biogenic compounds. Based on encouraging ^{34}S data from a comparative study of hair collected from 35 residents of the United Kingdom and individuals from Australia, Canada and Chile (Bol, Marsh and Heaton, 2007), I have rated this potential high enough to propose in 2007 that it should be possible to extract useful information on the recent geographic movement and whereabouts of a person suspected to be involved in serious and organized crime, including terrorism, from as a little as a single strand of hair using LA-MC-ICP-MS. It remains to be seen if this hypothesis will receive the support required for it to be tested and validated.

Given the aforementioned differences between marine, freshwater and terrestrial δ^{34}S values, with the help of sulfur isotope analysis of animal or human tissue it is potentially possible to distinguish between freshwater/terrestrial or marine dietary sources (Privat, O'Connell and Hedges, 2007; Richards *et al.*, 2003; Richards, Fuller and Hedges, 2001). However, similar to considerations mentioned for drawing conclusions on human dietary habits based on δ^{15}N values, δ^{34}S values should be interpreted in the context of δ^{15}N and δ^{13}C values obtained from the same sample material. Presumed marine diet-derived ^{34}S signatures may actually be the result of consuming plants and animals cultivated and reared in coastal regions. One possible explanation for this phenomenon of artefact marine ^{34}S signatures is the use of seaweed and kelp as natural fertilizer or indeed feed stock for domestic animals such as sheep and cattle (Balasse, Tresset and Ambrose, 2006). The other explanation is referred to as the 'sea spray effect' (i.e. a result of airborne deposition of marine sulfur particles from the ocean to the coast), although the extent of this mode of marine sulfur transfer on to coastal regions has not been established yet. Having said that, the 'sea spray effect' may permit plausible interpretation of δ^{34}S values in terms of geographic provenance with relatively high ^{34}S signatures being taken to intimate coastal residency, even when δ^{13}C and δ^{15}N values indicate a terrestrial rather than marine-based diet.

Finally, a word of caution when it comes to interpreting δ^{34}S values of tissue from present-day people. These δ^{34}S values may also reflect the potential influence of ^{34}S isotopic composition of fossil fuels or from other anthropogenic sources contained in atmospheric emission and from there incorporated into food sources as well as into human tissue directly.

Chapter I.6
Stable Isotope Forensics in Everyday Life

To fill the preceding and occasionally somewhat dry chapters with life, what follows here is a compilation of stable isotope applications, albeit not always forensic applications in the true sense of the word. These are, however, important analytical applications nonetheless impacting as they do on our day-to-day life by making it directly or indirectly safer and more enjoyable through protection of our health and our environment.

In a report on *The Economic Impact of Counterfeiting and Piracy*, published in 2007 (http://www.oecd.org/dataoecd/13/12/38707619.pdf; last accessed 24 July 2009), the Organization for Economic Cooperation and Development (OECD)[1] stated that in 2005:

> ... the volume of tangible counterfeit and pirated products in international trade could be up to US$200 billion. This figure does not, however, include counterfeit and pirated products that are produced and consumed domestically, nor does it include the significant volume of pirated digital products that are being distributed via the Internet. If these items were added, the total magnitude of counterfeiting and piracy worldwide could well be several hundred billion dollars more.

In the same report, the OECD highlights the links between counterfeiting consumer goods and organized crime as well as international terrorism.

> The groups involved in counterfeiting and piracy include mafias in Europe and the Americas and Asian "triads", which are also involved in heroin trafficking,

---

[1] OECD member countries: Australia, Austria, Belgium, Canada, the Czech Republic, Denmark, Finland, France, Germany, Greece, Hungary, Iceland, Ireland, Italy, Japan, Korea, Luxembourg, Mexico, the Netherlands, New Zealand, Norway, Poland, Portugal, the Slovak Republic, Spain, Sweden, Switzerland, Turkey, the United Kingdom and the United States of America; the Commission of the European Communities takes part in the work of the OECD.

Stable Isotope Forensics: An Introduction to the Forensic Application of Stable Isotope Analysis Wolfram Meier-Augenstein
© 2010 John Wiley & Sons, Ltd

prostitution, gambling, extortion, money laundering and human trafficking. To address the situation, Interpol created an Intellectual Property Crime Action Group in July 2002, to help combat trans-national and organised intellectual property (IP) crime by facilitating and supporting cross-border operational partnerships. Some governments have also established bilateral operational partnerships in border enforcement and criminal investigations.

In addition to the established link between counterfeiting and piracy and organised crime, Interpol has highlighted a disturbing relationship of counterfeiting and piracy with terrorist financing, with IP crime said to be becoming the preferred method of financing for a number of terrorist groups. The links take two basic forms:

- *Direct involvement*, where the terrorist group is implicated in the production or sale of counterfeit goods and remit a significant portion of those funds for the activities of the group. Terrorist organisations with direct involvement include groups which resemble or behave like organised crime groups.

- *Indirect involvement*, where sympathisers involved in IP crime provide financial support to terrorist groups via third parties.

Counterfeit products such as counterfeit pharmaceuticals are often substandard or even dangerous, posing health and safety risks to consumers that range from mild to life threatening. Given the impact of counterfeit products on the economy as well as consumer health and safety, sadly very little about the impact applied stable isotope analysis has made on detecting and combating consumer good fraud seems to be known to the general public, which is surprising given that a review published in 2001 on the increasing importance stable isotope profiling has gained in authenticity control of food and food ingredients lists 184 publications for this field alone (Rossmann, 2001). To a degree this is a sad indictment of the way real-life benefits gained from advances in life sciences, in general, and from stable isotope research, in particular, are being disseminated to the public. Even more saddening is the thought that this failure to communicate state-of-the-art techniques as well as insights gained to the benefit of human everyday life as well as the advancement of science also extends to schools, universities and colleges if comments and opinions regarding stable isotope techniques voiced by a number of scientists and so-called expert advisers are anything to go by. Every reader, in particular every student reader, of this book is therefore strongly encouraged to avail themselves of the positive aspects of the Internet, and to search for further literature and publications on applied stable isotope analysis because even a specialist subject book such as this will never manage to capture all the published information out there (chiefly because behind this book is an author who is only human and, hence, fallible after all).

I.6.1 'Food Forensics'

It is a surprisingly little-known fact that food products such as wine, certain spirits, high-quality single-seed vegetable oils, natural flavourings and honey are all subject to regular stable isotope analysis (more often than not in conjunction with other analytical techniques) to determine/verify product authenticity or to combat fraudulent labelling and misrepresentation (Ogrinc *et al.*, 2003). The British Food Standards Agency (FSA) has supported the development and application of stable isotope analytical techniques for food authentication since the mid 1990s, and indeed for some applications stable isotope analysis has been declared the method of choice. Similarly, the European Office for Wine, Alcohol and Spirit Drinks (BEVABS) has been using stable isotope analytical techniques and data to combat major fraud in the beverage sector since 1997, although stable isotope analysis of beverages has been used as early as 1993 (Calderone, Guillou and Naulet, 2003; Rossmann *et al.*, 1996). BEVABS was established by the European Union in 1993 and is now part of the Food Products Unit of the Institute for Health and Consumer Protection (IHCP) at the European Commission's Joint Research Centre (JRC) in Ispra (Italy).

I.6.1.1 Authenticity and Provenance of Single-Seed Vegetable Oils

Stable isotope fingerprinting of food and food ingredients to detect adulteration or misrepresentation can be regarded as the first forensic application of this technology, although such data have not yet been used in conjunction with a legal case or prosecution for fraudulent food adulteration. That said, a single-seed vegetable oil survey of corn (maize) oil sold in the United Kingdom was carried out in 1995 on behalf of the FSA (or Ministry of Agriculture, Fisheries and Food as it then was) and showed that 35% of oils analysed for their ^{13}C isotopic composition were found to contain undeclared oils from different sources and were therefore not 100% corn oil as declared on the label (Ministry of Agriculture, Fisheries and Food, 1995). These results were published, and both retailers and producers were informed of the result of this survey. Given the legal framework at the time (Food Safety Act 1990) – the survey was carried out before the Food Labelling Regulations 1996 – this survey did not result in any legal action. However, a follow-up survey was commissioned by the FSA in 2001 in which 61 samples of corn oil were again analysed for five different parameters including ^{13}C isotopic composition and this time none of the samples were declared suspicious. Authenticity of the oils was assessed by comparing the results of analysis with a purity specification for corn oil. This specification is contained in the Codex Standard for Named Vegetable Oils, which was adopted formally by the Codex Alimentarius Commission in July 2001. In their report on the 2001 food authenticity survey the FSA concluded that corn oil now sold in the United Kingdom complied fully with the fatty acid and stable carbon isotope ratio specifications contained in the Codex standard and is named correctly on the label.

To detect adulteration of olive oils, Angerosa *et al.* compared $\delta^{13}C$ values of the aliphatic alcoholic oil fractions and found those of the adulterant pomace oil to be significantly more negative than those of virgin and refined olive oils (Angerosa *et al.*, 1997). In a subsequent study, Angerosa *et al.* employed both ^{13}C and ^{18}O isotope analysis to determine the geographical origin of olive oils according to climatic regions from different Mediterranean countries such as Greece, Morocco and Spain (Angerosa *et al.*, 1999). Apart from blending high-quality single-seed vegetable oils such as corn or

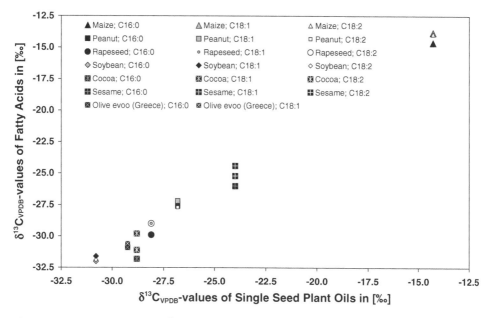

Figure I.9 Natural variation in ^{13}C isotopic composition of single-seed vegetable oils and selected fatty acids isolated from these oils. Figure is based on data published in Woodbury, S.E. et al., 1998 as well as author's own.

virgin olive oil with lower cost oils such as rapeseed oil (Kelly et al., 1997; Kelly and Rhodes, 2002; Woodbury et al., 1995; Woodbury, Evershed and Rossell, 1998), thermally induced degradation due to deodorization or steam washing can alter the ^{13}C isotopic signature of the whole oil and its key fatty acid components (Spangenberg, Macko and Hunziker, 1998). Other factors found to influence both the ^{13}C isotopic signature and relative fatty acid composition of olive oils are vintage (year of production), storage period (oxidation of unsaturated fatty acids), botanical species and maturity at harvest (Royer et al., 1999; Spangenberg and Ogrinc, 2001). See Figure I.9.

I.6.1.2 Authenticity and Provenance of Beverages

Another section of food products prone to misrepresentation, mislabelling and fraudulent use of adulterants are beverages. The seemingly non-descript term beverages is used here deliberately since the spectrum covered by accidental or fraudulent misrepresentation up to the production and sale of counterfeit brand products ranges from sparkling water to fruit juices and from beer and wine to white spirits and whisky. Authentic sparkling beverages usually contain only the CO_2 generated by fermentation while authentic sparkling mineral waters only contain the CO_2 naturally present in the spring. A much cheaper and easier method for carbonation is to saturate the beverage with industrial CO_2 from a gas tank or cylinder. Stable isotope analysis can detect the different modes of carbonation (Calderone et al., 2007; Calderone and Guillou, 2008) by comparing carbon and oxygen isotopic composition of natural CO_2 from authentic sources against industrial CO_2, which typically exhibits $\delta^{13}C$ and $\delta^{18}O$ values of below -33 and $-36‰$, respectively.

By equilibrating wine water from retail wines with CO_2 of known isotopic composition, $\delta^{18}O$ values of wine water were correlated with spatial climate and precipitation $\delta^{18}O$ patterns across the wine grape-growing regions of Washington, Oregon and California based on a spatial join with continuous GIS maps of these parameters. A regression model was implemented spatially in a GIS with which it was possible to predict wine $\delta^{18}O$ values for all vintages and generally reflected the consistent enrichment of wine from Napa relative to Livermore. GIS models of wine water $\delta^{18}O$ values could therefore become useful tools for independently verifying claims of regional origin and vintage (West, Ehleringer and Cerling, 2007).

Detecting alien sugar (i.e. natural sugar from a different plant source) in fruit juice and wine was a fairly simple task in the early days of food control using stable isotope analytical techniques since mainly cheap corn syrup (*Zea mays* L.; a C_4-plant) was predominantly used to boost sugar and/or ethanol content, respectively (Brooks *et al.*, 2002; Martinelli *et al.*, 2003). Determining $\delta^{13}C$ values of glucose or bulk carbon was sufficient to prove adulteration. However, addition of small amounts of C_4-plant sugars (10% or less) to C_3-plant products such as wine, fruit juice and honey or the addition of sugars from other C_3-plants (sugar beet, concentrated and deflavourized grape juice) could no longer be detected by these measurements, and more sophisticated analytical techniques had to be developed based on high-precision CSIA to detect fraudulent addition of sugars (Dennis, Massey and Bigwood, 1994) and even vitamin C from alien sources (Schmidt *et al.*, 1993). This second generation of 'forensic' stable isotope analytical techniques exploited the fact that in authentic fruit juices the $\delta^{13}C$ values of biogenetically related compounds such as L-ascorbic acid, L-malic acid and L-tartaric acid were strongly correlated with the $\delta^{13}C$ values of their parent sugars (Gensler, Rossmann and Schmidt, 1995; Gensler and Schmidt, 1994; Jamin *et al.*, 1997; Rossmann *et al.*, 1997; Weber *et al.*, 1997; Weber, Gensler and Schmidt, 1997). For example, there is an isotopic fractionation of 4.8‰ between the $\delta^{13}C$ values of L-ascorbic acid and its precursor glucose. This enrichment is position specific, mainly located in the C-1 position of biogenetically authentic L-ascorbic acid, and would appear to be the result of plant-specific kinetic isotope effects during biosynthesis, whereas L-ascorbic acid of biotechnological origin preserves the ^{13}C signature and distribution of glucose.

The research group led by Schmidt discovered position-specific ^{13}C depletion at C-1 in glycerol originating from natural sources and they suggested this unique feature might be used as a means to test for illegal addition of synthetic glycerol to wines (Weber, Kexel and Schmidt, 1997). They also found a consistent $\Delta\delta^{13}C$ fractionation between ethanol and citric acid of 2.4‰ in addition to the known $\Delta\delta^{13}C$ fractionation between fermented sugar and ethanol of $-1.7‰$ (Weber *et al.*, 1997).

The advent of on-line continuous flow ^2H stable isotope analysis offered a new avenue for control and authentication of food and food additives (Figure I.10). In order to authenticate fruit juices the ^2H isotopic composition of citric acid was studied comparing commercial citric acids with citric acid extracted from fruit juices. The developed method's ability to detect an addition of exogenous citric acid was successfully

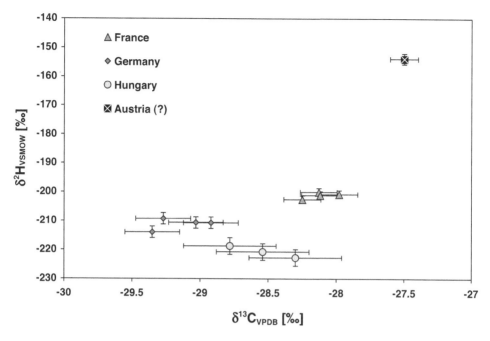

Figure I.10 Bivariate plot of δ^2H and δ^{13}C values of ethanol from a selection of European white wines, including one suspect sample of wine labelled as vintage Austrian wine.

tested by identifying an orange juice sample that had been spiked with commercially produced citric acid (Jamin et al., 2005).

Using a single-isotope, yet multicomponent approach to compound-specific isotope analysis (CSIA) of a best-selling blended whisky, radar diagrams (a.k.a. spider web plots) graphically representing δ^{13}C values of volatile whisky congeners acetaldehyde, ethyl acetate, n-propanol, isobutanol and amyl alcohol were constructed based on an analysis of eight product samples collected over 2 years of production to generate a fingerprint and its variability range of the authentic product. On the basis of this multivariate approach it was possible to discriminate two other whisky samples from the authentic popular Scotch whisky brand (Parker et al., 1998).

I.6.1.3 Authenticity and Provenance of other Premium Foods

An area where compound-specific isotope fingerprinting has become the method of choice is authenticity control of flavours, fragrances and essential oils. Substituting synthetic or 'nature identical' for natural flavours or fragrances is an all too easy way to defraud consumers by, for example, selling a product containing synthetic flavour or fragrance as the pure natural product or a product based on priced natural ingredients only. Natural lavender oil, for example, contains 30–60% linalyl acetate and retails at about £450/l. Synthetic linalyl acetate (97%) costs £108/l while natural linalyl acetate (around 80%) costs £180/l. Similarly, an extract of natural γ-decalactone, a flavour compound contained in peach and apricots, costs approximately £480/kg while the synthetic compound only costs £60/kg. By combining CSIA via gas chromatography (GC)-IRMS with hyphenated techniques such as enantioselective capillary GC (cCG) and multidimensional cGC it is possible to use stable isotope data for authenticity control of flavours, fragrances and essential oils (Juchelka *et al.*, 1998; Mosandl, 1995, 1997; Mosandl *et al.*, 1995; Nitz *et al.*, 1991; Nitz, Weinreich and Drawert, 1992). The analytical power of isotope analysis for discriminating between natural and nature-identical hemisynthetic or fully synthetic flavours and fragrances has resulted in their inclusion as part of the authentication process for a product to gain Natural Certification in the European Union and/or United States. To maintain EU Natural Certified status as defined in European Communities (88/388/EEC), Article 9 (2), each new batch of a natural product has to undergo this analysis. The corresponding US Natural Certificate status is defined in the US Federal Food, Drug, and Cosmetic Act, and the US Code of Federal Regulations, Title 21, Sections 101.22(a) (3) and 101.22(i) (4).

Other fascinating examples of the power of multivariate analysis including bulk-specific isotope analysis (BSIA) and CSIA are provenancing the geographic origin of premium long-grain rice (Kelly *et al.*, 2002), analysing vegetables and grains with the aim of distinguishing between conventional and organic methods of production (Schmidt *et al.*, 2005), determining the geographic authenticity of Emmental cheese (Pillonel *et al.*, 2003; Pillonel *et al.*, 2005), revealing the environmental and geographic history of fish based on the ^{2}H isotopic signature of fish muscle and otolith microchemistry (Whitledge, Johnson and Martinez, 2006), and determining the difference between farmed and wild caught fish such as salmon (Aursand, Mabon and Martin, 2000), sea bass (Bell *et al.*, 2007) and sea bream (Morrison *et al.*, 2007). Further information on the increasing use of natural abundance isotope variation and elemental concentrations as geographic 'tracers' to determine the provenance of food, which reflects consumers' increasing concern about the origin of the foods they eat, can be found in a review article published in 2005 (Kelly, Heaton and Hoogewerff, 2005).

I.6.2 Counterfeit Pharmaceuticals

The circulation of counterfeit pharmaceuticals is a problem affecting consumers worldwide and is a problem that is thought to be escalating. Counterfeit drugs range from products with the correct ingredients but not manufactured to the standards of the pharmaceutical industry such as Good Manufacturing Practice (GMP), products with the wrong ingredients or not declared ingredients, products containing insufficient amounts of the active pharmaceutical ingredient (API) and products containing no API whatsoever. The World Health Organization (WHO) estimates that counterfeit medicines currently account for approximately 10% of the global pharmaceutical drug market with a cost to the pharmaceutical industry of US$46 billion per year. The WHO expects this figure to increase to US$75 billion by 2010 (http://www.who.int/medicines/services/counterfeit/impact/ImpactF_S/en/). The cost to patients' health resulting from receiving counterfeit pharmaceuticals instead of the genuine product is difficult to quantify but could be devastating. Even though not a case of counterfeit medicine, the case of 1235 Chinese babies suffering ill health and two babies dying from consuming milk powder feed to which melamine had been fraudulently added to increase nitrogen content, thus making the product appear to be rich in protein, is a stark example of the human cost counterfeit products extract from the public.

In their 2007 report, the OECD had the following to say on the subject of counterfeit pharmaceuticals.

> In the case of *pharmaceuticals*, trademark-infringing products may include correct ingredients in incorrect quantities or may be composed according to a wrong formula. Products can furthermore contain non-active or even toxic ingredients. Ailments which could be remedied by genuine products may go untreated or worsen; in some cases this may lead to death. Most purchasers of counterfeit pharmaceuticals are likely to be completely unaware that they have been victimised.

In other words, counterfeit pharmaceuticals cannot only be substandard, they can even be dangerous, posing health and safety risks to consumers that range from mild to life threatening.

In light of the potential, yet unquantifiable human cost and the quantifiable economical cost to the pharmaceutical industry it is rather surprising a finding that published articles on the subject of detecting and combating counterfeit pharmaceuticals are few and far between. However, all these publications agree on multivariate analysis including stable isotope analysis of one or more isotopes to be the method of choice (Jasper *et al.*, 2004; Santamaria-Fernandez, Hearn and Wolff, 2008; Wokovich *et al.*, 2005).

Detection of counterfeit pharmaceuticals and authentication of genuine products are two sides of the same coin. Product authentication seems a logical and integral part of combating fraud and counterfeit goods (*cf.* Figures I.11 and I.12). It would also seem the logical thing to do for manufacturers of genuine products to protect themselves from unjustified claims for damages caused by taking a counterfeit medicine that may match the genuine medicine in terms of ingredients including the API but may have been manufactured to lesser standards and, hence, may be contaminated. Given that both the European Union and the United States have already introduced Natural Certification based on isotope analysis to guarantee authenticity of natural food and cosmetics ingredients, one should imagine the pharmaceutical industry will soon adopt a similar approach or scheme.

I.6.3 Environmental Forensics

Environmental forensics is another example of stable isotope forensics having a tangible impact on everyday life and public health by helping to enforce legislation such as the European Environmental Liability Directive 2004 (2004/35/EC) – the first EC

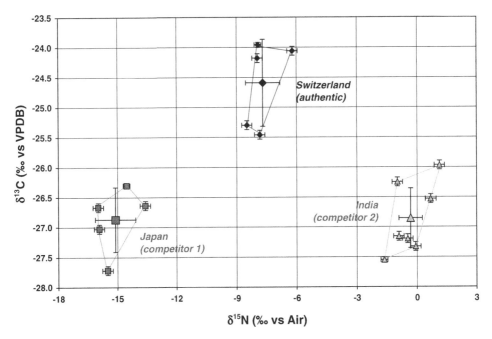

Figure I.11 Isotopic bivariate plot of $\delta^{13}C$ and $\delta^{15}N$ values of the API folic acid from three different manufacturers at three different locations; mean $\pm 1\sigma$ of each cluster is shown in the middle of the cluster. © 2007 J.P. Jasper, Nature's Fingerprint™/Molecular Isotope Technologies, LLC, Niantic, CT 06357-1815, USA, reproduced with permission.

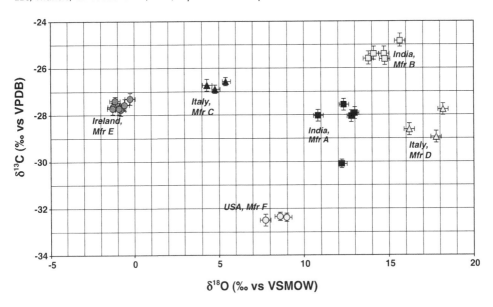

Figure I.12 Isotopic bivariate plot of $\delta^{13}C$ and $\delta^{18}O$ values of the API naproxen from six different manufacturers at four different locations. © 2007 J.P. Jasper, Nature's Fingerprint™/ Molecular Isotope Technologies, LLC, Niantic, CT 06357-1815, USA, reproduced with permission.

legislation to include the 'polluter pays' principle. Of similar importance in this respect is its forerunner, the European Water Framework Directive (2000/60/EC), that came into force in December 2000 and became part of UK law in December 2003 (http://www.defra.gov.uk/environment/quality/water/waterquality/index.htm).

Many of the methods employed today towards elucidating chronology of an environmental pollution event, its point of origin and ultimately the polluter owe their origin to the desire of many a geochemist (and archaeologist) to extract as much information as possible contained in fossil biomarkers, such as sedimentary long-chain fatty alcohols and sterols (Jones et al., 1991; Mudge, Belanger and Nielsen, 2008), triterpene-derived hydrocarbons (Hauke et al., 1992), neutral monosaccharides (Moers et al., 1993), long-chain alkanes (Bakel, Ostrom and Ostrom, 1994; Bjoroy, Hall and Moe, 1994; Ficken, Barber and Eglinton, 1998; Huang et al., 1996; Ishiwatari, Uzaki and Yamada, 1994), alkanes, polycyclic aromatic hydrocarbons (PAHs) and isoprenoids (Lichtfouse et al., 1997; Lichtfouse, 2000; Wilhelms, Larter and Hall, 1994). The common denominator for this explosive advancement in answering geochemical questions was the commercial availability of CF-IRMS systems coupled to a gas chromatograph via an on-line combustion interface (GC/C-IRMS) thus facilitating ^{13}C (or ^{15}N)-CSIA of individual compounds contained in a complex sample mixture.

Monitoring atmospheric gases such as CH_4, CO and CO_2 linked to environmental and climate changes and differentiating between natural (e.g. bacterial) and anthropogenic sources by measuring δ^{13}C values of these gases had traditionally been carried out using dual-inlet IRMS systems. These measurements required time-consuming sample preparation steps involving large gas sample volumes. The high abundance sensitivity of GC/C-IRMS in conjunction with low volumetric flow rates opened up the use of PLOT (Porous-Layer Open-Tubular) fused silica capillary columns for routine GC analysis of highly volatile organic compounds (HVOCs) and permanent gases. ^{13}C-CSIA of HVOCs and permanent gases has now become the method of choice for scientists as air samples between 50 µl and 5 ml can be analysed on-line without any prior sample preparation (Archbold et al., 2005; Baylis, Hall and Jumeau, 1994; Clayton et al., 1997; Davis et al., 2005; Harper et al., 2001, 2003; Kalin et al., 2001; Merritt, Hayes and Marias, 1995; Waldron et al., 1998; Zeng et al., 1994).

High-precision CSIA by GC/C-IRMS is now a standard analytical tool used in conjunction with traditional techniques such as GC-flame ionization detection and GC-MS to determine point of origin and identify sources of oil spills and oil pollution including release of petrol and diesel fuels (Boyd et al., 2006; Davis et al., 2005; Hough et al., 2006; Mansuy, Philp and Allen, 1997; Oudijk, 2005; Philp, 2007; Philp, Allen and Kuder, 2002; Uzaki, Yamada and Ishiwatari, 1993), ocean transported bitumen (Dowling et al., 1995), characterization of refractory wastes at heavy-oil contaminated sites (Whittaker et al., 1996; Whittaker, Pollard and Fallick, 1995) and to trace the sources of PAHs in the environment (McRae et al., 1996; Murphy and Brown, 2005; Pies, Ternes and Hofmann, 2008; Saber, Mauro and Sirivedhin, 2005, 2006; Sun et al., 2003).

There is growing evidence that a chemical hailed as the saviour of the environment from lead pollution through car exhaust emissions of anti-knocking fuel additives will only create a new environmental problem instead. This chemical is methyl *tert*-butyl ether (MTBE); similar to benzene, it is quite resilient to weathering and biotic breakdown, be it in soil or in groundwater. The only remediation treatment that seems to work relatively efficiently is abiotic breakdown by a strong oxidizing agent such as potassium permanganate ($KMnO_4$). Not surprisingly therefore MTBE has become the focus of environmental forensic research and environmental forensics application exploiting its ^{13}C signature and ^{13}C isotopic fractionation associated with its degradation to determine 'Earliest Demonstrable Inception Date' and 'Latest Possible Initiation Date' when investigating groundwater plumes and environmental contamination resulting from accidental or deliberate fuel spills (Elsner *et al.*, 2007; Oudijk, 2005, 2008; Schmidt, 2003; Schmidt *et al.*, 2004; Smallwood *et al.*, 2001; Smallwood, Philp and Allen, 2002).

Applying stable isotope techniques to a field of analytical science deals with detection, identification, spatial distribution, ageing and natural attenuation as well as degradation of pollutants or xenobiotics in a complex environment is of course not always straightforward and can be fraught with difficulties. As already mentioned in previous chapters, considering potential pitfalls and contextual interpretation of data are two of the most essential guiding principles when undertaking such work (Blessing, Jochmann and Schmidt, 2008; Ehleringer *et al.*, 2008b; Schmidt *et al.*, 2004).

I.6.4 Wildlife Forensics

Stable isotope wildlife forensics (i.e. tracing the origin of animals) can be considered the flip-side of the stable isotope ecology coin whose aim is, amongst others, to elucidate food webs and migration patterns of wild animals (Fry, 2006; Hobson and Wassenaar, 1999; Thompson *et al.*, 2005; West *et al.*, 2006). Biological sciences have benefited greatly from applied stable isotope techniques, and the study of the dietary and migration patterns of elephants by Cerling *et al.* (2006) is a good example. This project used sequential (i.e. growth rate-related) time-resolved ^{13}C and ^{15}N isotope analysis of elephants' tail hair to generate a chronological history of the elephants' eating habits and even their feeding locations. Averaged $\delta^{13}C$ and $\delta^{15}N$ values from all elephants in the study showed gradual changes from season to season. The results also showed that isotopic make-up of an individual elephant's hair could be significantly different from the control group if the elephant in question was involved in night-time crop raiding. Obviously determining such differences is only possible if the crop's isotopic composition is significantly different from the control group's staple diet.

Staying with the example of elephants, carbon, nitrogen and strontium isotope compositions of elephants in Amboseli Park, Kenya, were measured using a multivariate isotope analytical approach to examine changes in diet and habitat use since the 1960s (Koch *et al.*, 1995). Obviously a tool enabling scientists to determine changes in diet and feeding locations based on measurable and quantifiable parameters rather than based on observational and anecdotal evidence alone can also be used to determine if ivory comes from controlled and regulated sources or from poached animals. Similarly, stable isotope techniques applied to unravel diet change and migrationary patterns of wild birds (Kelly, 2006; Kelly, Ruegg and Smith, 2005) can be employed to determine if a bird sold as bred in captivity came indeed from a controlled and licensed breeder was illegally caught in the wild and illegally imported (Bearhop *et al.*, 2003; Bowen, Wassenaar and Hobson, 2005; Braune, Hobson and Malone, 2005; Fox and Bearhop, 2008; Hobson *et al.*, 2004; Hobson, 2005; Kelly, Thompson and Newton, 2009; Wassenaar and Hobson, 2000, 2006).

I.6.5 Anti-Doping Control

This area of applied stable isotope techniques does not really come under the forensic science umbrella since in most countries taking performance-enhancing drugs is not a criminal offence. Furthermore, in challenges of failed doping tests the burden of proof rests with the athlete. In a reversal of the normal process of law an athlete who has failed a drug test is presumed guilty until proven innocent. One could argue it would be desirable for this situation to change even if would be just for the reason to ensure that laboratories engaged in traditional and stable isotope analytical techniques to detect and provide evidence of a doping offence have to conform to the level of standard that forensic science service providers and their work have to comply with when presenting forensic evidence in a court of law. We will revisit this matter in the Appendix III.A of Part III of this book where we will apply the knowledge we have gained and consider what lessons should be learned for real stable isotope forensics from hopefully unintended side-effects of ill-advised, ill-considered and ill-applied stable isotope techniques.

However much justifiable criticism one could voice concerning the methods employed by anti-doping agencies or practices in use in some anti-doping laboratories, one has to face the undeniable fact that the use of performance-enhancing drugs in most sports has been a problem for decades and continues to be so. As methods of cheating in competition and gaining an unfair advantage over honest, hard-working athletes who devote their lives to their chosen career and to representing their country in sporting events have become increasingly sophisticated, analytical techniques employed to detect doping offences must similarly develop. Since the use of synthetic steroids became easier to detect due to increasing sensitivity of analytical instruments such as desktop quadrupole and time-of-flight mass spectrometers, use of performance-enhancing substances moved on to natural or nature-identical steroids and hormones. Given what we have learned thus far from the preceding chapters it will come as no surprise that ^{13}C-CSIA was seen as the solution to the task of detecting natural or nature-identical steroid or hormone abuse (Abramson, Osborn and Teffera, 1996; Ueki, 1998, 2001; Ueki and Okano, 1999).

It was not until the 1998 Winter Olympic Games in Nagano and the 2000 Olympic Games in Sydney that stable isotope analysis as a method to establish the presence of exogenous testosterone was mentioned in anti-doping rules of sport and in fact applied. However, the suggestion that ^{13}C-CSIA using GC/C-IRMS could be used to detect doping with steroids goes back to 1994 (Becchi *et al.*, 1994), and names such as Aguilera, Becchi, Catlin, Flenker, Shackleton and Schänzer are firmly linked with research and development into doping control by means of ^{13}C-CSIA and, equally as important, with population studies to determine confidence limits for δ^{13}C values of steroids and steroid metabolites reflecting intra- and inter-individual natural variability (Aguilera *et al.*, 1996a, 1996b, 1997, 1999, 2001; Aguilera, Chapman and Catlin, 2000; Bourgogne *et al.*, 2000; Buisson *et al.*, 2005; Flenker *et al.*, 2007; Hebestreit *et al.*, 2006). In the meantime advances in CSIA analytical techniques enabled Schänzer's group in Cologne to investigate the potential of improving on ^{13}C-CSIA of testosterone metabolites by measuring differences in ^{2}H isotopic composition of testosterone metabolites caused by pathway-specific isotope fractionation, particularly during the hydrogenation of testosterone to 5β-androstanediol and 5α-androstanediol. First reported differences between $δ^2$H values for testosterone metabolites [5β-androstanediol – 5α-androstanediol] and the subsequent metabolite pair [androsterone – etiocholanolone] were 12 and 21‰, respectively (Piper *et al.*, 2009).

Chapter I.7
Summary of Part I

The sheer number of applications of stable isotope analytical techniques presented in the preceding chapters as well as their wide spectrum clearly demonstrates that this technique is a very powerful tool, providing quantitative and qualitative information that cannot be obtained by any other means. Government agencies such as the British FSA whose remit is authenticity of food and food ingredients, BEVABS at the European Commission JRC's IHCP that aims to ensure correct implementation of EU wine quality legislation and was set up to combat major frauds in this area or sports bodies such as the World Anti-Doping Agency have already declared stable isotope analysis a method of choice and an indispensible tool in combating counterfeit or mislabelled food stuffs or combating doping in sports.

However, no matter what the application, one underlying principle of stable isotope forensics is already emerging. Similar to the tenet that meaningful forensic interpretation of evidence and analytical data has to be contextual and should be supported by corroborating information obtained from independent techniques, stable isotope forensic examination of evidence should also be based on a multivariate approach examining information based on independent variables. This may already be achieved through multi-isotope profiling where single isotopic signatures are independent variables, which when combined will work like a multitumbler combination lock with only one solution in X^Y of possible combinations with X being the number of positions per tumbler and Y being the number of tumblers. However, given what we have learned about the drivers that account for the differences in isotopic composition of two otherwise chemically indistinguishable compounds it is clear that even multivariate stable isotope profiles should be combined with analytical data from independent analytical techniques to yield the maximum possible level of discrimination. That being said, based on what we have learned thus far it seems clear that multivariate stable isotope profiles of natural compounds and materials hold the potential to serve as a powerful screening tool and to provide real-time forensic information so as to help directing resources in criminal investigations.

Stable Isotope Forensics: An Introduction to the Forensic Application of Stable Isotope Analysis Wolfram Meier-Augenstein
© 2010 John Wiley & Sons, Ltd

Chapter I.8
Set Problems

(1) Almost all natural chemical elements occur in subtly different forms called isotopes.
True ❏/False ❏.

(2) All isotopes are radioactive and, hence, decay.
True ❏/False ❏.

(3) What are the two main differences between isotopes of the same chemical element?

(4) Natural variations in isotopic abundance of ^{13}C can typically be of the order of:

   a) 1.0 atom% ❏

   b) 0.11 atom% ❏

   c) 0.01 atom% ❏

   d) 0.001 atom% ❏

   e) 0.0006 atom% ❏

(5) Calculate the bond length for ^1H–^{35}Cl and ^2H–^{35}Cl from values for rotational constant B given in Section I.4.1.

(6) Imagine the omnivore twin brother of the vegan James McDoe in the trophic level shift example of Section I.5.4.1 would not have lived in Scotland, but would be an expat living in South Africa. Using Figure I.8 as a reference point, where would you expect the $\delta^{15}N/\delta^{13}C$ signature of his hair to appear in such a bivariate plot?

References Part I

Abramson, F.P., Osborn, B.L. and Teffera, Y. (1996) Isotopic differences in human growth hormone preparations. *Analytical Chemistry*, **68**, 1971–1972.

Aguilera, R., Becchi, M., Casabianca, H., Hatton, C.K., Catlin, D.H., Starcevic, B. and Pope, H.G. (1996a) Improved method of detection of testosterone abuse by gas chromatography/combustion/isotope ratio mass spectrometry analysis of urinary steroids. *Journal of Mass Spectrometry*, **31**, 169–176.

Aguilera, R., Becchi, M., Grenot, C., Casabianca, H. and Hatton, C.K. (1996b) Detection of testosterone misuse: comparison of two chromatographic sample preparation methods for gas chromatographic-combustion/isotope ratio mass spectrometric analysis. *Journal of Chromatography B*, **687**, 43–53.

Aguilera, R., Becchi, M., Mateus, L., Popot, M.A., Bonnaire, Y., Casabianca, H. and Hatton, C.K. (1997) Detection of exogenous hydrocortisone in horse urine by gas chromatography combustion carbon isotope ratio mass spectrometry. *Journal of Chromatography B*, **702**, 85–91.

Aguilera, R., Catlin, D.H., Becchi, M., Phillips, A., Wang, C., Swerdloff, R.S., Pope, H.G. and Hatton, C.K. (1999) Screening urine for exogenous testosterone by isotope ratio mass spectrometric analysis of one pregnanediol and two androstanediols. *Journal of Chromatography B*, **727**, 95–105.

Aguilera, R., Chapman, T.E. and Catlin, D.H. (2000) A rapid screening assay for measuring urinary androsterone and etiocholanolone delta C-13 (parts per thousand) values by gas chromatography/combustion/isotope ratio mass spectrometry. *Rapid Communications in Mass Spectrometry*, **14**, 2294–2299.

Aguilera, R., Chapman, T.E., Starcevic, B., Hatton, C.K. and Catlin, D.H. (2001) Performance characteristics of a carbon isotope ratio method for detecting doping with testosterone based on urine diols: controls and athletes with elevated testosterone/epitestosterone ratios. *Clinical Chemistry*, **47**, 292–300.

Ambler, R.P., Macko, S.A., Sykes, B., Griffiths, J.B., Bada, J. and Eglinton, G. (1999) Documenting the diet in ancient human populations through stable isotope analysis of hair – discussion. *Philosophical Transactions of the Royal Society of London Series B Biological Sciences*, **354**, 75–76.

Ambrose, S.H. and Krigbaum, J. (2003) Bone chemistry and bioarchaeology. *Journal of Anthropological Archaeology*, **22**, 193–199.

Angerosa, F., Camera, L., Cumitini, S., Gleixner, G. and Reniero, F. (1997) Carbon stable isotopes and olive oil adulteration with pomace oil. *Journal of Agricultural and Food Chemistry*, **45**, 3044–3048.

Angerosa, F., Bréas, O., Contento, S., Guillou, C., Reniero, F. and Sada, E. (1999) Application of stable isotope ratio analysis to the characterization of the geographical origin of olive oils. *Journal of Agricultural and Food Chemistry*, **47**, 1013–1017.

Archbold, M.E., Redeker, K.R., Davies, S., Elliot, T. and Kalin, R.M. (2005) A method for carbon stable isotope analysis of methyl halides and chlorofluorocarbons at pptv concentrations. *Rapid Communications in Mass Spectrometry*, **19**, 337–342.

Aursand, M., Mabon, F. and Martin, G.J. (2000) Characterization of farmed and wild salmon (*Salmo salar*) by a combined use of compositional and isotopic analyses. *Journal of the American Oil Chemists' Society*, **77**, 659–666.

Bakel, A.J., Ostrom, P.H. and Ostrom, N.E. (1994) Carbon isotopic analysis of individual n-alkanes – evaluation of accuracy and application to marine particulate organic material. *Organic Geochemistry*, **21**, 595–602.

Balasse, M., Bocherens, H. and Mariotti, A. (1999) Intra-bone variability of collagen and apatite isotopic composition used as evidence of a change of diet. *Journal of Archaeological Science*, **26**, 593–598.

Balasse, M., Tresset, A. and Ambrose, S.H. (2006) Stable isotope evidence (delta C-13, delta O-18) for winter feeding on seaweed by Neolithic sheep of Scotland. *Journal of Zoology*, **270**, 170–176.

Balzer, A., Gleixner, G., Grupe, G., Schmidt, H.L., Schramm, S. and TurbanJust, S. (1997) In vitro decomposition of bone collagen by soil bacteria: the implications for stable isotope analysis in archaeometry. *Archaeometry*, **39**, 415–429.

Baylis, S.A., Hall, K. and Jumeau, E.J. (1994) The analysis of the C1–C5 components of natural gas samples using gas chromatography-combustion-isotope ratio mass spectrometry. *Organic Geochemistry*, **21**, 777–785.

Bearhop, S., Furness, R.W., Hilton, G.M., Votier, S.C. and Waldron, S. (2003) A forensic approach to understanding diet and habitat use from stable isotope analysis of (avian) claw material. *Functional Ecology*, **17**, 270–275.

Becchi, M., Aguilera, R., Farizon, Y., Flament, M.M., Casabianca, H. and James, P. (1994) Gas chromatography combustion isotope ratio mass spectrometry analysis of urinary steroids to detect misuse of testosterone in sport. *Rapid Communications in Mass Spectrometry*, **8**, 304–308.

Bell, J.G., Preston, T., Henderson, R.J., Strachan, F., Bron, J.E., Cooper, K. and Morrison, D.J. (2007) Discrimination of wild and cultured European sea bass (*Dicentrarchus labrax*) using chemical and isotopic analyses. *Journal of Agricultural and Food Chemistry*, **55**, 5934–5941.

Bell, L.S., Cox, G. and Sealy, J. (2001) Determining isotopic life history trajectories using bone density fractionation and stable isotope measurements: a new approach. *American Journal of Physical Anthropology*, **116**, 66–79.

Birchall, J., O'Connell, T.C., Heaton, T.H.E. and Hedges, R.E.M. (2005) Hydrogen isotope ratios in animal body protein reflect trophic level. *Journal of Animal Ecology*, **74**, 877–881.

Bjoroy, M., Hall, P.B. and Moe, R.P. (1994) Stable carbon isotope variation of *n*-alkanes in Central Graben oils. *Organic Geochemistry*, **22**, 355–381.

Blessing, M., Jochmann, M.A. and Schmidt, T.C. (2008) Pitfalls in compound-specific isotope analysis of environmental samples. *Analytical and Bioanalytical Chemistry*, **390**, 591–603.

Bol, R., Marsh, J. and Heaton, T.H.E. (2007) Multiple stable isotope (O-18, C-13, N-15 and S-34) analysis of human hair to identify the recent migrants in a rural community in SW England. *Rapid Communications in Mass Spectrometry*, **21**, 2951–2954.

Bourgogne, E., Herrou, V., Mathurin, J.C., Becchi, M. and de Ceaurriz, J. (2000) Detection of exogenous intake of natural corticosteroids by gas chromatography/combustion/isotope ratio mass spectrometry: application to misuse in sport. *Rapid Communications in Mass Spectrometry*, **14**, 2343–2347.

Bowen, G.J., Wassenaar, L.I. and Hobson, K.A. (2005) Global application of stable hydrogen and oxygen isotopes to wildlife forensics. *Oecologia*, **143**, 337–348.

Boyd, T.J., Osburn, C.L., Johnson, K.J., Birgl, K.B. and Coffin, R.B. (2006) Compound-specific isotope analysis coupled with multivariate statistics to source-apportion hydrocarbon mixtures. *Environmental Science and Technology*, **40**, 1916–1924.

Braune, B.M., Hobson, K.A. and Malone, B.J. (2005) Regional differences in collagen stable isotope and tissue trace element profiles in populations of long-tailed duck breeding in the Canadian Arctic. *Science of the Total Environment*, **346**, 156–168.

Brooks, J.R., Buchmann, N., Phillips, S., Ehleringer, B., Evans, R.D., Lott, M., Martinelli, L.A., Pockman, W.T., Sandquist, D., Sparks, J.P., Sperry, L., Williams, D. and Ehleringer, J.R. (2002) Heavy and light beer: a carbon isotope approach to detect C-4 carbon in beers of different origins, styles, and prices. *Journal of Agricultural and Food Chemistry*, **50**, 6413–6418.

Buisson, C., Hebestreit, M., Weigert, A.P., Heinrich, K., Fry, H., Flenker, U., Banneke, S., Prevost, S., Andre, F., Schaenzer, W., Houghton, E. and Le Bizec, B. (2005) Application of stable carbon isotope analysis to the detection of 17 beta-estradiol administration to cattle. *Journal of Chromatography A*, **1093**, 69–80.

Calderone, G. and Guillou, C. (2008) Analysis of isotopic ratios for the detection of illegal watering of beverages. *Food Chemistry*, **106**, 1399–1405.

Calderone, G., Guillou, C. and Naulet, N. (2003) [Official methods based on stable isotope techniques for analysis of food. Ten years' of European experience]. *L'Actualité Chimique*, (8/9), 22–24.

Calderone, G., Guillou, C., Reniero, F. and Naulet, N. (2007) Helping to authenticate sparkling drinks with C-13/C-12 of CO_2 by gas chromatography-isotope ratio mass spectrometry. *Food Research International*, **40**, 324–331.

Cerling, T.E., Wittemyer, G., Rasmussen, H.B., Vollrath, F., Cerling, C.E., Robinson, T.J. and Douglas-Hamilton, I. (2006) Stable isotopes in elephant hair document migration patterns and diet changes. *Proceedings of the National Academy of Sciences of the United States of America*, **103**, 371–373.

Clayton, C.J., Hay, S.J., Baylis, S.A. and Dipper, B. (1997) Alteration of natural gas during leakage from a North Sea salt diapir field. *Marine Geology*, **137**, 69–80.

Coplen, T.B. (1994) Reporting of stable hydrogen, carbon, and oxygen isotopic abundances. *Pure and Applied Chemistry*, **66**, 273–276.

Coplen, T.B. (1996) More uncertainty than necessary. *Paleoceanography*, **11**, 369–370.

Coplen, T.B., Brand, W.A., Gehre, M., Groning, M., Meijer, H.A.J., Toman, B. and Verkouteren, R.M. (2006a) After two decades a second anchor for the VPDB delta C-13 scale. *Rapid Communications in Mass Spectrometry*, **20**, 3165–3166.

Coplen, T.B., Brand, W.A., Gehre, M., Groning, M., Meijer, H.A.J., Toman, B. and Verkouteren, R.M. (2006b) New guidelines for delta C-13 measurements. *Analytical Chemistry*, **78**, 2439–2441.

Craig, H. (1961) Isotopic variation in meteoric waters. *Science*, **133**, 1702–1703.

D'Angela, D. and Longinelli, A. (1993) Oxygen isotopic composition of fossil mammal bones of Holocene age – paleoclimatological considerations. *Chemical Geology*, **103**, 171–179.

Daux, V., Lecuyer, C., Heran, M.A., Amiot, R., Simon, L., Fourel, F., Martineau, F., Lynnerup, N., Reychler, H. and Escarguel, G. (2008) Oxygen isotope fractionation between human phosphate and water revisited. *Journal of Human Evolution*, **55**, 1138–1147.

Davis, A., Howe, B., Nicholson, A., McCaffery, S. and Hoenke, K.A. (2005) Use of geochemical forensics to determine release eras of petrochemicals to groundwater, Whitehorse, Yukon. *Environmental Forensics*, **6**, 253–271.

DeNiro, M.J. (1985) Postmortem preservation and alteration of *in vivo* bone-collagen isotope ratios in relation to paleodietary reconstruction. *Nature*, **317**, 806–809.

Dennis, M.J., Massey, R.C. and Bigwood, T. (1994) Investigation of the sorbitol content of wines and an assessment of its authenticity using stable-isotope ratio mass-spectrometry. *Analyst*, **119**, 2057–2060.

Dobberstein, R.C., Huppertz, J., von Wurmb-Schwark, N. and Ritz-Timme, S. (2008) Degradation of biomolecules in artificially and naturally aged teeth: implications for age estimation based on aspartic acid racemization and DNA analysis. *Forensic Science International*, **179**, 181–191.

Dowling, L.M., Boreham, C.J., Hope, J.M., Murray, A.P. and Summons, R.E. (1995) Carbon isotopic composition of hydrocarbons in ocean-transported bitumens from the coastline of Australia. *Organic Geochemistry*, **23**, 729–737.

Ehleringer, J.R., Bowen, G.J., Chesson, L.A., West, A.G., Podlesak, D.W. and Cerling, T.E. (2008a) Hydrogen and oxygen isotope ratios in human hair are related to geography. *Proceedings of the National Academy of Sciences of the United States of America*, **105**, 2788–2793.

Ehleringer, J.R., Cerling, T.E., West, J.B., Podlesak, D.W., Chesson, L.A. and Bowen, G.J. (2008b) Spatial considerations of stable isotope analyses in environmental forensics. *Issues in Environmental Science and Technology* **26**, 36–53.

Elsner, M., Mckelvie, J., Couloume, G.L. and Lollar, B.S. (2007) Insight into methyl *tert*-butyl ether (MTBE) stable isotope fractionation from abiotic reference experiments. *Environmental Science and Technology*, **41**, 5693–5700.

Ficken, K.J., Barber, K.E. and Eglinton, G. (1998) Lipid biomarker, delta C-13 and plant macrofossil stratigraphy of a Scottish montane peat bog over the last two millennia. *Organic Geochemistry*, **28**, 217–237.

Flenker, U., Hebestreit, M., Piper, T., Hulsemann, F. and Schanzer, W. (2007) Improved performance and maintenance in gas chromatography/isotope ratio mass spectrometry by precolumn solvent removal. *Analytical Chemistry*, **79**, 4162–4168.

Fox, T. and Bearhop, S. (2008) The use of stable-isotope ratios in ornithology. *British Birds*, **101**, 112–130.

Friedman, I. and O'Neil, J.R. (1977) *Compilation of Stable Isotope Fractionation Factors of Geochemical Interest*, Vol. **440**-KK, US Geological Survey, Reston, VA.

Frost, H.M. (1990) Skeletal structural adaptations to mechanical usage (SATMU). 1. Redefining Wolff's law – the bone modeling problem. *Anatomical Record*, **226**, 403–413.

Fry, B. (2006) *Stable Isotope Ecology*, Springer, New York.

Fuller, B.T., Richards, M.P. and Mays, S. (2001) Stable isotopes from bone and tooth collagen reveal dietary patterns and the age of weaning at the medieval village site of Wharram Percy. *Journal of Bone and Mineral Research*, **16**, 41.

Fuller, B.T., Richards, M.P. and Mays, S.A. (2003) Stable carbon and nitrogen isotope variations in tooth dentine serial sections from Wharram Percy. *Journal of Archaeological Science*, **30**, 1673–1684.

Fuller, B.T., Fuller, J.L., Sage, N.E., Harris, D.A., O'Connell, T.C. and Hedges, R.E.M. (2004) Nitrogen balance and delta N-15: why you're not what you eat during pregnancy. *Rapid Communications in Mass Spectrometry*, **18**, 2889–2896.

Fuller, B.T., Fuller, J.L., Sage, N.E., Harris, D.A., O'Connell, T.C. and Hedges, R.E.M. (2005) Nitrogen balance and delta N-15: why you're not what you eat during nutritional stress. *Rapid Communications in Mass Spectrometry*, **19**, 2497–2506.

Gensler, M. and Schmidt, H.L. (1994) Isolation of the main organic acids from fruit juices and nectars for carbon isotope ratio measurements. *Analytica Chimica Acta*, **299**, 231–237.

Gensler, M., Rossmann, A. and Schmidt, H.L. (1995) Detection of added L-ascorbic acid in fruit juices by isotope ratio mass spectrometry. *Journal of Agricultural and Food Chemistry*, **43**, 2662–2666.

Handley, L.L. and Scrimgeour, C. (1997) Terrestrial plant ecology and ^{15}N natural abundance: the present limits to interpretation for uncultivated systems with original data from a Scottish old field. *Advances in Ecological Research*, **27**, 134–213.

Harbeck, M., Dobberstein, R., Ritz-Timme, S., Schroder, I. and Grupe, G. (2006) Degradation of biomolecules in bones: effect on the biological trace analysis taking the example of stable isotope ratios in collagen. *Anthropologischer Anzeiger*, **64**, 273–282.

Harper, D.B., Kalin, R.M., Hamilton, J.T.G. and Lamb, C. (2001) Carbon isotope ratios for chloromethane of biological origin: potential tool in determining biological emissions. *Environmental Science and Technology*, **35**, 3616–3619.

Harper, D.B., Hamilton, J.T.G., Ducrocq, V., Kennedy, J.T., Downey, A. and Kalin, R.M. (2003) The distinctive isotopic signature of plant-derived chloromethane: possible application in constraining the atmospheric chloromethane budget. *Chemosphere*, **52**, 433–436.

Hauke, V., Graff, R., Wehrung, P., Trendel, J.M., Albrecht, P., Riva, A., Hopfgartner, G., Gulacar, F.O., Buchs, A. and Eakin, P.A. (1992) Novel triterpene-derived hydrocarbons of the arborane/fernane series in sediments. 2. *Geochimica et Cosmochimica Acta*, **56**, 3595–3602.

Hebestreit, M., Flenker, U., Fussholler, G., Geyer, H., Guntner, U., Mareck, U., Piper, T., Thevis, M., Ayotte, C. and Schanzer, W. (2006) Determination of the origin of urinary norandrosterone traces by gas chromatography combustion isotope ratio mass spectrometry. *Analyst*, **131**, 1021–1026.

Hedges, R.E.M. (2002) Bone diagenesis: an overview of processes. *Archaeometry*, **44**, 319–328.

Hedges, R.E.M. and Reynard, L.M. (2007) Nitrogen isotopes and the trophic level of humans in archaeology. *Journal of Archaeological Science*, **34**, 1240–1251.

Hedges, R.E.M., Clement, J.G., Thomas, C.D.L. and O'Connell, T.C. (2007) Collagen turnover in the adult femoral mid-shaft: modeled from anthropogenic radiocarbon tracer measurements. *American Journal of Physical Anthropology*, **133**, 808–816.

Henton, E., Meier-Augenstein, W. and Kemp, H.F. (2009) The use of oxygen isotopes in sheep molars to investigate past herding practices at the Neolithic settlement of Çatalhöyük, Central Anatolia. *Archaeometry*, doi: 10.1111/j.1475-4754.2009.00492.x.

Hobbie, E.A. and Werner, R.A. (2004) Intramolecular, compound-specific, and bulk carbon isotope patterns in C-3 and C-4 plants: a review and synthesis. *New Phytologist*, **161**, 371–385.

Hobson, K.A. (2005) Using stable isotopes to trace long-distance dispersal in birds and other taxa. *Diversity and Distributions*, **11**, 157–164.

Hobson, K.A. and Wassenaar, L.I. (1999) Stable isotope ecology: an introduction. *Oecologia*, **120**, 312–313.

Hobson, K.A., Bowen, G.J., Wassenaar, L.I., Ferrand, Y. and Lormee, H. (2004) Using stable hydrogen and oxygen isotope measurements of feathers to infer geographical origins of migrating European birds. *Oecologia*, **141**, 477–488.

Hoefs, J. (2009) *Stable Isotope Geochemistry*, 6th edn, Springer, Berlin.

Hough, R.L., Whittaker, M., Fallick, A.E., Preston, T., Farmer, J.G. and Pollard, S.J.T. (2006) Identifying source correlation parameters for hydrocarbon wastes using compound-specific isotope analysis. *Environmental Pollution*, **143**, 489–498.

Huang, Y.S., Bol, R., Harkness, D.D., Ineson, P. and Eglinton, G. (1996) Postglacial variations in distributions, C-13 and C-14 contents of aliphatic hydrocarbons and bulk organic matter in 3 types of British acid upland soils. *Organic Geochemistry*, **24**, 273–287.

Iacumin, P., Bocherens, H., Mariotti, A. and Longinelli, A. (1996) An isotopic palaeoenvironmental study of human skeletal remains from the Nile Valley. *Palaeogeography Palaeoclimatology Palaeoecology*, **126**, 15–30.

Ishiwatari, R., Uzaki, M. and Yamada, K. (1994) Carbon isotope composition of individual *n*-alkanes in recent sediments. *Organic Geochemistry*, **21**, 801–808.

Jamin, E., Gonzalez, J., Remaud, G., Naulet, N., Martin, G.G., Weber, D., Rossmann, A. and Schmidt, H.L. (1997) Improved detection of sugar addition to apple juices and concentrates using internal standard C-13 IRMS. *Analytica Chimica Acta*, **347**, 359–368.

Jamin, E., Martin, F., Santamaria-Fernandez, R. and Lees, M. (2005) Detection of exogenous citric acid in fruit juices by stable isotope ratio analysis. *Journal of Agricultural and Food Chemistry*, **53**, 5130–5133.

Jasper, J.P., Westenberger, B.J., Spencer, J.A., Buhse, L.F. and Nasr, M. (2004) Stable isotopic characterization of active pharmaceutical ingredients. *Journal of Pharmaceutical and Biomedical Analysis*, **35**, 21–30.

Jim, S., Ambrose, S.H. and Evershed, R.P. (2004) Stable carbon isotopic evidence for differences in the dietary origin of bone cholesterol, collagen and apatite: implications for their use in palaeodietary reconstruction. *Geochimica et Cosmochimica Acta*, **68**, 61–72.

Jones, A.M., O'Connell, T.C., Young, E.D., Scott, K., Buckingham, C.M., Iacumin, P. and Brasier, M.D. (2001) Biogeochemical data from well preserved 200 ka collagen and skeletal remains. *Earth and Planetary Science Letters*, **193**, 143–149.

Jones, D.M., Carter, J.F., Eglinton, G., Jumeau, E.J. and Fenwick, C.S. (1991) Determination of d^{13}C values of sedimentary straight chain and cyclic alcohols by gas chromatography/isotope ratio mass spectrometry. *Biological Mass Spectrometry*, **20**, 641–646.

Juchelka, D., Beck, T., Hener, U., Dettmar, F. and Mosandl, A. (1998) Multidimensional gas chromatography coupled on-line with isotope ratio mass spectrometry (MDGC-IRMS): progress in the analytical authentication of genuine flavor components. *Journal of High Resolution Chromatography*, **21**, 145–151.

Kalin, R.M., Hamilton, J.T.G., Harper, D.B., Miller, L.G., Lamb, C., Kennedy, J.T., Downey, A., McCauley, S. and Goldstein, A.H. (2001) Continuous flow stable isotope methods for study of delta C-13 fractionation during halomethane production and degradation. *Rapid Communications in Mass Spectrometry*, **15**, 357–363.

Kelly, A., Thompson, R. and Newton, J. (2009) Stable hydrogen isotope analysis as a method to identify illegally trapped songbirds. *Science & Justice*, **48**, 67–70.

Kelly, J.F. (2000) Stable isotopes of carbon and nitrogen in the study of avian and mammalian trophic ecology. *Canadian Journal of Zoology/Revue Canadienne de Zoologie*, **78**, 1–27.

Kelly, J.F. (2006) Stable isotope evidence links breeding geography and migration timing in wood warblers (Parulidae). *Auk*, **123**, 431–437.

Kelly, J.F., Ruegg, K.C. and Smith, T.B. (2005) Combining isotopic and genetic markers to identify breeding origins of migrant birds. *Ecological Applications*, **15**, 1487–1494.

Kelly, S., Parker, I., Sharman, M., Dennis, J. and Goodall, I. (1997) Assessing the authenticity of single seed vegetable oils using fatty acid stable carbon isotope ratios (C-13/C-12). *Food Chemistry*, **59**, 181–186.

Kelly, S., Baxter, M., Chapman, S., Rhodes, C., Dennis, J. and Brereton, P. (2002) The application of isotopic and elemental analysis to determine the geographical origin of premium long grain rice. *European Food Research and Technology*, **214**, 72–78.

Kelly, S., Heaton, K. and Hoogewerff, J. (2005) Tracing the geographical origin of food: the application of multi-element and multi-isotope analysis. *Trends in Food Science and Technology*, **16**, 555–567.

Kelly, S.D. and Rhodes, C. (2002) Emerging techniques in vegetable oil analysis using stable isotope ratio mass spectrometry. *Grasas y Aceites*, **53**, 34–44.

Keppler, F. and Rockmann, T. (2007) Methane, plants and climate change. *Scientific American*, **296**, 52–57.

Keppler, F., Kalin, R.M., Harper, D.B., McRoberts, W.C. and Hamilton, J.T.G. (2004) Carbon isotope anomaly in the major plant C-1 pool and its global biogeochemical implications. *Biogeosciences*, **1**, 123–131.

Keppler, F., Hamilton, J.T.G., Brass, M. and Rockmann, T. (2006) Methane emissions from terrestrial plants under aerobic conditions. *Nature*, **439**, 187–191.

Keppler, F., Harper, D.B., Kalin, R.M., Meier-Augenstein, W., Farmer, N., Davis, S., Schmidt, H.L., Brown, D.M. and Hamilton, J.T.G. (2007) Stable hydrogen isotope ratios of lignin methoxyl groups as a paleoclimate proxy and constraint of the geographical origin of wood. *New Phytologist*, **176**, 600–609.

Keppler, F., Hamilton, J.T.G., McRoberts, W.C., Vigano, I., Brass, M. and Rockmann, T. (2008) Methoxyl groups of plant pectin as a precursor of atmospheric methane: evidence from deuterium labelling studies. *New Phytologist*, **178**, 808–814.

Koch, P.L., Fisher, D.C. and Dettman, D. (1989) Oxygen isotope variation in the tusks of extinct proboscideans – a measure of season of death and seasonality. *Geology*, **17**, 515–519.

Koch, P.L., Heisinger, J., Moss, C., Carlson, R.W., Fogel, M.L. and Behrensmeyer, A.K. (1995) Isotopic tracking of change in diet and habitat use in African elephants. *Science*, **267**, 1340–1343.

Koch, P.L., Tuross, N. and Fogel, M.L. (1997) The effects of sample treatment and diagenesis on the isotopic integrity of carbonate in biogenic hydroxylapatite. *Journal of Archaeological Science*, **24**, 417–429.

Lee-Thorp, J. and Sponheimer, M. (2003) Tooth enamel remains a virtually closed system for stable light isotope and trace element archives in fossils [Abstract]. *American Journal of Physical Anthropology*, **120** (Suppl. 36), 138.

Lee-Thorp, J.A. (2008) On isotopes and old bones. *Archaeometry*, **50**, 925–950.

Lichtfouse, E. (2000) Compound-specific isotope analysis. Application to archaeology, biomedical sciences, biosynthesis, environment, extraterrestrial chemistry, food science, forensic science, humic substances, microbiology, organic geochemistry, soil science and sport. *Rapid Communications in Mass Spectrometry*, **14**, 1337–1344.

Lichtfouse, E., Budzinski, H., Garrigues, P. and Eglinton, T.I. (1997) Ancient polycyclic aromatic hydrocarbons in modern soils: C-13, C-14 and biomarker evidence. *Organic Geochemistry*, **26**, 353–359.

Lock, C.M. and Meier-Augenstein, W. (2008) Investigation of isotopic linkage between precursor and product in the synthesis of a high explosive. *Forensic Science International*, **179**, 157–162.

Longinelli, A. (1984) Oxygen isotopes in mammal bone phosphate – a new tool for paleohydrological and paleoclimatological research. *Geochimica et Cosmochimica Acta*, **48**, 385–390.

Luz, B., Kolodny, Y. and Horowitz, M. (1984) Fractionation of oxygen isotopes between mammalian bone-phosphate and environmental drinking water. *Geochimica et Cosmochimica Acta*, **48**, 1689–1693.

Macko, S.A., Engel, M.H., Andrusevich, V., Lubec, G., O'Connell, T.C. and Hedges, R.E.M. (1999a) Documenting the diet in ancient human populations through stable isotope analysis of hair. *Philosophical Transactions of the Royal Society of London Series B Biological Sciences*, **354**, 65–75.

Macko, S.A., Lubec, G., Teschler-Nicola, M., Andrusevich, V. and Engel, M.H. (1999b) The Ice Man's diet as reflected by the stable nitrogen and carbon isotopic composition of his hair. *FASEB Journal*, **13**, 559–562.

Majoube, M. (1971) Oxygen-18 and deuterium fractionation between water and steam. *Journal de Chimie Physique et de Physico-Chimie Biologique*, **68**, 1423–1436.

Mansuy, L., Philp, R.P. and Allen, J. (1997) Source identification of oil spills based on the isotopic composition of individual components in weathered oil samples. *Environmental Science and Technology*, **31**, 3417–3425.

Martinelli, L.A., Moreira, M.Z., Ometto, J.P.H.B., Alcarde, A.R., Rizzon, L.A., Stange, E. and Ehleringer, J.R. (2003) Stable carbon isotopic composition of the wine and CO_2 bubbles of sparkling wines: detecting C-4 sugar additions. *Journal of Agricultural and Food Chemistry*, **51**, 2625–2631.

Mays, S.A., Richards, M.P. and Fuller, B.T. (2002) Bone stable isotope evidence for infant feeding in mediaeval England. *Antiquity*, **76**, 654–656.

McRae, C., Love, G.D., Murray, I.P., Snape, C.E. and Fallick, A.E. (1996) Potential of gas chromatography isotope ratio mass spectrometry to source polycyclic aromatic hydrocarbon emissions. *Analytical Communications*, **33**, 331–333.

Meier-Augenstein, W. (1999) Applied gas chromatography coupled to isotope ratio mass spectrometry. *Journal of Chromatography A*, **842**, 351–371.

Meier-Augenstein, W. and Fraser, I. (2008) Forensic isotope analysis leads to identification of a mutilated murder victim. *Science & Justice*, **48**, 153–159.

Melander, L. and Saunders, W.H. (1980) *Reaction Rates of Isotopic Molecules*, John Wiley & Sons, Inc., New York

Merritt, D.A., Hayes, J.M. and Marias, D.J.D. (1995) Carbon isotopic analysis of atmospheric methane by isotope-ratio-monitoring gas chromatography-mass spectrometry. *Journal of Geophysical Research – Atmospheres*, **100**, 1317–1326.

Minagawa, M. and Wada, E. (1984) Stepwise enrichment of N-15 along food-chains – further evidence and the relation between delta-N-15 and animal age. *Geochimica et Cosmochimica Acta*, **48**, 1135–1140.

Moers, M.E.C., Jones, D.M., Eakin, P.A., Fallick, A.E., Griffiths, H. and Larter, S.R. (1993) Carbohydrate diagenesis in hypersaline environments – application of GC-IRMS to the stable-isotope analysis of derivatized saccharides from surficial and buried sediments. *Organic Geochemistry*, **20**, 927–933.

Mook, W.G. (ed.) (2000) *Environmental Isotopes in the Hydrological Cycle – Principles and Applications, Volume I: Introduction – Theory, Methods, Review*, International Atomic Energy Agency, Vienna.

Morrison, D.J., Dodson, B., Slater, C. and Preston, T. (2000) C-13 natural abundance in the British diet: implications for C-13 breath tests. *Rapid Communications in Mass Spectrometry*, **14**, 1321–1324.

Morrison, D.J., Preston, T., Bron, J.E., Hemderson, R.J., Cooper, K., Strachan, F. and Bell, J.G. (2007) Authenticating production origin of gilthead sea bream (*Sparus aurata*) by chemical and isotopic fingerprinting. *Lipids*, **42**, 537–545.

Mosandl, A. (1995) Enantioselective capillary gas chromatography and stable isotope ratio mass spectrometry in the authenticity control of flavors and essential oils. *Food Reviews International*, **11**, 597–664.

Mosandl, A. (1997) Progress in the authenticity assessment of wines and spirits. *Analusis*, **25**, M31–M38.

Mosandl, A., Braunsdorf, R., Bruche, G., Dietrich, A., Hener, U., Karl, V., Kopke, T., Kreis, P., Lehmann, D. and Maas, B. (1995) New methods to assess authenticity of natural flavors and essential oils. *ACS Symposium Series*, **596**, 94–112.

Mudge, S.M., Belanger, S.E. and Nielsen, A.M. (2008) *Fatty Alcohols: Anthropogenic and Natural Occurrence in the Environment*, RSC Publishing, Cambridge.

Murphy, B.L. and Brown, J. (2005) Environmental forensics aspects of PAHs from wood treatment with creosote compounds. *Environmental Forensics*, **6**, 151–159.

Nardoto, G.B., Silva, S., Kendall, C., Ehleringer, J.R., Chesson, L.A., Ferraz, E.S.B., Moreira, M.Z., Ometto, J.P.H.B. and Martinelli, L.A. (2006) Geographical patterns of human diet derived from stable-isotope analysis of fingernails. *American Journal of Physical Anthropology*, **131**, 137–146.

Nitz, S., Kollmannsberger, H., Weinreich, B. and Drawert, F. (1991) Enantiomeric distribution and C-13/C-12 isotope ratio determination of gamma-lactones – appropriate methods for the differentiation between natural and nonnatural flavors. *Journal of Chromatography A*, **557**, 187–197.

Nitz, S., Weinreich, B. and Drawert, F. (1992) Multidimensional gas chromatography - isotope ratio mass spectrometry (MDGC-IRMS). A. System description and technical requirements. *Journal of High Resolution Chromatography*, **15**, 387–391.

O'Connell, T.C. and Hedges, R.E.M. (1999) Investigations into the effect of diet on modern human hair isotopic values. *American Journal of Physical Anthropology*, **108**, 409–425.

O'Connell, T.C., Hedges, R.E.M., Healey, M.A. and Simpson, A.H.R. (2001) Isotopic comparison of hair, nail and bone: modern analyses. *Journal of Archaeological Science*, **28**, 1247–1255.

Ogrinc, N., Kosir, I.J., Spangenberg, J.E. and Kidric, J. (2003) The application of NMR and MS methods for detection of adulteration of wine, fruit juices, and olive oil. A review. *Analytical and Bioanalytical Chemistry*, **376**, 424–430.

Oudijk, G. (2005) Fingerprinting and age-dating of gasoline releases – a case study. *Environmental Forensics*, **6**, 91–99.

Oudijk, G. (2008) Compound-specific stable carbon isotope analysis of MTBE in groundwater contamination fingerprinting studies: the use of hydrogeologic principles to assess its validity. *Environmental Forensics*, **9**, 40–54.

Parker, I.G., Kelly, S.D., Sharman, M., Dennis, M.J. and Howie, D. (1998) Investigation into the use of carbon isotope ratios (C-13/C-12) of Scotch whisky congeners to establish brand authenticity using gas chromatography combustion isotope ratio mass spectrometry. *Food Chemistry*, **63**, 423–428.

Pearson, O.M. and Lieberman, D.E. (2004) The aging of Wolff's 'Law': ontogeny and responses to mechanical loading in cortical bone. *Yearbook of Physical Anthropology*, **47**, 63–99.

Petzke, K.J., Boeing, H. and Metges, C.C. (2005) Choice of dietary protein of vegetarians and omnivores is reflected in their hair protein C-13 and N-15 abundance. *Rapid Communications in Mass Spectrometry*, **19**, 1392–1400.

Petzke, K.J., Boeing, H., Klaus, S. and Metges, C.C. (2005) Carbon and nitrogen stable isotopic composition of hair protein and amino acids can be used as biomarkers for animal-derived dietary protein intake in humans. *Journal of Nutrition*, **135**, 1515–1520.

Philp, R.P. (2007) The emergence of stable isotopes in environmental and forensic geochemistry studies: a review. *Environmental Chemistry Letters*, **5**, 57–66.

Philp, R.P., Allen, J. and Kuder, T. (2002) The use of the isotopic composition of individual compounds for correlating spilled oils and refined products in the environment with suspected sources. *Environmental Forensics*, **3**, 341–348.

Pies, C., Ternes, T.A. and Hofmann, T. (2008) Identifying sources of polycyclic aromatic hydrocarbons (PAHs) in soils: distinguishing point and non-point sources using an extended PAH spectrum and *n*-alkanes. *Journal of Soils and Sediments*, **8**, 312–322.

Pillonel, L., Badertscher, R., Froidevaux, P., Haberhauer, G., Holzl, S., Horn, P., Jakob, A., Pfammatter, E., Piantini, U., Rossmann, A., Tabacchi, R. and Bosset, J.O. (2003) Stable isotope ratios, major, trace and radioactive elements in Emmental cheeses of different origins. *Lebensmittel-Wissenschaft und -Technologie*, **36**, 615–623.

Pillonel, L., Badertscher, R., Casey, M., Meyer, J., Rossmann, A., Schlichtherle-Cerny, H., Tabacchi, R. and Bosset, J.O. (2005) Geographic origin of European Emmental cheese: characterisation and descriptive statistics. *International Dairy Journal*, **15**, 547–556.

Piper, T., Thevis, M., Flenker, U. and Schanzer, W. (2009) Determination of the deuterium/hydrogen ratio of endogenous urinary steroids for doping control purposes. *Rapid Communications in Mass Spectrometry*, **23**, 1917–1926.

Privat, K.L., O'Connell, T.C. and Hedges, R.E.M. (2007) The distinction between freshwater- and terrestrial-based diets: methodological concerns and archaeological applications of sulphur stable isotope analysis. *Journal of Archaeological Science*, **34**, 1197–1204.

Reynard, L.M. and Hedges, R.E.M. (2008) Stable hydrogen isotopes of bone collagen in palaeodietary and palaeoenvironmental reconstruction. *Journal of Archaeological Science*, **35**, 1934–1942.

Richards, M.P., Fuller, B.T. and Hedges, R.E.M. (2001) Sulphur isotopic variation in ancient bone collagen from Europe: implications for human palaeodiet, residence mobility, and modern pollutant studies. *Earth and Planetary Science Letters*, **191**, 185–190.

Richards, M.P., Mays, S. and Fuller, B.T. (2002) Stable carbon and nitrogen isotope values of bone and teeth reflect weaning age at the Medieval Wharram Percy site, Yorkshire, UK. *American Journal of Physical Anthropology*, **119**, 205–210.

Richards, M.P., Fuller, B.T., Sponheimer, M., Robinson, T. and Ayliffe, L. (2003) Sulphur isotopes in palaeodietary studies: a review and results from a controlled feeding experiment. *International Journal of Osteoarchaeology*, **13**, 37–45.

Rieley, G. (1994) Derivatization of organic compounds prior to gas chromatographic combustion isotope ratio mass spectrometric analysis – identification of isotope fractionation processes. *Analyst*, **119**, 915–919.

Ritz-Timme, S., Laumeier, I. and Collins, M.J. (2003) Aspartic acid racemization: evidence for marked longevity of elastin in human skin. *British Journal of Dermatology*, **149**, 951–959.

Rossmann, A. (2001) Determination of stable isotope ratios in food analysis. *Food Reviews International*, **17**, 347–381.

Rossmann, A., Schmidt, H.L., Reniero, F., Versini, G., Moussa, I. and Merle, M.H. (1996) Stable carbon isotope content in ethanol of EC data bank wines from Italy, France and Germany. *Zeitschrift fur Lebensmittel-Untersuchung und -Forschung*, **203**, 293–301.

Rossmann, A., Koziet, J., Martin, G.J. and Dennis, M.J. (1997) Determination of the carbon-13 content of sugars and pulp from fruit juices by isotope ratio mass spectrometry (internal reference method) – a European interlaboratory comparison. *Analytica Chimica Acta*, **340**, 21–29.

Royer, A., Gerard, C., Naulet, N., Lees, M. and Martin, G.J. (1999) Stable isotope characterization of olive oils. I – compositional and carbon-13 profiles of fatty acids. *Journal of the American Oil Chemists' Society*, **76**, 357–363.

Saber, D., Mauro, D. and Sirivedhin, T. (2006) Environmental forensics investigation in sediments near a former manufactured gas plant site. *Environmental Forensics*, **7**, 65–75.

Saber, D.L., Mauro, D. and Sirivedhin, T. (2005) Applications of forensic chemistry to environmental work. *Journal of Industrial Microbiology and Biotechnology*, **32**, 665–668.

Santamaria-Fernandez, R. and Hearn, R. (2008) Systematic comparison of delta S-34 measurements by multicollector inductively coupled plasma mass spectrometry and evaluation of full uncertainty budget using two different metrological approaches. *Rapid Communications in Mass Spectrometry*, **22**, 401–408.

Santamaria-Fernandez, R., Hearn, R. and Wolff, J.C. (2008) Detection of counterfeit tablets of an antiviral drug using delta S-34 measurements by MC-ICP-MS and confirmation by LA-MC-ICP-MS and HPLC-MC-ICP-MS. *Journal of Analytical Atomic Spectrometry*, **23**, 1294–1299.

Schmidt, H.L., Butzenlechner, M., Rossmann, A., Schwarz, S., Kexel, H. and Kempe, K. (1993) Intermolecular and intramolecular isotope correlations in organic compounds as a criterion for authenticity identification and origin assignment. *Zeitschrift fur Lebensmittel-Untersuchung und -Forschung*, **196**, 105–110.

Schmidt, H.L., Rossmann, A., Voerkelius, S., Schnitzler, W.H., Georgi, M., Grassmann, J., Zimmermann, G. and Winkler, R. (2005) Isotope characteristics of vegetables and wheat from conventional and organic production. *Isotopes in Environmental and Health Studies*, **41**, 223–228.

Schmidt, T.C. (2003) Analysis of methyl *tert*-butyl ether (MTBE) and *tert*-butyl alcohol (TBA) in ground and surface water. *Trends in Analytical Chemistry*, **22**, 776–784.

Schmidt, T.C., Zwank, L., Elsner, M., Berg, M., Meckenstock, R.U. and Haderlein, S.B. (2004) Compound-specific stable isotope analysis of organic contaminants in natural environments: a critical review of the state of the art, prospects, and future challenges. *Analytical and Bioanalytical Chemistry*, **378**, 283–300.

Schoeninger, M.J. and DeNiro, M.J. (1982) Diagenetic effects on stable isotope ratios in bone apatite and collagen. *American Journal of Physical Anthropology*, **57**, 225.

Schoeninger, M.J. and DeNiro, M.J. (1984) Nitrogen and carbon isotopic composition of bone collagen from marine and terrestrial animals. *Geochimica et Cosmochimica Acta*, **48**, 625–639.

Schoeninger, M.J., DeNiro, M.J. and Tauber, H. (1983) Stable nitrogen isotope ratios of bone collagen reflect marine and terrestrial components of prehistoric human diet. *Science*, **220**, 1381–1383.

Senbayram, M., Dixon, L., Goulding, K.W.T. and Bol, R. (2008) Long-term influence of manure and mineral nitrogen applications on plant and soil ^{15}N and ^{13}C values from the Broadbalk Wheat Experiment. *Rapid Communications in Mass Spectrometry*, **22**, 1735–1740.

Sharp, Z.D. (2007) *Principles of Stable Isotope Geochemistry*, Pearson Prentice Hall, Upper Saddle River, NJ.

Sharp, Z.D., Atudorei, V. and Furrer, H. (2000) The effect of diagenesis on oxygen isotope ratios of biogenic phosphates. *American Journal of Science*, **300**, 222–237.

Shemesh, A., Kolodny, Y. and Luz, B. (1983) Oxygen isotope variations in phosphate of biogenic apatites. 2. Phosphorite rocks. *Earth and Planetary Science Letters*, **64**, 405–416.

Smallwood, B.J., Philp, R.P., Burgoyne, T.W. and Allen, J.D. (2001) The use of stable isotopes to differentiate specific source markers for MTBE. *Environmental Forensics*, **2**, 215–221.

Smallwood, B.J., Philp, R.P. and Allen, J.D. (2002) Stable carbon isotopic composition of gasolines determined by isotope ratio monitoring gas chromatography mass spectrometry. *Organic Geochemistry*, **33**, 149–159.

Spangenberg, J.E. and Ogrinc, N. (2001) Authentication of vegetable oils by bulk and molecular carbon isotope analyses with emphasis on olive oil and pumpkin seed oil. *Journal of Agricultural and Food Chemistry*, **49**, 1534–1540.

Spangenberg, J.E., Macko, S.A. and Hunziker, J. (1998) Characterization of olive oil by carbon isotope analysis of individual fatty acids: Implications for authentication. *Journal of Agricultural and Food Chemistry*, **46**, 4179–4184.

Sun, C.G., Snape, C.E., McRae, C. and Fallick, A.E. (2003) Resolving coal and petroleum-derived polycyclic aromatic hydrocarbons (PAHs) in some contaminated land samples using compound-specific stable carbon isotope ratio measurements in conjunction with molecular fingerprints. *Fuel*, **82**, 2017–2023.

Thompson, D.R., Bury, S.J., Hobson, K.A., Wassenaar, L.I. and Shannon, J.P. (2005) Stable isotopes in ecological studies. *Oecologia*, **144**, 517–519.

Ueki, M. (1998) Doping control in sports and the current procedures for its detection. *Japanese Journal of Toxicology and Environmental Health*, **44**, 75–82.

Ueki, M. (2001) Doping analysis for steroid and differentiation of their origins (Review). *Bunseki Kagaku*, **50**, 287–300.

Ueki, M. and Okano, M. (1999) Analysis of exogenous dehydroepiandrosterone excretion in urine by gas chromatography/combustion/isotope ratio mass spectrometry. *Rapid Communications in Mass Spectrometry*, **13**, 2237–2243.

Uzaki, M., Yamada, K. and Ishiwatari, R. (1993) Carbon-isotope evidence for oil pollution in long-chain normal alkanes in Tokyo Bay sediments. *Geochemical Journal*, **27**, 385–389.

Waldron, S., Watson-Craik, I.A., Hall, A.J. and Fallick, A.E. (1998) The carbon and hydrogen stable isotope composition of bacteriogenic methane: a laboratory study using a landfill inoculum. *Geomicrobiology Journal*, **15**, 157–169.

Wassenaar, L.I. and Hobson, K.A. (2000) Stable-carbon and hydrogen isotope ratios reveal breeding origins of red-winged blackbirds. *Ecological Applications*, **10**, 911–916.

Wassenaar, L.I. and Hobson, K.A. (2006) Stable-hydrogen isotope heterogeneity in keratinous materials: mass spectrometry and migratory wildlife tissue subsampling strategies. *Rapid Communications in Mass Spectrometry*, **20**, 2505–2510.

Weber, D., Gensler, M. and Schmidt, H.L. (1997) Metabolic and isotopic correlations between D-glucose, L-ascorbic acid and L-tartaric acid. *Isotopes in Environmental and Health Studies*, **33**, 151–155.

Weber, D., Kexel, H. and Schmidt, H.L. (1997) C-13 pattern of natural glycerol: origin and practical importance. *Journal of Agricultural and Food Chemistry*, **45**, 2042–2046.

Weber, D., Rossmann, A., Schwarz, S. and Schmidt, H.L. (1997) Correlations of carbon isotope ratios of wine ingredients for the improved detection of adulterations. 1. Organic acids and ethanol. *Zeitschrift fur Lebensmittel-Untersuchung und -Forschung A: Food Research And Technology*, **205**, 158–164.

West, J.B., Bowen, G.J., Cerling, T.E. and Ehleringer, J.R. (2006) Stable isotopes as one of nature's ecological recorders. *Trends in Ecology and Evolution*, **21**, 408–414.

West, J.B., Ehleringer, J.R. and Cerling, T.E. (2007) Geography and vintage predicted by a novel GIS model of wine delta O-18. *Journal of Agricultural and Food Chemistry*, **55**, 7075–7083.

Whittaker, M., Pollard, S.J.T. and Fallick, T.E. (1995) Characterization of refractory wastes at heavy oil-contaminated sites – a review of conventional and novel analytical methods. *Environmental Technology*, **16**, 1009–1033.

Whittaker, M., Pollard, S.J.T., Fallick, A.E. and Preston, T. (1996) Characterisation of refractory wastes at hydrocarbon-contaminated sites. 2. Screening of reference oils by stable carbon isotope fingerprinting. *Environmental Pollution*, **94**, 195–203.

Whitledge, G.W., Johnson, B.M. and Martinez, P.J. (2006) Stable hydrogen isotopic composition of fishes reflects that of their environment. *Canadian Journal of Fisheries and Aquatic Sciences*, **63**, 1746–1751.

Wilhelms, A., Larter, S.R. and Hall, K. (1994) A comparative-study of the stable carbon isotopic composition of crude-oil alkanes and associated crude-oil asphaltene pyrolysate alkanes. *Organic Geochemistry*, **21**, 751–760.

Wilson, A.S., Taylor, T., Ceruti, M.C., Chavez, J.A., Reinhard, J., Grimes, V., Meier-Augenstein, W., Cartmell, L., Stern, B., Richards, M.P., Worobey, M., Barnes, I. and Gilbert, M.T.P. (2007) Stable isotope and DNA evidence for ritual sequences in Inca child sacrifice. *Proceedings of the National Academy of Sciences of the United States of America*, **104**, 16456–16461.

Wokovich, A.M., Spencer, J.A., Westenberger, B.J., Buhse, L.F. and Jasper, J.P. (2005) Stable isotopic composition of the active pharmaceutical ingredient (API) naproxen. *Journal of Pharmaceutical and Biomedical Analysis*, **38**, 781–784.

Woodbury, S.E., Evershed, R.P., Rossell, J.B., Griffith, R.E. and Farnell, P. (1995) Detection of vegetable oil adulteration using gas chromatography combustion isotope ratio mass spectrometry. *Analytical Chemistry*, **67**, 2685–2690.

Woodbury, S.E., Evershed, R.P. and Rossell, J.B. (1998) Purity assessments of major vegetable oils based on delta C-13 values of individual fatty acids. *Journal of the American Oil Chemists' Society*, **75**, 371–379.

Zeng, Y.Q., Mukai, H., Bandow, H. and Nojiri, Y. (1994) Application of gas chromatography combustion isotope ratio mass spectrometry to carbon isotopic analysis of methane and carbon monoxide in environmental samples. *Analytica Chimica Acta*, **289**, 195–204.

Part II
Instrumentation and Analytical Techniques

Chapter II.1
Mass Spectrometry versus Isotope Ratio Mass Spectrometry

For a number of reasons it is not possible to detect in a quantitative manner the differences in isotopic composition at natural abundance levels using a molecular scanning or 'organic' mass spectrometer, typically a quadrupole mass spectrometer equipped with an electron multiplier. These reasons chiefly comprise ion statistics of a standard electron impact source, detector characteristics and detector statistics of an electron multiplier, and the fact that molecular mass spectrometry (MS) uses a single detector and therefore cannot continuously detect any particular isotopologue ion pair (or triplet, etc.) of X^+ and $(X+1)^+$ (and $X+2^+$, etc.) simultaneously the whole time. Even though isotope ratio measurement by means of selected ion monitoring (SIM) mode is possible in principle, it is limited by the fact that statistically 50% of all detectable ions of either mass X^+ or $X+1^+$ are lost to detection and, hence, to the data integration process. For the case of observing X^+, $X+1^+$ and $X+2^+$, theoretically only one-third of all ions for each mass are detected. Using cascading electron multipliers as universal detectors limits the reproducibility of ion current measurements due to the variability of the cascading signal amplification process. For these reasons, under optimized conditions the achievable precision of isotope abundance measurement using molecular MS in SIM mode can only be as good as 0.05 atom%; realistically, it will be of the order of 0.1–0.5 atom% (Table II.1). The resulting limitations on accuracy of isotope abundance measurements in SIM mode therefore only permit the detection of changes in isotopic abundance of the order of 0.5 atom% excess (APE) (Meier-Augenstein, 1999b; Preston and Slater, 1994; Rennie *et al.*, 1996); in the case of ^{13}C (overall natural abundance 1.108 atom%) this means the isotopic abundance of one carbon in every molecule of a given organic compound has to increase either to 1.61 atom% (+0.5 APE) or to decrease to 0.61 atom% (−0.5 APE) with neither value being anywhere near typical natural abundance levels for ^{13}C. In other words, scanning MS cannot provide reliable

Table II.1 Comparison of MS and IRMS system performance when applied to stable isotope analysis at near-natural abundance levels.

	Molecular or organic GC-MS	EA-IRMS	GC/C(TC)-IRMS
Sample introduction	injection of liquid (or gaseous) sample matrix on to GC column	solid sample in tin or silver capsules; liquid sample by direct injection	injection of liquid (or gaseous) sample matrix on to GC column
Sample separation	yes, by GC	no	yes, by GC
Sample manipulation prior to MS analysis	none	combustion/reduction (or pyrolysis) of sample into CO_2 and N_2 (or H_2 and CO) in the elemental analyser	combustion/reduction (or thermal conversion) of compounds into CO_2 and N_2 (or H_2 and CO) in the interface
Interface	heated transfer capillary directly connected to ion source	transfer capillary with open split	capillary, incorporating wide bore combustion/reduction furnaces
Mass analyser	quadrupole	magnet	magnet
Detector	one electron multiplier	multiple FC collector	multiple FC collector
Mode of charged mass detection	SIM, e.g. switching between M^+ and $(M+1)^+$ $((M+1)^+/M^+$ ratios are calculated on the basis of measured ion current intensities)	simultaneous detection of particles with three adjacent masses (e.g. m/z 44, 45 and 46 for $^{12}C^{16}O_2$, $^{13}C^{16}O_2$ and $^{12}C^{18}O^{16}O$, respectively)	simultaneous detection of particles with three adjacent masses (e.g. m/z 28, 29 and 30 for $^{14}N^{14}N$, $^{14}N^{15}N$ and $^{15}N^{15}N$, respectively)
Measurable enrichment range (APE)	+0.02 to +100[a]	−0.7 to +2.0[b]	−0.7 to +2.0[b]
Sample size requirement	≤1 pmol	5.5 nmol to 5 μmol[c]	0.05–5 nmol[d]
CSIA	yes, but only with a precision of 0.05 atom% at best[a]	no[e]	yes, with a precision of 0.0001 atom%

[a]Using a multiple-labelled tracer (e.g. [$^2H_5$]phenylalanine) and measuring $(M+n)^+/M^+$ ratios (e.g. $(M+5)^+/M^+$), enrichments down to 0.2 APE can be reliably detected, while some researchers claim detection limits of +0.02 APE in organic compounds are achievable with modern organic MS.

[b]The range given here is based on the author's own observations and measurements of ^{13}C isotopic composition of natural and tracer compounds using a standard triple collector with a fixed amplifier set-up optimized for measurements at natural abundance level. The quoted upper limit of 2 APE is not a physical but a practical limit constrained by isotope linearity considerations and desirable accuracy and precision.

[c]The required sample size depends on the type of isotope analysis as well as on the type and make of CF-IRMS instrument. Analysis of ^{13}C abundance requires only small amounts of material because of the high abundance of carbon in organic compounds. The opposite is true for analysis of ^{15}N or ^{34}S abundance, due to the low abundance of nitrogen or sulfur in organic compounds. Furthermore, for compounds containing only one atom of nitrogen, 2 mol equiv. of compound have to be combusted to generate 1 mol equiv. of N_2. At the other end of the scale, ^2H and ^{18}O isotope analysis of water on a modern High temperature thermal conversion elemental analysis (TC/EA)-IRMS requires only 0.1 μl or 5.5 nmol of water.

[d]In addition to the considerations mentioned under footnote 'c', it should be noted that depending on relative element abundance in a given organic compound CSIA by GC/C-IRMS requires injection of 0.05–5 nmol per individual compound *on column*.

[e]Using elemental analysis (EA)-IRMS, CSIA is only possible for off-line isolated and purified compounds. This is an updated version of a table originally published by the author (Meier-Augenstein, 1999b); © 1999 Lippincott Williams & Wilkins, reproduced in part with permission.

quantitative information on differences in isotopic composition of organic compounds at natural abundance levels caused by the kinetic or thermodynamic isotope effects mentioned above. For this reason, multicollector isotope ratio mass spectrometers have become the instruments of choice to determine the origin or history of a given organic compound by measuring its characteristic isotope 'signature'.

However, one has to acknowledge that the first attempt to measure isotope ratios of organic compounds by standard single-detector MS came up with the solution to convert organic compounds into simple, permanent gases such as CO_2 to improve on the performance of SIM by jumping from M^+ to $(M+1)^+$ and formed the basis of future continuous flow isotope ratio mass spectrometry (CF-IRMS) systems. Matthews and Hayes (1978) reported for their system that a level of enrichment as low as +0.004 APE could be detected (for definition of APE see Chapter II.2).

If one considers the situation of natural ^{13}C isotope abundance, one is looking at a spectrum ranging from 1.1441 to 1.0342 atom% (+0.033 to −0.077 APE, i.e. a range of 0.1099 atom%) with most organic compounds falling into a range from 1.1099 to 1.0539 atom% (−0.001 to −0.057 APE, i.e. a range of 0.056 atom%), thus being somewhat depleted in ^{13}C compared to the natural abundance value for the international reference material VPDB. For example, ^{13}C abundance in petrochemicals such as benzene or toluene is typically 1.078 atom%, while sugar from sugar beet or sugar cane has a ^{13}C composition of 1.083 or 1.099 atom%, respectively. Reliable detection of differences on such a small scale requires isotope ratio measurement that is accurate to at least 0.001 atom% and precise to 0.0001 atom%. This means even when operated in SIM mode, with an uncertainty of measurement of 0.1 atom% at best, scanning MS cannot provide reliable quantitative information on differences in isotopic composition at natural abundance levels of below 0.2 atom% caused by kinetic or thermodynamic isotope effects (Meier-Augenstein, 1999a; Rieley, 1994). For this reason, IRMS has become the method of choice to determine subtle changes in isotopic composition at natural abundance levels caused by isotopic fractionation during (bio)chemical and physicochemical processes (Brand, 1996; Brenna et al., 1997; Meier-Augenstein, 1999a).

IRMS is based on non-scanning magnetic sector instruments in which mass-filtered ions are stigmatically focused onto dedicated Faraday cup (FC) collectors, positioned along the focal plane of the full deflection radius r (see Equation II.1) in such a way that each ion beam comprised of ions of a particular mass (represented by the mass/charge ratio m/z) falls into its appropriate cup for continuous and simultaneous ion recording of all ions and, hence, complete data integration:

$$r^2 = \frac{2V}{B} \times \frac{m}{z} \qquad (II.1)$$

where r is the deflection radius, V is the accelerating voltage, B is the magnetic field, m is the fragment mass and z is the charge (usually $=1$).

Due to this design IRMS instruments achieve highly accurate and precise measurement of isotopic abundance at the expense of the flexibility of scanning MS. IRMS systems have been specifically designed to measure isotopic composition at low enrichment and natural abundance levels with a high degree of accuracy and precision. To this end, any complex material or compound subject to stable isotope analysis has to be converted into a permanent gas such as CO_2 or N_2 that must be isotopically representative of the parent material. The sample gas is subsequently introduced into the ion source of the IRMS instrument. Inside the IRMS system, molecular CO_2^+ and N_2^+ ions are formed in and emerge from the ion source, are accelerated by a fixed accelerating voltage V and are separated by a magnetic sector set to a fixed magnetic field strength B throughout the experiment. Mass-filtered ions are focused onto dedicated FC detectors positioned specifically for the masses of interest; see Figure II.1. Unlike electron

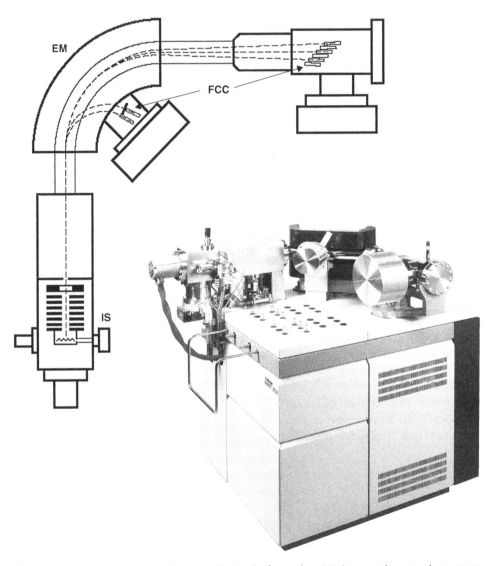

Figure II.1 Schematic (top) and picture (bottom) of a modern IRMS magnetic sector instrument with a multicollector analyser (IS, ion source; EM, electromagnet; FCC, Faraday cup collectors). © 2008 Thermo-Fisher Scientific, reproduced with permission.

multipliers used in scanning organic mass spectrometers where a hit by a fragment ion leads to an electron cascade, FC detectors are actual ion counters. For example, IRMS instruments equipped to determine CO_2 for ^{13}C abundance analysis have three FCs for measurement of masses m/z 44, 45 and 46, positioned so that the ion beam of each mass falls simultaneously into the appropriate cup. The masses m/z 44, 45 and 46 correspond to the CO_2 isotopologues $^{12}C^{16}O_2$, $^{13}C^{16}O_2$ and $^{12}C^{18}O^{16}O$, respectively.

High-precision stable isotope analysis of ^{13}C isotopic abundance at both natural abundance and low enrichment levels can yield measurements of ^{13}C abundance with a typical precision of 0.0001 atom% or 0.1‰. Thanks to this high precision, even minute changes in ^{13}C isotopic abundance of 0.0003 atom% or 0.25‰ can be reliably detected, irrespective of whether these minute changes have been caused by kinetic isotope effects associated with enzyme-mediated biochemical reactions or batch-to-batch variations of reaction conditions during chemical synthesis.

Chapter II.2
Instrumentation and δ Notation

IRMS systems have been designed to measure isotopic composition at low enrichment and natural abundance levels. In an IRMS system, molecular ions emerge from the ion source and are separated by a magnetic sector set to a single field strength throughout the experiment. No energy filtering is used so as to optimize transmission near unity, giving a typical absolute sensitivity of 10^{-3} (about 800–1400 molecules enter the ion source per ion detected) and better for top-end instruments such as the MAT 253 (ThermoFisher Scientific, Bremen, Germany). Mass-filtered ions are stigmatically focused onto dedicated FC detectors positioned specifically for the masses of interest. As mentioned before, IRMS instruments equipped to determine CO_2 have three FCs for measurement of *m/z* 44, 45 and 46, positioned so that the ion beam of each mass falls simultaneously in the appropriate cup. Each FC has a dedicated amplifier mounted on the vacuum housing to minimize noise pick-up, and dedicated counters/recorders for continuous and simultaneous recording of all relevant ion beams and their ratios. Each amplifier has a different resistor setting to account for differences in natural abundance of the various isotopes and, hence, resulting differences in ion currents. While absolute values for resistors used by different manufacturers may vary, the ratio between resistor values in the three amplifiers of a triple FC collector array is set to $1:100:300$ for cup 1 : cup 2 : cup 3, respectively, thus making cup 2 100 times more sensitive than cup 1. In older instruments, resistors of fixed values were hardwired into the printed circuit boards of the amplifiers. Today's state-of-the-art instruments are equipped with electronically controlled resistors whose values can be changed via proprietary software.

FCs are the detectors of choice for IRMS for two major considerations. (i) FCs are highly stable and rugged, and rarely need to be replaced, unlike electron multipliers whose sensitivity degrades with use and age. (ii) The absolute precision required for IRMS determinations is at least 10^{-4}, which is attainable based on counting statistics with at least 10^8 particles detected. Ion currents that achieve these levels are well within the range detectable by FCs. For instance, the ion beam currents in atmospheric CO_2 analysis will typically be around 10 nA, 100 pA and 30–40 pA for *m/z* 44, 45 and 46, respectively.

Owing to the selectively amplified multicollector set-up minute variations in very small amounts of the heavier isotope can be detected in the presence of large amounts of the lighter isotope. The abundance A_S of the heavier isotope n_2 in a sample is defined as:

$$A_S = \frac{R_S}{(1 + R_S)} \times 100 \text{ (atom\%)} \qquad (\text{II.2})$$

where R_S is the ratio n_2/n_1 of the two isotopes for the sample. The enrichment of an isotope in a sample as compared to a standard value (A_{STD}) is given in APE:

$$\text{APE} = A_S - A_{STD} \qquad (\text{II.3})$$

Since changes in isotopic composition at natural abundance levels are invariably very small, measured sample isotope ratios (R_S) are related to the measured isotope ratio of a reference material or standard (R_{STD}) of known isotopic composition and the result of this operation is reported in the δ notation already introduced in Part I (Equation I.1):

$$\delta_S = \left(\frac{[R_S - R_{STD}]}{R_{STD}} \right) \times 1000 \text{ (\textperthousand)} \qquad (\text{I.1})$$

By virtue of Equations I.1 and II.1, δ_S values can be converted into abundance values A_S:

$$A_S = \frac{100}{\frac{1}{\left(\frac{\delta_S}{1000}+1\right) R_{STD}} + 1} \text{ (atom\%)} \qquad (\text{II.4})$$

Apart from its ease of use (i.e. the convenience of having minute absolute differences in isotopic composition expanded into much larger numbers typically falling in the -100 to $+100$‰ range), there is another reason why the δ notation is a useful and appropriate way to express results of isotope analysis at natural abundance levels.

Mass spectrometric measurements are more precise when monitoring a ratio of ion currents or ion counts for two different mass ion beams, rather than just monitoring individual absolute ion beams. Influences detrimental to measurement precision such as instrument noise and fluctuating ion source and analyser conditions will affect all ion beams, and will therefore cancel out when ion beam or ion count ratios are monitored, and this is indeed what happens in an isotope ratio mass spectrometer.

As given by its definition in Equation I.1, the δ value of a sample is calculated from two separate ratios (R_S and R_{STD}) and a ratio-of-ratios (R_S/R_{STD}). Relating the measured heavy/light isotope ratio of the sample to that of an internationally recognized calibration material or international reference material normalizes or anchors measured isotope ratios to a defined reference point so results are reported on a standardized scale and can be compared between different laboratories and instruments from different manufacturers. Another benefit of using the ratio-of-ratios approach based on concurrently measured heavy/light isotope ratios of a reference material is its way of normalizing both the heavy and light isotope (fractions) in the sample to the respective contents of the reference material irrespective of individual IRMS experimental conditions. A final remark on terminology as used and promoted by the IAEA for standard materials – a calibration material is automatically also reference material, but a reference material is not a calibration material. For example, VSMOW is a calibration material because it anchors the ^{2}H isotope scale; while SLAP, although used as a second end-member for scale correction (*cf*. Section II.6.1), is a reference material but not a calibration material (see Chapter I.2 for scale definitions).

II.2.1 Dual-Inlet Isotope Ratio Mass Spectrometry

Until the commercial availability of CF-IRMS systems enabling on-line isotope analysis of organic compounds in the late 1980s, initial attempts to exploit the information locked into stable isotope ratios at natural abundance levels for forensic purposes were confined to employing dual-inlet IRMS systems. Typically, off-line preparation involves multiple steps on custom-designed vacuum lines equipped with high-vacuum and sample-compression pumps, concentrators using cryogenic or chemical traps, reactions in furnaces using catalysts or true reagents, and micro-distillation steps. Contamination and isotopic fractionation are a constant threat at any step, and manual off-line methods are generally slow and tedious, usually requiring large sample amounts. The quality of the results depends considerably on operator skill and dedication.

The dual adjustable-volume inlet system facilitates sample/standard comparison under almost nearly identical circumstances. The highest precision is necessary to compensate for (i) normal fluctuations in instrument response and (ii) ion source non-linearity, which produces isotope ratios that depend on the source gas pressure. The latter phenomenon can be understood from consideration of the physical arrangement of an electron impact ion source. The source assembly includes magnets that collimate the electron flow from the filament through the ion box. Once formed, ions are accelerated across magnetic field lines, which induce minor but measurable mass selectivity. As the mean free path of ions depends on the relatively high pressure in the closed ion source, the mean position from which ions escape the ion box depends on pressure as well as on mass. This translates into a subtle dependence of isotope ratio on pressure, which becomes apparent in the precision of these measurements. The severity of this effect in any particular source defines the relative importance of sample/standard pressure matching. CF-IRMS instruments have only become available because modern ion sources have been designed for maximum linearity, and because the helium carrier gas flow maintains the ion source pressure nearly constant and independent of the level of sample gas flowing into the source.

II.2.2 Continuous Flow Isotope Ratio Mass Spectrometry

In the IRMS inlet system referred to as 'continuous flow', a helium carrier gas passes continuously into the ion source and sweeps bands (or peaks) of analyte gas into the source for analysis. This approach overcomes the sample size requirements for viscous flow because the helium carrier gas stream maintains viscous conditions independent of sample size. Although originally developed for on-line coupling of a gas chromatograph to an IRMS system, this approach has also been successfully used to connect IRMS systems to a variety of automated sample processing devices such as an elemental analyser.

Since the inception of the first CF interface to couple an elemental analyser to a multicollector IRMS in 1983 (Preston and Owens, 1983), CF-IRMS instruments have become an indispensable tool for stable isotope analysis (*cf*. Figure II.2). Another benefit of the CF set-up is that an isotope reference gas can be introduced directly into the ion

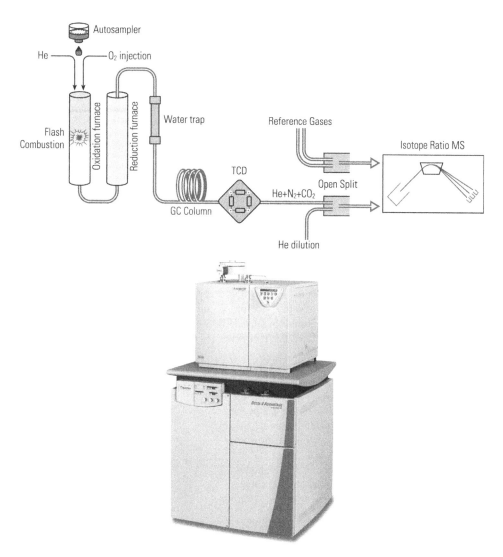

Figure II.2 Schematic (top) and picture (bottom) of a typical EA-IRMS system. © 2008 Thermo-Fisher Scientific, reproduced with permission.

source at virtually any time during an analysis, thus bracketing the sample peak between multiple reference gas peaks of known isotopic composition (Meier-Augenstein, 1997; Werner and Brand, 2001).

Around the same time that Preston and Owens (1983) built the first elemental analysis (EA)-IRMS system, Barrie *et al.* built the first compound-specific IRMS system by coupling a dual-collector mass spectrometer to a gas chromatograph via an on-line combustion interface (C) thus permitting continuous recording of $(M+1)^+/M^+$ isotope ratios (Barrie, Bricout and Koziet, 1984). Their instrument produced isotope ratio values that were an order of magnitude more precise than those obtained from an optimized single-collector GC/C-MS system operated in SIM mode (Matthews and Hayes, 1978) on which their design was based. Key dates in instrument research and development influencing the design and evolution of commercially available CF-IRMS systems are given in Table II.2.

Table II.2 Key dates in instrument research and development influencing design and evolution of commercially available CF-IRMS systems.

Year		Reference
1983	First CF-IRMS system comprising an elemental analyser coupled to an isotope ratio mass spectrometer (EA-IRMS)	Preston and Owens (1983)
1984	First ^{13}C compound-specific CF-IRMS system coupling a gas chromatograph to an isotope ratio mass spectrometer via a combustion interface	Barrie, Bricout and Koziet (1984); Preston and Owens (1985)
1992	Software algorithms for isotopic calibration, background correction and peak deconvolution in CSIA	Tegtmeyer *et al.* (1992); Ricci *et al.* (1994)
1993	First GC(-MS)/C-IRMS hybrid system permitting simultaneous CSIA and compound identification	Meier-Augenstein *et al.* (1994); Meier-Augenstein (1995)
1994	Universal combustion/reduction interface for ^{13}C- and ^{15}N-CSIA of organic compounds by GC/C-IRMS	Merritt *et al.* (1995)
1994	First pyrolysis interface for ^{18}O-CSIA of organic compounds by GC/Py-IRMS	Brand, Tegtmeyer and Hilkert (1994)
1996	First on-line ^{2}H isotope analysis by CF-IRMS of liquid samples using on-line sample reduction and a novel high-dispersion IRMS	Prosser and Scrimgeour (1995); Begley and Scrimgeour (1996)
1996	A reference gas inlet module for sample peak size and shape matched isotopic calibration in CF-IRMS	Meier-Augenstein (1997)
1997	First on-line system for position specific isotope analysis (PSIA)	Corso and Brenna (1997)
2001	On-line ^{2}H and ^{18}O isotope analysis of liquid (including water) and solid samples using high-temperature conversion on glassy carbon coupled to IRMS (TC/EA-IRMS)	Sharp, Atudorei and Durakiewicz (2001)
2004	Artefact and fractionation-free coupling of high performance liquid chromatography (HPLC) to IRMS	Krummen *et al.* (2004)

II.2.3 Bulk Material Stable Isotope Analysis

Modern commercially available elemental analysers provide an automated means for on-line high-precision stable isotope analysis of bulk or single-component materials (bulk stable isotope analysis (BSIA)). Samples are placed in a capsule, typically silver or tin, and loaded into a carousel for automated analysis. From there, the sample is dropped into a heated reactor that contains an oxidant, such as copper and chromium oxide for carbon or sulfur analysis, where combustion takes place in a helium atmosphere with an excess of oxygen. Combustion products such as N_2, NO_x, CO_2 and water are transported in a carrier gas stream of helium through a reduction furnace for removal of excess oxygen and conversion of nitrous oxides into N_2. A drying tube is used to remove all traces of water in the system. The gas-phase products are separated on a GC column under isothermal conditions and detected non-destructively before introduction to the IRMS system. Precision of measurement is typically 1σ ($\delta^{13}C$) $< 0.15‰$ and 1σ ($\delta^{15}N$) $< 0.3‰$ (Brand, 1996; Hofmann and Brand, 1996).

Oxygen and hydrogen are the two most recent elements for which CF-IRMS data of bulk or individual compounds have been reported. Oxygen-containing samples are converted on-line to CO by pyrolytic reaction with carbon (the 'Unterzaucher reaction') (Brand, Tegtmeyer and Hilkert, 1994). Several other reports subsequently appeared extending this principle to include bulk isotope analysis of hydrogen (Begley and Scrimgeour, 1997), although the most widely sold CF-IRMS system for ^2H and ^{18}O isotope analysis is based on a system operating in the presence of glassy carbon and at high temperatures in excess of $1400\,°C$ to suppress formation of CH_4, thus ensuring quantitative sample conversion of organic samples as well as water into H_2 (Sharp, Atudorei and Durakiewicz, 2001). In the case of nitrogen-containing compounds, CO is separated from N_2 using a GC column packed with a 5 Å molecular sieve. Precision of measurement (1σ) for samples such as water and organic compounds is typically better than $±2$ and $±0.3‰$ for reported δ^2H and $\delta^{18}O$ values, respectively.

Nowadays, thanks to advances in electronics and instrument design, dual-isotope analysis of ^{15}N and ^{13}C is possible from the same sample in one analytical run on a modern combustion/reduction elemental analysis (EA)-IRMS system. Provided good chromatographic separation of the N_2 peak from the CO_2 peak can be achieved, a high-precision magnetic field or accelerating voltage jump from ^{15}N to ^{13}C mode can be performed, resulting in a total analysis time of 7 min per sample. Similarly, dual-isotope analysis for ^2H and ^{18}O from the same sample can be carried out using high-temperature thermal conversion elemental analysis (TC/EA). Here, high-temperature conversion of the sample on glassy carbon at 1400–$1450\,°C$ is used to generate H_2 and CO (see Figure II.3). Both solid and liquid samples, the latter by means of a special liquid injector, can be analysed for ^2H and ^{18}O simultaneously with the total analysis time being as fast as 6 min per sample. In conclusion, EA-IRMS and TC/EA-IRMS would appear to be the method of choice for many forensic applications (drugs, explosives, hair, fingernails, etc.), either as a quickly performed initial measurement helping to focus efforts and resources or as a means to provide additional intelligence on samples for which conventional analytical methods do not yield sufficient information on similarity and provenance.

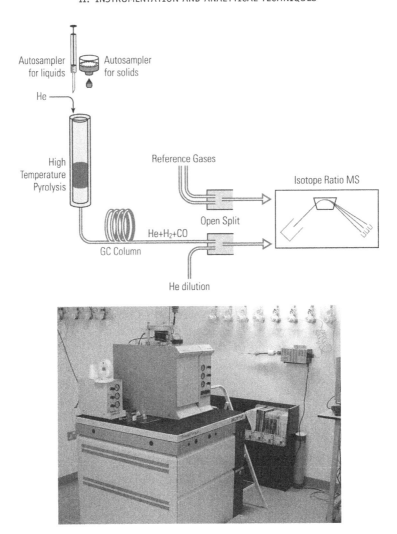

Figure II.3 Schematic (top) and picture (bottom) of a TC/EA-IRMS system. © 2008 Thermo-Fisher Scientific, reproduced with permission.

II.2.4 Compound-Specific Stable Isotope Analysis

Accurate and precise stable isotope analysis depends on careful sample manipulation from initial sampling through to final sample preparation. In addition, in the case of on-line compound-specific isotope analysis (CSIA) of individual constituents in a complex sample matrix via GC/C-IRMS, high-precision CSIA very much depends on high-resolution capillary GC (HRcGC) including multidimensional GC (MDGC) by carefully matching column properties such as stationary phase polarity, film thickness and column dimensions to the separation task at hand (Juchelka *et al.*, 1998; Meier-Augenstein, 1999a, 2002, 2004; Nitz, Weinreich and Drawert, 1992). Demands on sample size, sample derivatization, quality of GC separation, interface design and isotopic calibration have been discussed in a number of reviews and original research articles (Brand, 1996; Brenna, 1994; Ellis and Fincannon, 1998; Meier-Augenstein, 1999b; Metges and Petzke, 1999; Pietzsch et al., 1995).

The need for sample conversion into simple analyte gases has prompted the design of a special interface adapted to the particular sample size and carrier gas flow requirements of cGC. In a set-up for either ^{13}C- or ^{15}N-CSIA, the GC effluent is fed into a combustion reactor (C), either a quartz glass or ceramic tube, filled with oxidised copper/platinum (CuO/Pt) or oxidised copper/nickel/platinum (CuO/NiO/Pt) wires and maintained at a temperature of approximately 820 or 940 °C, respectively (Merritt et al., 1995; Rautenschlein, Habfast and Brand, 1990) (cf. Figure II.4). Users of this technique need to be aware of the influence of combustion tube packing on the analytical performance of GC/C-IRMS (Eakin, Fallick and Gerc, 1992). Similar to the set-up discussed for BSIA above, for conversion of NO_x generated alongside N_2 during the combustion, a reduction tube, filled with copper and held at 600 °C, is positioned behind the combustion reactor. Next in line is a water trap to remove water vapour generated during combustion and most instrument manufacturers employ a Nafion^{TM} tube for this purpose that acts as a semipermeable membrane through which water passes freely while all the other combustion products are retained in the carrier gas stream. Quantitative water removal prior to admitting the combustion gases into the ion source is essential because any water residue would lead to protonation of, for example, CO_2 to produce HCO_2^+ (m/z 45), which interferes with the isotopic analysis of $^{13}CO_2$ (isobaric interference) (Leckrone and Hayes, 1998).

II.2.4.1 CSIA and Compound Identification

Even more important, especially from a forensic point of view, is the following problem. Since organic compounds have to be converted into simple analyte gases, isotopically representative of the parent material, naturally all structural information that could otherwise be used to confirm the identity of the organic compound whose isotopic signature has been measured is lost. One potential 'solution' to overcome this dilemma is to split the sample and analyse another aliquot of the same sample on a scanning GC-MS system employing the same yet not identical chromatographic conditions that were used for CSIA on the GC/C-IRMS system. By 'the same' we mean in this context the same type of GC column (stationary phase; film thickness; column dimensions), same carrier gas management and same temperature programme. Given the variability in system performance due to inter- and intra-batch differences for GC columns and gas chromatographs (even from the same manufacturer) it is nigh impossible to achieve *identical* chromatographic conditions in two separate GC systems. Prior to the advent of MS/IRMS hybrid systems (see below) peak *identification* in GC/C-IRMS relied on peak matching by comparing peak retention parameters such as relative retention or linear retention indices against retention parameters established on another GC system equipped with a mass-selective detector such as a scanning mass spectrometer for compound peak identification.

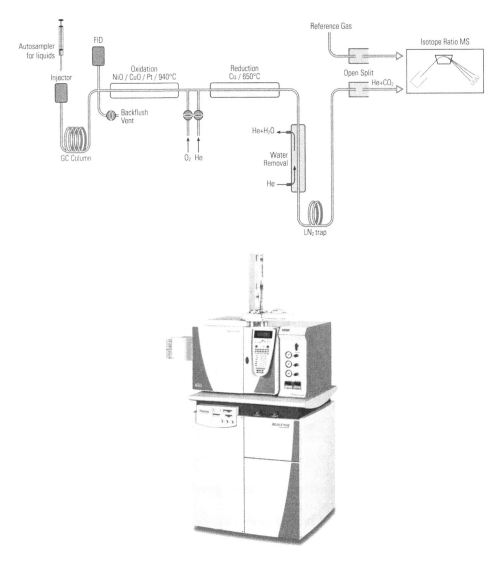

Figure II.4 Schematic (top) and picture (bottom) of a GC/C-IRMS system. © 2008 Thermo-Fisher Scientific, reproduced with permission.

This approach relies ultimately on a mere comparison of gas chromatographic properties of a given organic compound such as its relative retention or its linear retention index – a *modus operandi* that could cause severe problems if such results were put forward as evidence in a court of law, especially if the underlying demands on gas chromatographic conditions to this end were not properly met (*cf*. Appendix III.A in Part III). Compound 'identification' by mapping or matching peaks from chromatograms obtained on separate instruments is fraught with a number of problems. While approved methods such as comparison of relative retention or linear retention indices do exist to compensate for slight differences in GC column length, carrier gas flow or, in the case of linear retention indices, even the temperature programme, these methods yield valid results only if GC columns of the same internal diameter, same film thickness and same stationary-phase polarity are being used in either system. However, even strict observance of these rules does not safeguard against the eventuality that due to natural variability in sample matrix a highly symmetrical and perfectly shaped GC peak can be caused by more than just one compound (just consider the wide range of compounds present in a complex sample matrix such as urine (Meier-Augenstein *et al.*, 1993)).

This problem was recognized by me in 1993 and, in collaboration with Willi Brand, a hyphenated mass spectrometric hybrid system was developed that enables CSIA while at the same time recording a conventional mass spectrum of the peak generated by the presumed organic target compound to aid its unambiguous identification as well as demonstrating peak purity (Meier-Augenstein *et al.*, 1994). To this end, a GC/C-IRMS systems was interfaced with an ion trap mass spectrometer splitting the GC effluent to the conversion interface to facilitate simultaneous admission to the ion source of the organic MS without incurring isotopic fractionation (Meier-Augenstein *et al.*, 1995; Meier-Augenstein, 1995) (*cf.* Figure II.5). This particular set-up, namely the use of an ion trap mass spectrometer, not only enables recording of a standard mass spectrum, but also permits multiple MS characterization (MS/MS or MS^n) of a compound, thus meeting the forensic requirement of MS^2 characterization of two or more parent/daughter fragment ion pairs.

II.2.4.2 Position-Specific Isotope Analysis

If we were to compare IRMS to microscopy, we could say moving from BSIA of, for example, a complex protein such as collagen (which is made up of three individual protein strands that in turn are made up of about 1000 individual amino acids each) to CSIA is like increasing the level of magnification on a microscope by a factor of 1000. This move to CSIA yields much more detailed information since isotopic shifts or isotopic fractionation will affect non-essential amino acids in a different way compared to essential amino acids, to stay with the example of protein. However, isotopic signatures such as the $\delta^{13}C$ value of a particular compound are a composite signal of all the carbons comprising this compound. In other words, the compound-specific $\delta^{13}C$ value is an averaged signal similar to a bulk $\delta^{13}C$ value of a material being an averaged signal of all the carbon comprising that material. Sometimes, not even a compound-specific stable isotope profile will provide the necessary degree of discriminatory power and a more sophisticated approach is required.

Measurement of intra-molecular variations or differences in isotopic abundance due to kinetic isotope effects during biosynthesis enables us to extend the scope of GC-IRMS. In 1997, an on-line pyrolysis system for position-specific isotope analysis (PSIA) of selected compounds from a complex mixture was described in detail for the first time (Corso and Brenna, 1997). Corso and Brenna coupled a GC (GC-1) for sample separation prior to pyrolysis to a second GC (GC-2) for separation of pyrolytic products of the selected sample compound. Furthermore, they installed a valve into GC-2

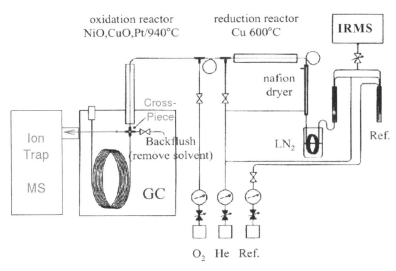

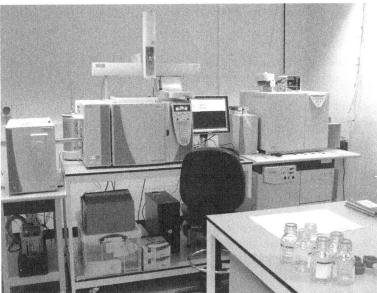

Figure II.5 Schematic (top) and picture (bottom) of a GC(-MS)/C-IRMS hybrid system (showing from left to right: mass spectrometer, gas chromatograph with combustion reactor and autosampler, and the interface unit to the IRMS system; the IRMS system itself is partially obscured by the table in the front right hand corner) based on the author's original design (Meier-Augenstein et al., 1994; Meier-Augenstein, 1995).

to permit separated pyrolysis fragments to be admitted to an organic MS for structure analysis and fragment identification. As we have already seen in Part I (Section I.6.1.2), CSIA is a very powerful technique in food authenticity control where it exploits the fact that biogenetically related compounds are linked by position-specific ^{13}C isotopic shifts in their carbon skeleton. Addition of chemically identical and nature-identical but exogenous flavour enhancers such as citric acid to a premium single-source fruit juice can thus be identified since the exogenous additive will have a different position-specific ^{13}C signature compared to the corresponding expected ^{13}C signature of the exclusively endogenous compound (Weber, Gensler and Schmidt, 1997).

II.2.4.3 CSIA of Polar, Non-Volatile Organic Compounds

Even though GC-IRMS applications account for the vast majority of compound-specific isotope analyses, this does not mean this technique is not without its challenges. The major prerequisite for any type of GC analysis is for the compounds of interest to be sufficiently volatile. The two standard remedies for situations when dealing with compounds of medium or low volatility such as long-chain fatty acids, steroids or amino acids are high-temperature cGC (HTcGC) or derivatization (Meier-Augenstein, 1999a, 2002, 2004). The latter, however, invariably involves the introduction of additional carbon in the form of, for example, methyl or acetyl groups leading to a change in the ^{13}C isotopic composition of the target compound. Due to the petrochemical origin of most derivatization agents this usually means observed or measured δ^{13}C values for the derivatized compound are lower (more negative) than the true δ^{13}C value of the underivatized parent material. In addition, depending on the reaction mechanism associated with derivatization, isotopic fractionation typically affects the ^{13}C signature of the derivatizing agent and, hence, the carbon(s) added during this reaction (Rieley, 1994). This of course makes calculation of the true δ^{13}C value of the parent compound difficult since correction based on a simple mass balance equation does not account for this effect.

The obvious solution is therefore to avoid derivatization altogether and to aim for direct stable isotope analysis of the compounds of interest. While HTcGC offers a feasible alternative to derivatization for compound classes such as long-chain fatty acids (Evershed, 1996; Woodbury, Evershed and Rossell, 1998), long-chain fatty alcohols and steroids, this technique cannot be employed for extremely polar compounds such as amino acids and sugars. However, chromatographic separation of free (i.e. underivatized) amino acids and sugars is easily achievable by high-performance liquid chromatography.

Mirroring the instrumental evolution seen in organic MS where for similar consideration GC-MS instruments were soon followed by mass spectrometer coupled to liquid chromatography systems (LC-MS), research into devising an LC-IRMS system began as early as 1993 (Caimi and Brenna, 1993). Despite several successful application demonstrations of this system to protein, lipids and lipophilic vitamins (Caimi and Brenna, 1995a, 1995b, 1996) it was never adopted by any of the instrument manufacturers. While Caimi and Brenna pursued the moving-wire (or moving-belt) interface for LC-IRMS, an alternative interface in the form of a chemical reaction interface was presented in 1996 (Teffera, Kusmierz and Abramson, 1996). However, it took another 8 years before an LC-IRMS was described that would be adopted by manufacturers and end-users alike (Krummen *et al.*, 2004). Since it is based on chemical oxidation of organic compounds by peroxodisulfate and subsequent removal of the CO_2 formed from the mobile phase via a semipermeable membrane this interface imposes restrictions on mobile-phase composition (only aqueous phases with inorganic pH modifiers can be used) and on the type of isotope analysis (^{13}C only). Despite these restrictions, however, LC-IRMS will no doubt make a major contribution to the compound-specific analysis of free amino acids and amino acids laid down in important structural proteins such as bone collagen (McCullagh, Juchelka and Hedges, 2006).

Chapter II.3
Isotopic Calibration and Quality Control in Continuous Flow Isotope Ratio Mass Spectrometry

Stable isotope data for any light element stable isotope must be reported in the literature on the scale defined by the calibration material for the corresponding isotope (e.g. ^{13}C data are reported on the VPDB scale, whereas ^{2}H data are reported on the VSMOW scale). All modern CF-IRMS systems come equipped with an inlet permitting mass-discrimination-free introduction of a suitable reference gas into the ion source. This approach safeguards against day-to-day variation in IRMS performance, but does not address important issues such as the 'principle of identical treatment' of sample and reference materials for meaningful correction and quality control of analytical results (Bowen *et al.*, 2005; Meier-Augenstein, 1997; Meier-Augenstein, Watt and Langhans, 1996; Werner and Brand, 2001). Since the nature of BSIA by EA-IRMS prevents us from mixing the sample with a standard, good practice demands bracketing a batch of samples by a set of two standards of known, yet different isotopic composition (ideally two international reference materials; see Chapter I.2 for details of CIAAW, NIST and IAEA web sites). In addition, system performance should be cross-checked by including at least one reference material or an in-house standard treated as unknown sample as part of a batch run to act as an acquisition quality control sample. Typically, a batch analysis may comprise four or five actual samples run at least in triplicate with standards (in triplicate) on either side for subsequent end-member correction (*cf*. Section II.3.1). An example of such a quality control procedure is illustrated in Table II.3 showing the scenario for a sample batch sequence comprising a set of samples enveloped by two reference materials. For reasons of statistical analysis samples should be analysed at least in quadruplet ($n = 4$). If one would wish to apply likelihood ratios to come to a conclusion about the likely match between a control and recovered (seized) specimen

---

Stable Isotope Forensics: An Introduction to the Forensic Application of Stable Isotope Analysis   Wolfram Meier-Augenstein
© 2010 John Wiley & Sons, Ltd

Table II.3 Sample batch sequence composition in BSIA favouring high sample throughput under stable experimental conditions using ^{2}H isotope analysis of water as an example.

Sample ID[a]	δ^2H (‰)			δ^2H$_{VSMOW}$ (‰) Accepted
	Measured	Mean	SD	
BLANK				
RM1 (VSMOW)				
RM1 (VSMOW)				
RM1 (VSMOW)				0
RM1 (VSMOW)				
RM1 (VSMOW)				
DUMMY (Tap)				
Samples				
AQC (GISP)				
AQC (GISP)				
AQC (GISP)				−189.73
AQC (GISP)				
AQC (GISP)				
Samples				
DUMMY (Tap)				
RM2 (SLAP)				
RM2 (SLAP)				
RM2 (SLAP)				−428
RM2 (SLAP)				
RM2 (SLAP)				
BLANK				

[a]RM = international reference material (here VSMOW and SLAP) used to anchor (standardize) sample results on the VSMOW scale. AQC = acquisition quality control sample (here GISP) used to control quality of measurement results and standardization.

samples (Evett *et al.*, 2000) then standards, controls or case samples should be analysed in even numbered multiples (i.e. $n = 4$, 6, etc.); clearly, the higher the number of replicates the longer the run time and the slower the turnaround time.

If conditions in the laboratory make instrument drift likely, a sample batch sequence should definitely be preceded and followed by a set of two standards run at least in triplicate each to permit drift correction (i.e. account for a systematic difference in scale adjustment between the start and the end of a sample batch sequence). Finally, two empty tin or silver capsules should be placed at either end of a complete sample set to act as blanks. Raw sample data can thus be blank corrected and calibrated against the reference gas peaks by the proprietary instrument software. Sample δ values thus obtained are then stretch-and-shift corrected to adjust measured isotope data to internationally accepted references scales. We will learn how to do this in the following chapter.

II.3.1 Two-Point or End-member Scale Correction

In Part I of this book we have already learned about how and why the measured isotope ratio value of a given substance is related or referred to the measured isotope ratio value of an international reference material with an accepted (true) isotope abundance δ value. In the following we will learn how measured δ values are adjusted to internationally accepted reference scales using ^{2}H and ^{13}C bulk isotope analysis as examples.

II.3.1.1 Scale Correction of Measured δ^2H Values

For a number of system-inherent reasons the measured difference between two compounds of different isotopic composition tends to be smaller than the actual difference. This effect becomes more pronounced and, hence, more noticeable the larger the difference in isotopic composition typically is. We call this effect the dynamic scale compression and it is at its most severe for the isotope with the lowest natural abundance, namely ^{2}H. For reported δ^2H values to be in line with accepted IAEA guidelines they must (i) be anchored or calibrated against the calibration material VSMOW (δ^2H = 0‰), and (ii) the difference in δ^2H values between the reference materials VSMOW and SLAP *must be* 428‰. Should this difference for measured δ^2H values of VSMOW and SLAP be less than 428‰ then a Z-factor (Coplen, 1988) or stretching factor (Sharp, Atudorei and Durakiewicz, 2001) has to be applied to the data. Stretched or scale-adjusted data require to be shifted by a constant off-set to bring them in line with the VSMOW scale. This constant is related to fractionation occurring at the open split prior to the ion source inlet of the IRMS (Coplen, 1988; Nelson, 2000).

In the example shown in Table II.4, averaged measured δ^2H values for VSMOW and SLAP are +1.95 and −411.75‰. To bring the measured difference of 413.7‰ between VSMOW and SLAP in line with the accepted scale difference of 428‰,

Table II.4 Sample VSMOW – SLAP δ^2H scale correction.

						Mean	1σ
VSMOW$_{measured}$	δ^2H	+1.89	+2.28	+1.87	+1.77	+1.95	0.2
SLAP$_{measured}$	δ^2H	−411.56	−410.88	−411.52	−413.05	−411.75	0.8
GISP$_{measured}$	δ^2H	−180.80	−183.11	−182.13	−181.96	−182.0	0.8
VSMOW$_m$ − SLAP$_m$	$\Delta\delta^2$H$_m$					413.7	
VSMOW$_{accepted}$	δ^2H	0					
SLAP$_{accepted}$	δ^2H	−428					
VSMOW$_{ac}$ − SLAP$_{ac}$	$\Delta\delta^2$H$_a$	428					

			stretch	stretched	shift	adjusted
$\Delta\delta^2$H$_{ac}$/$\Delta\delta^2$H$_m$		428/413.7 =	1.035			
VSMOW$_{stretched}$	δ^2H			2.02		0
SLAP$_{stretched}$	δ^2H			−425.98		−428
VSMOW$_{ac}$ − VSMOW$_s$		0 − 2.02 =			−2.02	
SLAP$_{ac}$ − SLAP$_s$		−428 − (−413.7) =			−2.02	
GISP$_{adjusted}$	δ^2H			−188.4		−190.4

measured δ^2H values have to be stretched by 3.5% (i.e. a stretch factor of 1.035 has to be applied). This value we obtain from the ratio of [VSMOW−SLAP]$_{measured}$/ [VSMOW−SLAP]$_{accepted}$. Scale-adjusting (stretching) the measured δ^2H values for VSMOW and SLAP in this example leads to two new values, which are now used to determine the off-set or shift to bring measured δ^2H values finally in line with the VSMOW reference framework. For both reference materials the difference of [$\delta^2H_{accepted} - \delta^2H_{stretched}$] amounts to −2.02. The full correction equation for this example can now be obtained by combining the two terms we have determined to perform a VSMOW/SLAP scale correction for δ^2H values of samples run as part of a batch sequence (e.g. as illustrated in Table II.4):

$$\delta^2H_{adjusted} = 1.035\ \delta^2H_{measured} - 2.02$$

It is good practice to quality control the performance of the correction equation and to this end one should always include a third compound of known isotopic composition (ideally another international reference material) into a sample batch run to be analysed contemporaneously. Preferably this third reference material should have an isotopic composition that lies numerically halfway between the two end-members used for scale correction. For the δ^2H scale we have such a material meeting both conditions with GISP, whose δ^2H value on the VSMOW scale is defined as −189.73‰. In our example, the scale correction of the measured δ^2H value for GISP yields −190.4‰, which is in good agreement with its accepted δ^2H value, given the typical measurement precision for δ^2H values is ±1–2‰.

II.3.1.2 Scale Correction of Measured $\delta^{13}C$ Values

If one considers ^{13}C to be approximately 70–80 times more abundant in nature than ^2H one could be forgiven for thinking that scale compression may be a minor concern in ^{13}C isotope analysis. While such an assumption would be a recipe for disaster in, for example, geological and geochemical applications where ^{13}C analyses of carbonates on dual-inlet instruments yield a precision of typically 0.01‰, it would still have to be considered unsafe even for continuous flow applications where typically a precision of 0.1‰ can be achieved.

The need for a second anchor for the VPDB scale was addressed in 2006 when LSVEC lithium carbonate was proposed as a lower end anchor for the VPDB scale with an assigned $\delta^{13}C$ value of −46.60‰ opposite NBS-19 calcium carbonate with an assigned $\delta^{13}C$ value of +1.95‰ (Coplen et al., 2006a). Obviously, these are reference materials that are not immediately useful when analysing organic materials and compounds; however, applying the same principle, IAEA-certified organic reference materials such as IAEA-CH6 and IAEA-600 or USGS 40 and USGS 41 (see Table II.5) can be chosen to serve the same purpose.

The figures shown in Table II.6 aptly illustrate the importance of appropriate scale correction even for ^{13}C isotope analysis. The upper part of Table II.6 shows the results of a two-end-member correction where end-members with vastly different $\delta^{13}C_{VPDB}$

Table II.5 Organic ^{13}C reference materials available from the IAEA.

International reference material	Code	$\delta^{13}C_{VPDB}$ (‰)[a]
Sucrose	IAEA-CH-6	−10.449
Polyethylene	IAEA-CH-7	−32.151
Wood	IAEA-C4	−24.0
Wood	IAEA-C5	−25.5
Sucrose	IAEA-C6	−10.8
Oxalic acid	IAEA-C7	−14.5
Oxalic acid	IAEA-C8	−18.3
Caffeine	IAEA-600	−27.771
L-Glutamic acid	USGS 40	−26.389
L-Glutamic acid	USGS 41	+37.626
Cellulose	IAEA-CH-3	−24.724
Wood	IAEA-C9	−23.9

[a] Note the δ^{13}C values given here are the latest values published by the IAEA as of 30 November 2006.
An expanded list of international reference materials is given in Table I.2.

Table II.6 Sample two-end-member VPDB δ^{13}C scale corrections showing the effect on appropriate and inappropriate choice of end-members (shown in italic type).

Sample ID[a]	δ^{13}C measured	δ^{13}C measured mean ± 1σ	δ^{13}C expected (VPDB)[b]	δ^{13}C adjusted mean ± 1σ
Appropriate				
RM IAEA-CH7	*−31.63*			
RM IAEA-CH7	*−31.58*	*−31.70 ± 0.17*	*−31.80*	*−31.80 ± 0.18*
RM IAEA-CH7	*−31.90*			
RM IAEA-CH6	*−10.81*			
RM IAEA-CH6	*−10.73*	*−10.80 ± 0.07*	*−10.40*	*−10.40 ± 0.07*
RM IAEA-CH6	*−10.86*			
AQC1-Leu	−30.40			
AQC1-Leu	−30.51	−30.47 ± 0.06	30.52	−30.53 ± 0.06
AQC1-Leu	−30.49			
Inappropriate				
RM IAEA-CH7	−31.63			
RM IAEA-CH7	−31.58	−31.70 ± 0.17	−31.80	−31.80 ± 0.18
RM IAEA-CH7	−31.90			
RM IAEA-CH6	−10.81			
RM IAEA-CH6	−10.73	−10.80 ± 0.07	−10.40	−10.17 ± 0.07
RM IAEA-CH6	−10.86			
AQC1-Leu	*−30.40*			
AQC1-Leu	*−30.51*	*−30.47 ± 0.06*	*30.52*	*−30.52 ± 0.07*
AQC1-Leu	*−30.49*			

[a] RM = international reference material (here IAEA-CH-7 and IAEA-CH-6) used to anchor (standardize) sample results on the VSMOW scale. AQC = acquisition quality control sample (here leucine) used to control quality of measurement results and standardization

[b] The expected δ^{13}C values for IAEA-CH7 and IAEA-CH6 used here reflect those published by the IAEA prior to 30 November 2006.

values have been chosen, thus bracketing the typical range of sample $\delta^{13}C$ values. To demonstrate how crucial meeting this condition is, the same $\delta^{13}C$ values have been used in the lower part of Table II.6, but this time choosing an end-member pair, IAEA-CH7 and QC1-Leu, whose $\delta^{13}C$ values are far too close to each other. As a result the 'corrected' $\delta^{13}C$ value of the international reference material IAEA-CH6 becomes $-10.17‰$ (i.e. 0.23‰ too positive), with this difference being three times the calculated error of measurement. For a measured $\delta^{13}C$ value of $-1.0‰$ the results between [IAEA-CH7 – IAEA-CH6] and [IAEA-CH7 – QC1-Leu] end-member pair correction would increase from -0.47 to $-0.04‰$ (i.e. a difference of 0.33‰). For a technique where a difference of 0.5‰ is regarded as significant such a scaling error cannot be tolerated.

Chapter II.4
Statistical Analysis of Stable Isotope Data within a Forensic~Context

II.4.1 Chemometric Analysis

Owing to the fact that in looking at a bivariate or trivariate plot comprised of two or three different stable isotope signatures, most of the time one looks at a multivariate plot of independent variables and graphs based on raw data (i.e. observed data can convey already a level of information and discrimination not too dissimilar to that gleaned from graphs derived from chemometric data analysis tools such as multivariate statistical analytical methods). However, merely plotting measured data may not always reveal shared similarities that may exist between samples made from identical source materials but in different batches with batch-to-batch differences due to variation in, for example, reaction temperature not affecting all isotopes to the same degree. Chemometric techniques for analysis are probably the most widely used techniques by analytical chemists due to the inherently multivariate nature of measurements in analytical chemistry (Brereton, 2006).

Principal component analysis (PCA) essentially reduces an array of measured and potentially correlated variables (relative abundance of key organic compounds, trace metal content, ^{13}C isotope abundance, etc.) into a small number of uncorrelated vectors. This method can be very useful when analysing datasets comprised of results from complementary analytical methods with the aim of determining whether general relationships between data exist. In other words, the aim of this particular chemometric method is to determine if similarities exist between m objects characterized by n analytical parameters.

In contrast, multivariate statistical methods such as hierarchical cluster analysis (HCA) and discriminant analysis are used to answer the question if samples fall into groups. For this purpose objects m and measurements n are arranged in a matrix placing objects in rows and measurement in columns, and converted into a similarity matrix such as a correlation matrix or a Euclidean distance matrix. Similarity matrices are $m \times m$ object-related matrices. A correlation coefficient of 1 in a correlation matrix is meant to imply samples have indistinguishable characteristics, whereas in a Euclidean distance matrix the same is implied if the distance has a value of 0. In a subsequent step objects are linked by connecting single objects to each other in groups. This is an iterative process that continues until all objects have joined one large group. The result of HCA is presented in the form of a dendrogram with objects ordered according to their similarities.

When looking at the results from a blind analysis of drug samples that came from a series of seizures by police forces in the United Kingdom and Sweden, a combination of trivariate data plots and HCA yielded information in line with what was known about the exhibits at the time. A trivariate plot of stable isotope signatures of Ecstasy tablets seized in two different European countries already showed some separation of the isotope data (Figure II.6). The three data points shown encircled in Figure II.6 represent samples from three sets of Ecstasy tablets seized from the same person at the same time. Using statistical software package SYSTAT 11 (www.systat.com), HCA was performed employing the Euclidean distance measure and furthest-neighbour distance method (Figure II.7). At the similarity level the resulting dendrogram grouped exhibits case 2, case 3 and case 4; no other significant relationships were determined between the remaining seizures.

Chemometric methods will prove to be useful in cases where stable isotope data form part of a wider multivariate dataset complementary to analytical data such as

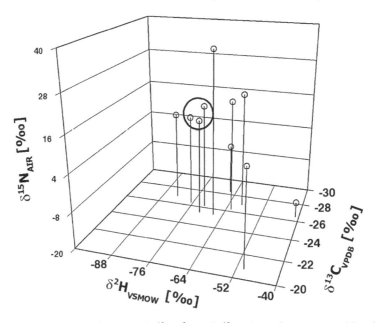

Figure II.6 Trivariate plots of measured $\delta^{13}C$, δ^2H and $\delta^{15}N$ values of 10 Ecstasy tablets from eight different seizures in two different European countries. A circle is drawn around the data points of tablets from the same seizure.

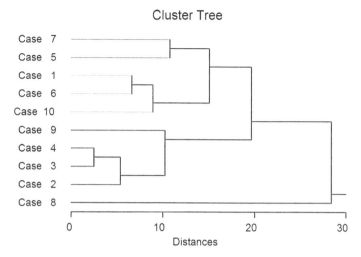

Figure II.7 HCA (furthest neighbour, Euclidean distance) using δ^2H, δ^{13}C, δ^{15}N and δ^{18}O values as well as MDMA content from 10 Ecstasy tablets from eight different seizures in two different European countries.

relative compound abundance, relative content of a particular class of compounds or elemental composition, to name but a few (Mudge, Belanger and Nielsen, 2008). Using such multivariate datasets in conjunction with PCA has been successfully applied to distinguish between wild sea bass and farmed sea bass, and to further distinguish between fish farmed at two different locations (Figure II.8). Here, PCA included

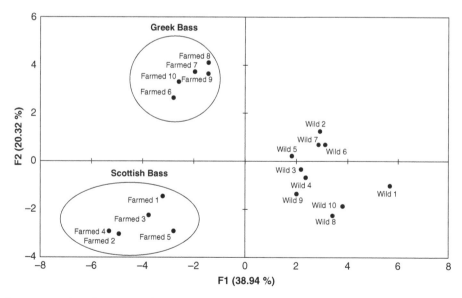

Figure II.8 Plot of PCA score factors for the first two principal components of multivariate data from farmed and wild European sea bass. © 2007 ACS Publications, reproduced with permission from Bell *et al.* (2007).

long-chain fatty acid abundance, $\delta^{13}C$ values of long-chain fatty acids, bulk ^{13}C isotopic composition, nitrogen elemental composition and lipid content (Bell *et al.*, 2007). Environmental forensics is another example of applying multiple analysis of variance (MANOVA), PCA and HCA to the problem of pollutant source differentiation in sites with groundwater contamination (Boyd *et al.*, 2006). Here, compound-specific ^{13}C isotope data of an array of aliphatic hydrocarbons and monocyclic aromatics (benzene, toluene, ethyl-benzene, xylene = BTEX) have been successfully used to determine if hydrocarbons in two samples were statistically different (MANOVA) and to determine how closely related the two samples were (PCA and HCA).

II.4.2 Bayesian Analysis

The Bayesian approach to the interpretation of evidence in forensic science, and in particular the use of likelihood ratios (Aitken *et al.*, 2006; Aitken and Lucy, 2002; Aitken, Zadora and Lucy, 2007), seems to be a very useful strategy to evaluating datasets comprised of stable isotope data alone with a view to testing the hypothesis if two otherwise indistinguishable compounds or materials are related or not. Probably the first application of the likelihood approach to evaluate forensic stable isotope data was reported in 2007 in the context of the forensic examination of Semtex samples (Pierrini *et al.*, 2007). In this particular application three different approaches to analysing data were compared: (i) multivariate likelihood ratio, which allows for between-group variance as well as within-group variance for multivariate normally distributed data, (ii) likelihood ratio using a kernel distribution for the distribution of between-group variability (Aitken and Lucy, 2004), and (iii) likelihood ratio using a multivariate Hotelling's T^2 distance and a kernel density estimate (Champod, Evett and Kuchler, 2001). A recent study of different likelihood ratio approaches to compare 3,4-methylenectioxymethamphetamine (MDMA) tablets compared distance methods where likelihood ratios were based on distance or similarity measures of corresponding properties or characteristics (e.g. Euclidean distance) with distribution methods where likelihood ratios were based on the distribution of the values describing the tablet characteristics (Bolck *et al.*, 2009).

As in the preceding chapter, I would like to use results from a research project that was carried out in the context of an actual case to answer the question if stable isotope fingerprinting could be used to distinguish between samples of white architectural paint that were otherwise indistinguishable (Farmer, Meier-Augenstein and Lucy, 2009). This example should very nicely illustrate the use of likelihood ratios and the power of this approach in conjunction with multivariate stable isotope signatures.

Finding oneself in a situation where one has to start from scratch because something like this had not been done before, one has to recruit or assemble a study cohort sufficiently strong in numbers to represent the entire population of white paints. To this end we first purchased 51 white architectural paints from a number of different retailers in the United Kingdom and the Republic of Ireland (Table II.7).

A range of 8‰ was observed for measured $\delta^{13}C$ values of all 51 paints, with the majority of paints falling within a narrow range for $\delta^{13}C$ −31 to −26‰, while some of

Table II.7 List of 51 white architectural paints from different sources.

Number	Batch	Make	Type	Colour/class
1	100H17929	Dulux Trade	eggshell low odour	pure brilliant white
2	0403H237763	Dulux Trade	eggshell quick drying	pure brilliant white
3	0503H2243264	Dulux	non-drip gloss	pure brilliant white
4	0605HH246829	Dulux Weathershield	exterior high gloss	pure brilliant white
5	06HH0306	Dulux Trade	weathershield, exterior high gloss	pure brilliant white
6	16H0303	Dulux Once	one-coat satinwood	pure brilliant white
7	24H0302557	Dulux	satinwood	pure brilliant white
8	19H0303	Dulux Once	one-coat gloss	pure brilliant white
9	D213796	Crown	non-drip gloss	pure brilliant white
10	D198566	Crown	liquid gloss	pure brilliant white
11	K168293	Crown	quick-dry gloss	pure brilliant white
12	D214287	Crown Solo	one-coat satin	pure brilliant white
13	K167245	Crown	non-drip satin breatheasy	pure brilliant white
14	D214711	Crown Solo	one-coat gloss	pure brilliant white
15	D215505	B&Q Colours	non-drip gloss	white
16	D206145	B&Q Outside	one-coat exterior gloss	pure white
17	D210717	B&Q Colours	liquid gloss	pure white
18	D215010	B&Q	one-coat high gloss	pure white
19	D206288	B&Q	liquid gloss	pure white
20	K15433	B&Q Everwhite	satinwood, low-odour, non-yellowing	white
21	K168012	B&Q Everwhite	gloss low-odour, non-yellowing quick dry	white
22	K168018	B&Q Everwhite	satin low-odour non-yellowing quick dry	white
23	32429701	B&Q Value	non-drip gloss	pure brilliant white
24	D210210	Santex	one-coat exterior gloss waterproof	pure brilliant white
25	DC5212UF/3313	International Ranch Paint	exterior gloss self-undercoating	brilliant white
26	P6703319/6693	Hammerite	satin	white
27	PC672021/5074	Hammerite	smooth	white
28	F134 (L663534095)	Ronseal	hammered	white
29	A234 (L661534066)	Ronseal	smooth	white
30	B05112	Johnstones	acrylic, low-odour egg shell	brilliant white
31	B7072	Johnstones	acrylic gloss	brilliant white
32	B14053	Johnstones	satinwood one coat	brilliant white
33	B01053	Johnstones	gloss	brilliant white
34	B25033	Johnstones	non-drip one-coat gloss	brilliant white

(Continued)

Table II.7 (*Continued*)

Number	Batch	Make	Type	Colour/class
35	B08053	Johnstones	satin finish	brilliant white
36	B19053	Johnstones	eggshell	brilliant white
37	B25023	Leyland	one-coat gloss	brilliant white
38	B19023	Leyland	eggshell	brilliant white
39	B20033	Leyland	non-drip gloss	brilliant white
40	B20013	Leyland	gloss	brilliant white
41	0303H228436	Dulux	satin, water based	pure brilliant white
42	1161C39	Dulux	professional liquid gloss	pure brilliant white
43	0203H228436	Dulux	water-based gloss	pure brilliant white
44	H127748	Crown	satin quick dry	pure brilliant white
45	B11063	Homebase	weathercoat	brilliant white
46	B29053	Homebase	satinwood	brilliant white
47		Homebase	non-drip gloss	pure white
48		Homebase	liquid gloss	pure white
49		Homebase	non-drip one-coat gloss	pure white
50	B09053	Homebase	quick dry gloss	pure white
51	D199652	Brolac	full gloss	brilliant white

All trademarks acknowledged. © 2009 Elsevier, reproduced with permission from Farmer, Meier-Augenstein and Lucy (2009).

the paints showed a slightly more negative $\delta^{13}C$ value of $-32‰$. Measured $\delta^{18}O$ values showed a range of 10‰ from about 10 to 20‰ with a more uniform distribution across this range as seen for the paints' ^{13}C isotopic composition. Measured $\delta^{2}H$ values ranged from -80 to $-130‰$. Compiling results from the stable isotope analysis of all 51 white paint samples yields a three-variable stable isotope signature using mean $\delta^{13}C$, $\delta^{18}O$ and $\delta^{2}H$ values, which is shown in Figure II.9.

For the purpose of such a study one needs to define what information the likelihood ratio is supposed to convey. Generally speaking, the likelihood ratio approach enables the forensic scientist to calculate the ratio of the probability of the outcome of the comparison between the control and the recovered samples. The likelihood ratio approach enables the forensic scientist to calculate the ratio of the probability of the outcome of the comparison between control and recovered samples based on a given set of measurable sample characteristics. In our case these characteristics were comprised of a set of isotope abundance measurements IA. The two possible outcomes of the comparison between control C and recovered samples R can be put into two competing propositions:

(i) The proposition of the prosecution H_P that control specimen C and recovered specimen R come from the same bulk source or share the same sample history.

(ii) The proposition of the defence H_D that control specimen C and recovered specimen R do not come from the same bulk source or do not share the same sample history.

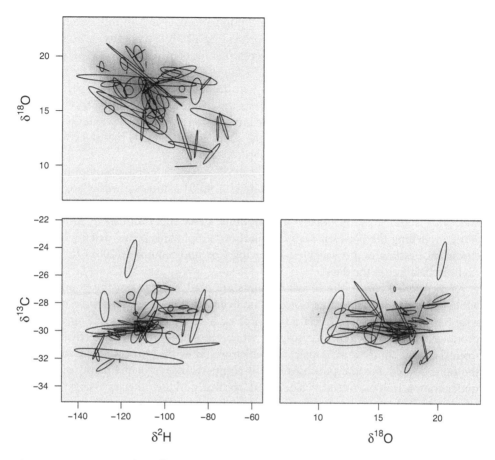

Figure II.9 Means of δ^2H, δ^{13}C and δ^{18}O observations for each of 51 samples of white paints plotted over the smoothed bivariate density (darker equates to higher density) for each variate pair. The items are represented as ellipses corresponding to 95% of the empirical distribution for that item.

The nature of the inference that can be made is determined in the main by three inter-related factors: the number of replicate measurements, intra (within)-source variation and inter (between)-source variation. The likelihood ratio (LR) can be taken to represent the aforementioned three factors and is written as:

$$\mathrm{LR} = \frac{\mathrm{Pr}(\mathrm{IA}|\mathrm{H_P})}{\mathrm{Pr}(\mathrm{IA}|\mathrm{H_D})} \qquad (\mathrm{II.5})$$

The term in the numerator of Equation II.5 represents the probability Pr of the observed degree of correspondence between recovered and control, taking into account intra-source variations (i.e. the variation observed for δ values obtained within replicate measurements from the same bulk white paint sample). The denominator term represents the probability Pr of the observed degree of correspondence taking into account inter-source variations (i.e. the variation observed for δ values obtained from isotopic measurements amongst a population of white paints of different provenance).

This approach requires data from relevant populations, which in this case have been represented by 51 white paint samples representing all white paints as a whole.

If LR < 1, the isotopic composition suggests that the control and recovered specimens are from different sources, and that the defence proposition is true. The closer the likelihood ratio is to zero, the more it suggests the control and recovered specimens are from different sources. If LR > 1, it suggests the control and recovered specimens are from the same source, and that the proposition of the prosecution is true. The greater the likelihood ratio becomes, the more the observation suggests that the control and recovered specimens are from the same source.

The next step is to calculate likelihood ratios for trivariate stable isotope signatures based on $\delta^{13}C$, $\delta^{18}O$ and $\delta^{2}H$ values obtained from 51 samples of white architectural paints following the calculation described by Aitken and Lucy (Aitken *et al.*, 2006; Aitken and Lucy, 2004), making normal assumptions about the within-item distributions while modelling the between-item distributions employing kernel density estimates. Covariance estimates are weighted following Cox and Solomon (2002) to allow for slight imbalances in the data.

As mentioned above, the 51 white paints are used to represent the population of white paints as a whole. Stable isotope data from all samples are compared on a pair-wise basis with the isotopic composition of each of the other paints. This approach produces a total of 1275 comparisons for which the paints are known to have not been from the same source and 51 pairs for which we know the paints to have come from the same source. For the purposes of this comparison one of the paints of the pair under comparison is arbitrarily labelled as 'control' and the other as 'recovered'. In an actual criminal investigation 'control' and 'recovered' would represent the sample of paint recovered from the scene of a crime and the sample recovered from the suspect, respectively.

On this basis, 27 (2.12%) of the calculated likelihood ratios of 1275 comparisons from paints known to originate from different sources gave LR > 1 (Table II.8). This indicates the paint samples are from the same source when in fact it is known the paints are from different sources. This would indicate a false-positive rate of about 2%. However, it should be noted that 11 of those 27 gave LR < 1×10^{1}. A likelihood ratio of this magnitude would provide very weak support (Evett *et al.*, 2000) for the proposition that the two paints shared a common source and a more realistic false-positive rate might be considered to be about 1.3% for which strong false-positive evidence existed, and which might actually mislead investigators.

Column 2 of Table II.8 gives the likelihood ratios calculated for comparisons made between the 51 cases where it is known that both samples are from the same source. For these calculations all three replicate observations from each paint have been used as both 'control' and 'recovered' samples. The likelihood ratios thus calculated are probably unrealistically high as the 'control' and 'recovered' samples matched exactly, and this is reflected in the zero percentage false-negative rate. This would not be expected in cases where it was genuinely uncertain whether two specimens of paint shared an origin.

To give a more realistic impression of the performance of IRMS observations in situations where two paint specimens are indeed from the same source, we conducted a second experiment using the first two replicates from each paint as the control item and

Table II.8 Percentage distributions for the likelihood ratios from each comparison.

Likelihood ratio	Different[a]	Same (all replicates)[b]	Same (two replicates)[c]
$\leq 10^{-6}$	90.04	0.00	0.00
10^{-6} to 10^{-5}	1.50	0.00	0.00
10^{-5} to 10^{-4}	1.25	0.00	0.00
10^{-4} to 10^{-3}	1.02	0.00	0.00
10^{-3} to 10^{-2}	1.41	0.00	0.00
10^{-2} to 10^{-1}	1.41	0.00	0.00
10^{-1} to 10^{0}	1.25	0.00	2.08
10^{0} to 10^{1}	0.86	0.00	2.08
10^{1} to 10^{2}	1.18	0.00	22.92
10^{2} to 10^{3}	0.08	58.82	68.75
10^{3} to 10^{4}	0.00	41.18	4.17
10^{4} to 10^{5}	0.00	0.00	0.00
10^{5} to 10^{6}	0.00	0.00	0.00
$\geq 10^{6}$	0.00	0.00	0.00
False positive/negative[d]	2.12	0.00	2.05

[a]There are 1275 comparisons between paint specimens known to originate from different sources. The percentage likelihood ratios for these are in the column 'Different'.
[b]There are 51 comparisons from paint specimens known to originate from the same source. All three replicates made for each specimen of paint were used to calculate the likelihood ratios comprising the column labelled 'Same (all replicates)'.
[c]In a second experiment, the first two replicates from each paint were used as the 'control' item and the second two replicates from each paint as the 'recovered' item. The percentage likelihood ratios are given in the column labelled 'Same (two replicates)'. Only 48 likelihood ratios could be calculated as three items had too few replicates observed for the calculations to be made.
[d]False-positive (between-source comparison giving a LR > 1) and false-negative (within-source comparison giving values of LR < 1) rates are given for each set of comparisons.
© 2009 Elsevier, reproduced with permission from Farmer, Meier-Augenstein and Lucy (2009).

the second two replicates from each paint as the recovered item. In this experiment the simulated control and recovered items shared only a single observation. The likelihood ratios from this second experiment are given in column 3 of Table II.8.

For the second experiment three likelihood ratios, those for items 41, 45 and 46, could not be calculated as too few replicates existed. From this second experiment only a single (around 2%) likelihood ratio proved to be less than one. If only strong misleading evidence is counted then this would reduce to a false-negative rate of zero.

It is highly promising for the use of stable isotope analysis as a forensic tool that the 51 white paint samples, which produce 1275 comparisons when using pair-wise comparison with likelihood ratios, produce a false-positive result in only around 2.6% of all hypothetical cases. One also has to consider that likelihood ratios have only been applied to the stable isotope signatures of the 51 white architectural paint samples and should of course also be applied to complementary data from other analytical techniques to determine the potential of these techniques for comparison.

Chapter II.5
Forensic Stable Isotope Analytical Procedures

It may, and probably will, surprise you, dear reader, to learn that as yet there are virtually no nationally or internationally agreed and validated Standard Operating Procedures in place for laboratories engaged in forensic stable isotope analysis to work towards. The two main reasons for this are (i) forensic stable isotope analysis is still a relatively young technique, and (ii) the chronic, almost systemic lack of funding to develop protocols and build databases that are fit-for-purpose.

Since November 2002, when funding of research into techniques and technology to combat crime and terrorism became part of the UK Research Councils' (RCUK) remit, only two projects focused on stable isotope techniques have received financial support. One was a network project to develop Forensic Applications of Stable Isotope Ratio Mass Spectrometry (through the Forensic Isotope Ratio Mass Spectrometry (FIRMS) Network), the other is a research project on Isotope Profiling of Drugs that has just been completed (cf. Chapter III.2.3). Given the urgent need by law enforcement and security agencies, and given the fact this technique was successfully used to aid identification of mutilated murder victims providing ample proof-of-concept (see Chapter III.5), it is hoped that much needed research into, for example, isotope profiling of human tissue in support of human provenancing to combat terrorism and people trafficking will receive the support and resources required to establish standardized protocols and procedures as well as the databases needed in support of forensic statistical analysis.

Potential future editions of this book, sufficient reader interest permitting, will hopefully see an expanded Part II containing information on internationally harmonized analytical protocols including a list of recommended matrix-matched reference materials with internationally agreed and accepted isotope abundance values. In the meantime, I have included three sample preparation protocols related to forensic stable isotope analysis of human remains in Appendix III.B in Part III of this book. These protocols have been scrutinized and reviewed by the FIRMS Network.

Stable Isotope Forensics: An Introduction to the Forensic Application of Stable Isotope Analysis Wolfram Meier-Augenstein
© 2010 John Wiley & Sons, Ltd

II.5.1 FIRMS Network

Despite the disappointing situation with regard to financial support for focused scientific research and development in this new and exciting field of forensic science, funded research projects outside the UK as well as privately financed research by enthusiastic individuals and individual research groups has already resulted in advances of our understanding of where and to what extent stable isotope signatures can make a valuable contribution to forensic science, in general, and criminal investigations, in particular. It is largely thanks to the FIRMS Network that received pump-priming support through a grant awarded by one of the UK's Research Councils as well as the subsequently formed FIRMS Steering Group, which received financial support from the Home Office after the Research Council grant had come to an end, that these efforts resulted in meaningful insights and information documented in over 50 peer-reviewed publications between 2002 and 2008, with 40 of these publications generated by members of the FIRMS Network.

Two key members of the FIRMS Network, both actively involved in the forensic application of stable isotope analysis in criminal investigations, are now ISO17025 accredited by the UK Accreditation Service for the particular type of stable isotope analysis they are engaged in. This has strengthened the FIRMS Network's knowledge base considerably and has not only led to an increase in high-profile oversees law enforcement organizations setting up stable isotope forensics laboratories of their own (see p. 253, 'Government agencies and institutes with dedicated stable isotope laboratories'), but has also attracted membership of end-users (i.e. police forces and other law enforcement agencies).

From its beginnings as a Research Council-funded network bringing together scientists interested and involved in stable isotope forensics, FIRMS (www.forensic-isotopes.org) has evolved into an organization with the objective of organizing and running regular inter-laboratory exercises, organizing international conferences, harmonizing isotope analytical procedures applied in a forensic context with a view to ensure best practice, identifying suitable matrix-matched reference materials for forensic isotope analysis and developing a forensic isotope database. In the meantime, certain government departments and various law enforcement agencies are already turning to FIRMS for advice on proficiency of stable isotope forensics service providers or quality assuring of government-funded projects involving forensic stable isotope analytical work. Therefore, anybody interested in becoming involved in forensic stable isotope analysis should consider joining FIRMS or, at the very least, should participate in International Laboratory Comparisons (ILCs) and other proficiency exercises organized by this organization.

Chapter II.6
Generic Considerations for Stable Isotope Analysis

Despite the current lack of nationally or internationally agreed/accepted guidelines and Standard Operation Procedures specifically created for forensic applications of stable isotope analytical techniques, generic guidelines based on principles of best practice can of course be given, and will be described and discussed in the following chapters. Absolutely essential to the entire analytical process is the aforementioned 'principle of identical treatment' of sample, acquisition quality control and reference materials to ensure meaningful correction and quality control of analytical results (Bowen *et al.*, 2005; Fraser and Meier-Augenstein, 2007; Meier-Augenstein, 1997; Meier-Augenstein, Watt and Langhans, 1996; Werner and Brand, 2001).

II.6.1 Generic Considerations for Sample Preparation

It may sound trivial, but it cannot be overemphasized enough that the quality of results of stable isotope analysis or indeed any analytical technique cannot be better than the quality of the sample analysed. In other words, sample preparation is a crucial part of the entire analytical process because it constitutes more often than not the performance-limiting step. For both BSIA and CSIA, close attention must be paid to the following points:

- Every step of the sample preparation procedure (sampling, sample storage, sample preparation, sample derivatization) must be scrutinized for its potential of mass-discriminatory effects.

- In addition to international reference materials for reference point calibration and scale adjustment, matrix-matched reference or perhaps better quality assurance materials with accepted δ values should be analysed concurrently.

- In the case of CSIA by GC/C-IRMS, an internal standard of a similar chemical nature and of known isotopic composition should be added to the sample *prior* to injection.

- In the case of BSIA by EA- or TC/EA-IRMS, a standard of a similar chemical nature and of known isotopic composition should be analysed concurrently (i.e. as part of the sample batch) (Werner and Brand, 2001).

- CSIA only: any derivatization agent and its isotopic signature must be kept consistent throughout the duration of an analytical project. This is best achieved by acquiring a large stock of the reagent from the same batch and storing it appropriately. The latter may include storage over drying agents, at low temperatures, under an inert gas or not exposed to light.

Due to the time-consuming nature of the process required to create a new international reference material, the rate with which organic international reference materials suitable for BSIA let alone CSIA have been made available by the IAEA has been slow, although there is an expectation that demands and active involvement from special user groups such as forensic laboratories and researchers will help to improve this situation. Apart from proficiency testing, ILCs organized by the FIRMS Network have the aim to identify compounds with the potential to become suitable matrix-matched secondary reference materials and to establish ILC consensus δ values for these. Any forensic laboratory planning to engage in forensic stable isotope analysis is strongly encouraged to contact the FIRMS Network for information on accreditation, ILCs or membership.

In the context of the efforts of quality control and quality assurance of CF-IRMS measurement, the names of a few individual scientists must be mentioned who have devoted their scientific life to these important areas of stable isotope analysis and who have produced a number of scientific publications and/or organic materials suitable for BSIA or CSIA and determined their isotopic composition against internationally accepted IAEA reference points. In collaboration with John M. Hayes (Woods Hole Oceanographic Institution) and A.L. Sessions (Caltech), Arndt Schimmelman at the University of Indiana has developed over 80 organic compounds as 'reference' materials for research purposes covering both EA-IRMS and GC/C-IRMS applications (http://mypage.iu.edu/~aschimme/hc.html). A complete listing of all secondary standards available from Arndt Schimmelman's laboratory is given in Table II.A.2 in Appendix II. Another scientist is Tom Brenna at Cornell University who developed a number of fatty acid standards specifically for CSIA by GC/C-IRMS (Caimi, Houghton and Brenna, 1994). Credit is also due to Willi Brand, Tyler Coplen, Manfred Gröning and Pier de Groot for their continued efforts and contributions to quality control and quality assurance of stable isotope analytical techniques (de Groot, 2004; Brand, 1996; Brand *et al.*, 2009a, 2009b; Brand and Coplen, 2001; Coplen, 1988, 1994, 1995, 1996; Coplen *et al.*, 2006a, 2006b; Coplen and Ramendik, 1998; Qi *et al.*, 2003; Werner and Brand, 2001).

II.6.2 Generic Considerations for BSIA

BSIA is considered to be one of the most robust of all the stable isotope analytical techniques. A well set-up and maintained IRMS system very rarely presents with problems not related to maintenance (e.g. filament change), so most of the few problems that occur are related to the sample side of the instrument. However, this should not lull us into a false sense of security. It pays to remember the following two homilies; 'what can go wrong will' and 'garbage in, garbage out'.

In addition, since the inception of high-temperature conversion-elemental analyser systems for ^{2}H- and ^{18}O-BSIA analysis the list of things that can go wrong in BSIA has become longer even though the sample side still accounts for the majority of the problems.

- Signal size and isotopic composition of the standard(s) should match or be at least close to those of the analyte(s) (Brenna *et al.*, 1997).

- Autosampler and autosampler action should be checked for leaks of carrier gas.

- Actual carrier gas flow rate should be monitored on a regular basis to detect early signs of 'ash' build-up in the reactor, thus causing a restriction that will result in peak shape and, hence, δ value deterioration.

- EA-IRMS only: water-trap should be inspected regularly for early signs of clogging.

- All GC parameters (selectivity of stationary phase, column length, temperature profile) should be exploited to their fullest potential to prevent peak overlap of sample conversion products (e.g. H_2 with N_2).

- GC column should be baked out on a regular basis to remove build-up of, for example, water traces that have become trapped on the column.

II.6.2.1 Isobaric Interference

Isobaric interference is a result of ions of equal mass but comprising isotopes of different elements being present in the ion source at the same time. In our case, the measurement of isotope ratios of light elements in permanent gases can be hampered by the presence of nitrogen containing gases such as nitrous oxide (N_2O; major abundant ion *m/z* 44) or nitrogen monoxide (NO; major abundant ion *m/z* 30) interfering with isotope analysis of ^{13}C (CO_2; major abundant ion *m/z* 44) or ^{18}O (CO; major abundant ion *m/z* 28, minor abundant ion *m/z* 30), respectively. Oxygen isotope analysis based on IRMS measurement of CO can also suffer from isobaric interference caused by concurrent presence of N_2 (major abundant ion *m/z* 28) in the ion source (Brand *et al.*, 2009a, 2009b).

There are two different causes for the above – chemical and chromatographic. The presence of nitrogen oxides is the result of non-quantitative conversion of nitrogen-containing compounds into N_2, which in a combustion/reduction interface, be it in an EA-IRMS or GC/C-IRMS system, is caused by an exhausted reduction reactor.

The presence of nitrogen, other than from atmospheric background, even in a system operating at a pressure of less than 2×10^{-6} mbar and thus considered leak tight, is caused by peak overlap of the N_2 peak with the subsequently eluting CO (or CO_2) peak and therefore indicative of a chromatographic problem. This requires a change of chromatographic conditions such as carrier gas flow, GC oven temperature, column selectivity or column dimensions to improve separation of N_2 from CO (or CO_2) and applies only to BSIA applications using TC/EA-IRMS (or EA-IRMS) instruments.

II.6.3 Particular Considerations for ^{2}H-BSIA

II.6.3.1 Keeping Your Powder Dry

Though seemingly obvious, it cannot be emphasized enough that samples destined for forensic analysis should be kept in sealed desiccators at all times to prevent contamination, if nothing else. This is even more important when planning to analyse solid compounds or materials for their ^{2}H (and/or ^{18}O) isotopic composition. In this case, samples must be stored in evacuated desiccators containing a desiccant such as self-indication phosphorus pentoxide to ensure samples are perfectly dry (i.e. do not contain any traces of moisture). Some compounds such as glucose, even when all free hydroxyl groups are derivatized, as in the case of glucose penta-acetate, are notoriously difficult to keep dry due to their ability to accommodate water molecules in their crystal lattice. In such an event, samples should be transferred from their desiccator to a drying oven prior to analysis (i.e. prior to being put on to the autosampler).

At this point I need to remind the reader that every compound, even if not regarded as hygroscopic, has its own particular equilibrium point for water or residual moisture content expressed as the water vapour pressure over or above a compound's surface. The analogy here is the case of apolar, hydrophilic organic solvents such as chloroform or toluene that are listed in any reference book as being immiscible with water. However, as every organic chemist knows this does not mean such solvents are absolute (i.e. completely water free), since depending on their particular equilibrium point for water adsorption they can contain up to 2% of water. Thus, if a particular chemical synthesis requires water-free conditions, solvents have to be distilled, often in the presence of a drying agent and the distillate, now an absolute solvent, has to be stored with a highly water-absorbent material added, typically zeolite molecular sieves, since otherwise the water-free solvent would soon absorb water from ambient atmospheric humidity until its particular equilibrium point has been reached once again.

The same is true for solid compounds and materials. A good experiment to demonstrate this visually can be carried out with copper sulfate ($CuSO_4$). To be more accurate, the blue copper sulfate crystals every chemist knows are actually comprised of

CuSO$_4$·5H$_2$O (i.e. per molecule of copper sulfate, this material contains five molecules of water integrated in its crystal lattice). If one dries this material in a drying oven one eventually obtains anhydrous (i.e. water-free) copper sulfate that is no longer blue but white. However, should this now water-free (anhydrous) material be removed from the drying oven and left on a work bench exposed to atmosphere, it will quickly start to absorb water from the surrounding atmospheric humidity and soon turn blue again. If we were interested in analysing the true ^{18}O isotopic composition of a particular copper sulfate sample (i.e. the ^{18}O signature of the sulfate group), analysing the blue copper sulfate would be wholly inappropriate since for every 2 mol equiv. of O$_2$ coming from the sulfate group the crystal water would contribute 2.5 mol equiv. of O$_2$.

The obvious solution is of course to analyse the dry, water-free copper sulfate. However, given how eager this material is to replace the water lost during the drying process, how do we ensure that it stays dry while it is sitting in the autosampler waiting to be analysed?

In order to avoid dried samples reabsorbing water from ambient humidity, the dry, water-free material has to be transferred into an autosampler that can be isolated from the surrounding atmosphere. At present the only type of autosampler meeting this requirement is the so-called 'Zero-Blank' autosampler (Costech Analytical Technologies, Valencia, CA, USA). Originally designed to eliminate nitrogen background (from atmosphere) when analysing samples for their elemental composition in an elemental analyser, its closed or sealed carousel design has proven to be the ideal solution to this problem since it permits purging atmosphere from the autosampler with dry helium and then sealing it (see Figure II.10). Once purged and sealed, dried samples sit in the autosampler tray in a protective atmosphere of dry helium and stay dry until analysed.

II.6.3.2 Total δ^2H versus True δ^2H Values

It could be argued that the point of total versus true δ^2H values of a given material or compound has already been covered in the previous chapter since the measured δ^2H value of a hygroscopic substance such as ammonium nitrate of unknown water content would be the total δ^2H value of this material, whereas the δ^2H value of an anhydrous, water-free sample of the same ammonium nitrate could be considered to be its true δ^2H value.

Sadly this is not the case and in the particular example of ammonium nitrate, matters are even more complicated. To study this problem further we first need to agree on terms – what we mean when we say this is the 'true' δ^2H value of a given compound. If we use the word 'true' to mean 'original' (i.e. reflecting that state of the compound or material when it was originally made), then we have to accept that only hydrogen-containing compounds with astronomically large pK_a values, which would only display their H$^+$-acidic character if given an aeon to do so, would qualify as having 'true' δ^2H values. In practical terms this means compounds containing only non-exchangeable hydrogen, such as most aliphatic and aromatic hydrocarbons with no relevant C–H acidic character e.g. the international reference material IAEA-CH7 (polyethylene foil) could be thought of as having a true, immutable δ^2H value.

Figure II.10 Costech Zero-Blank autosampler as used in our laboratory for BSIA by EA- or TC/EA-IRMS. The 1/16-inch stainless steel tubing above the isolation valve delivers the helium to the autosampler tray.

From this it is clear that a more appropriate term to use instead of 'true' would be 'non-exchangeable' δ^2H values, taken to mean the ^2H isotopic composition or signature of these hydrogen atoms in a molecule not able to act as H^+ acid within a relevant time frame. Conversely, hydrogen atoms contained in amino, carboxyl, hydroxyl and thiol groups or the ammonium ion, and including C–H acidic hydrogen atoms as present in acetonitrile or acetone, are classed as exchangeable hydrogen. Thus, even a dried and now water-free compound could still have two different δ^2H values associated with it – its total and its non-exchangeable δ^2H value. The problem associated with exchangeable hydrogen and, hence, the paramount importance of creating and agreeing on harmonized protocols for the ^2H isotope analysis of compounds and materials prone or liable to hydrogen exchange, has been highlighted by the outcomes of a series of ILCs organized by the FIRMS Network (Carter *et al.*, 2009).

Returning to the example of ammonium nitrate, it now becomes clear that the 'truest' δ^2H value ammonium nitrate may assume is a δ^2H value based on three out of four (75%) non-exchangeable hydrogen atoms since the ammonium ion (NH_4^+) has the potential to exchange completely at least one H^+ with water when treated with aqueous magnesium or calcium nitrate solutions and subsequently dried to prevent the ammonium nitrate grains or pellets from caking. However, this 75% scenario is just that – a scenario making the simplifying but incorrect assumption that one and only one hydrogen on the NH_4^+ molecule will exchange while the other three will stay fixed and remain unchanged. This assumption ignores the fact that NH_4^+ is a symmetrical molecule with four chemically equivalent hydrogen atoms. This means at any given moment any of the four hydrogen atoms can be the one whose N–H bond is broken to release a proton (H^+).

In reality, ammonium nitrate is so hygroscopic (Figure II.11) that it eventually becomes so wet it can react with various metals forming a variety of compounds, some of which are highly unstable (e.g. copper nitrate tetramine). Studies are currently under way to determine if and to what degree anti-caking-treated ammonium nitrate pellets are still able to exchange hydrogen. Preliminary findings suggest that ammonium nitrate as supplied by companies dealing with fine chemicals can exchange 1.2 of the four ammonium hydrogen atoms with surrounding atmospheric humidity in less than 5 days (see Section III.3.3).

II.6.3.2.1 ^2H Isotope Analysis of Human Hair

Another area of forensic interest where the difference between total versus non-exchangeable δ^2H values plays an important role is that of human provenancing (see Chapter III.5). Human provenancing, in contrast to human identification, does neither aim nor claim to identify a person, but rather aims to aid the identification process by providing information on a person's geographic provenance or life history. One potentially very useful source of information regarding a person's recent geographic life history is the isotopic signal recorded in human scalp hair and bone collagen (Bowen *et al.*, 2005; Ehleringer *et al.*, 2008; Fraser and Meier-Augenstein, 2007; Fraser, Meier-Augenstein and Kalin, 2006; Meier-Augenstein and Fraser, 2008; Reynard and Hedges, 2008; Sharp *et al.*, 2003).

II.6: GENERIC CONSIDERATIONS FOR STABLE ISOTOPE ANALYSIS

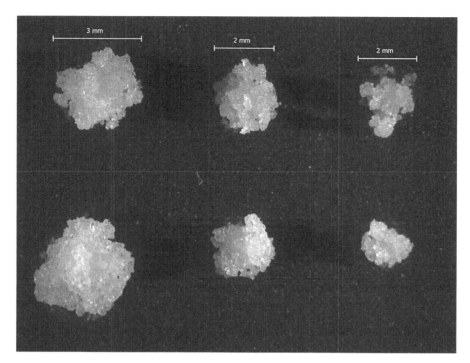

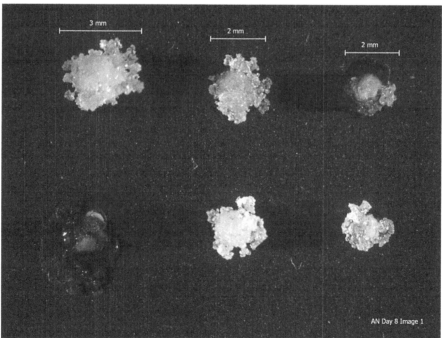

Figure II.11 (Top) Untreated ammonium nitrate from six different sources. Note the already 'wet' appearance of the sample in the top right corner. (Bottom) The same samples after an 8-day exposure to ambient atmosphere. Images provided courtesy of Sarah Benson (Forensic Operations Laboratory, Australian Federal Police, Canberra) (Benson, 2009).

The main constituent of hair is α-keratin (typically 91%) – a structural protein made up from 16 different amino acids with glycine being the most abundant. The presence of exchangeable hydrogen atoms in, for example, hydroxyl groups of amino acids such as threonine and tyrosine means hair is one of the materials for which a total and a non-exchangeable δ^2H value can be reported.

At this point it is necessary for us to consider what if anything influences the total δ^2H value of a given compound and thus necessitates the determination of the non-exchangeable δ^2H value of this material. We have just learned that even for a dry material free of any residual moisture or crystal water its total δ^2H value encompasses all chemically bound hydrogen while the non-exchangeable or true δ^2H value disregards the contribution of exchangeable hydrogen atoms present. Since hydrogen exchange can readily take place in a moderately humid atmosphere, we need to be prepared for the observation that total δ^2H values of a given material such as hair will be different if determined in two laboratories in locations where the isotopic composition of precipitation and, hence, local ambient humidity as well as local tap water is significantly different.

One work-around to this conundrum that is currently under consideration is for every laboratory engaged in the forensic isotope analysis of hair or similar materials to equilibrate the samples in a controlled environment with all laboratories using aliquots of the same standard waters of defined isotopic composition for this equilibration procedure. The other approach that is currently practiced is to equilibrate the sample together with at least two matrix-matched reference materials (e.g. keratin) of known 'true' ^2H isotopic composition under controlled conditions, e.g. in a sealed desiccator with water of known isotopic composition, to determine the exchange rate on the basis of the measured versus known δ^2H values of the reference materials and to correct the measured total δ^2H value of the sample accordingly to yield its non-exchangeable or true δ^2H value (Bowen *et al.*, 2005; Hobson *et al.*, 2004; Reynard and Hedges, 2008; Schimmelmann, 1991; Wassenaar and Hobson, 2003). The latter approach has been dubbed "comparative equilibration" (Kelly *et al.*, 2009).

A variation on the reference material approach is to equilibrate concurrently a matrix-matched reference material with the sample using water A of one particular known isotopic composition and to equilibrate in parallel another subsample of the reference material with water B of a known yet significantly different isotopic composition to water A (Bowen *et al.*, 2005; Chesson *et al.*, 2009; Fraser and Meier-Augenstein, 2007; Sharp *et al.*, 2003).

Irrespective of which of the aforementioned methods one chooses for determination of the 'true' δ^2H value of hair or other keratogenous material such as feathers, hair, hooves or whale baleen (Birchall *et al.*, 2005; Bowen, Wassenaar and Hobson, 2005; Hobson *et al.*, 2004; Hobson, Atwell and Wassenaar, 1999; Kelly, Ruegg and Smith, 2005; Kelly *et al.*, 2009; Meehan, Giermakowski and Cryan, 2004; Sharp *et al.*, 2003; Smith and Dufty, 2005; Wassenaar and Hobson, 2000, 2006), they all have two major features in common:

(i) Controlled equilibration of samples and/or matrix-matched standards and subsequent drying down.

(ii) Calculation of molar exchange fraction and of 'true' or non-exchangeable δ^2H values.

To afford a controlled and, hence, reproducible rate of hydrogen exchange, samples should be exposed to the same ambient environment such as water from a defined source in a hermetically sealed container for at least 4 days so labile hydrogen atoms prone to exchange will all reflect the same ^{2}H level as that of the water used for equilibration (Bowen *et al.*, 2005). Subsequent to equilibration, but prior to stable isotope analysis, samples must be dried down in an evacuated desiccator containing phosphorous pentoxide again for at least 7 days to remove residual moisture traces from the samples (Bowen *et al.*, 2005; Fraser and Meier-Augenstein, 2007).

The molar exchange fraction f_{Hxch} of exchangeable hydrogen can be calculated from the ratio of differences in observed δ^2H values for subsamples of hair equilibrated with two waters of different ^{2}H isotopic composition and the difference of the known δ^2H values for the two waters:

$$f_{Hxch} = \frac{\delta^2 H_{hair, waterA} - \delta^2 H_{hair, waterB}}{\delta^2 H_{waterA} - \delta^2 H_{waterB}} \quad (II.6)$$

With observed differences of typically 8.3‰ for $[\delta^2 H_{hair, waterA} - \delta^2 H_{hair, waterB}]$ and a determined difference of 87.7‰ for $[\delta^2 H_{waterA} - \delta^2 H_{waterB}]$ in our laboratory, calculated f_{Hxch} values for ground hair range from 0.083 to 0.105 (8.33 to 10.53%). Average f_{Hxch} values for ground hair and hair chopped into approximately 1 mm pieces were 0.094 and 0.135, respectively. These values are in very good agreement with published data where, dependent on the mechanical method of sample preparation, figures of 9.4 and 15.9% for pulverised and coarsely chopped hair, respectively, have been reported (Bowen *et al.*, 2005), which illustrates the need for a well-controlled method of sample preparation. However desirable a normalized method of ^{2}H isotope analysis followed by all laboratories engaged in forensic stable isotope analysis may be, inter-laboratory differences in f_{Hxch} do not necessarily invalidate results or the comparability as long as the 'principle of identical treatment' is adhered to (Bowen *et al.*, 2005; Chesson *et al.*, 2009; Fraser and Meier-Augenstein, 2007; Werner and Brand, 2001) and at least one matrix-matched reference material is taken through the equilibration procedure contemporaneously with the case samples to determine the actual f_{Hxch} associated with the particular sample preparation procedure. In doing so one will thus account for the variable nature of $\delta^2 H_{total}$ and f_{Hxch} that is influenced by various differences, such as how fine or coarse a sample was cut or ground on the day, to name but one. Even if a normalized protocol were to be adhered to by all laboratories across the globe, differences in parameters that are not easily controlled and kept constant such as barometric pressure, temperature and dew point would still result in minute inter- and intra-laboratory differences in f_{Hxch} and, hence, $\delta^2 H_{true}$; however, by controlling those parameters that can be controlled these differences can be kept to a minimum. In other words, since $\delta^2 H_{total}$ and f_{Hxch} are interdependent values, accounting for subtle inter- and intra-laboratory differences will ensure calculated values for $\delta^2 H_{true}$ will be identical within the analytical error of measurement.

It is, however, important to note that for a meaningful determination of true δ^2H values of materials such as hair or collagen using the comparative equilibration method (Kelly *et al.*, 2009; Wassenaar and Hobson, 2003, 2006) the absolute minimum requirement is the contemporaneous analysis of at least two reference materials for which

δ^2H_{true} and f_{Hxch} are known. Using a measured consensus value for δ^2H_{total} for a proteinogenous material comprising an undetermined mole fraction of exchangeable hydrogen is not acceptable as ILCs organized by the FIRMS Network have shown. ILC samples horse hair and paper that had been allowed to equilibrate with ambient humidity in the participating laboratories yielded standard deviations for the Huber estimate of the mean (H15 SD) for H15 mean values of δ^2H_{total} of ±11.05 and 7.35‰, respectively. Differences from the H15 mean values between lowest and highest δ^2H_{total} values for horse hair and paper were 44.3 and 21.3‰, respectively (Carter *et al.*, 2009). Considering that within-laboratory standard deviations for either material were typically 2.0 and 1.4‰, it is clear that external factors such as laboratory location-specific differences in ambient humidity level as well as its isotopic composition between the different geographic locations are among the factors contributing to these findings. This conclusion is supported by the ILC results for a material with no exchangeable hydrogen that were reported in the same paper. For polyethylene foil the values for H15 SD and within-laboratory standard deviation were 2.2 and 1.3‰, while the range around the H15 mean was 4.58‰.

On the basis that the overall hydrogen isotopic composition of hair is a mass balance of the non-exchangeable and exchangeable hydrogen in hair, and assuming that the isotope ratio of the exchangeable hydrogen is fixed for all samples (due to the 'principle of identical treatment'), one can describe the relation between $\delta^2H_{hair,\,true}$ ('true' = non-exchangeable), $\delta^2H_{hair,\,xch}$ ('xch' = exchangeable) and $\delta^2H_{hair,\,total}$ ('total' = measured) as:

$$\delta^2H_{hair,total} = (f_{Htrue} \times \delta^2H_{hair,true}) + (f_{Hxch} \times \delta^2H_{hair,xch}) \tag{II.7}$$

Since the mole fractions for the two contributing pools to the mass balance of total hydrogen (i.e. exchangeable hydrogen and non-exchangeable (true) hydrogen) add up to 1, we can substitute a term containing f_{Hxch} for f_{Htrue}:

$$f_{Htrue} + f_{Hxch} = 1 \tag{II.8}$$

and thus:

$$f_{Htrue} = 1 - f_{Hxch} \tag{II.9}$$

Using this relation between f_{Hxch} and f_{Htrue} we can rewrite Equation II.7:

$$\delta^2H_{hair,total} = ([1 - f_{Hxch}] \times \delta^2H_{hair,true}) + (f_{Hxch} \times \delta^2H_{hair,xch}) \tag{II.10}$$

Rearranging Equation II.10 yields:

$$\delta^2H_{hair,\,total} - (f_{Hxch} \times \delta^2H_{hair,\,xch}) = (1 - f_{Hxch}) \times \delta^2H_{hair,\,true} \tag{II.11}$$

To solve this mass balance equation for $\delta^2H_{hair,\,true}$ the assumption that the isotope ratio of the exchangeable hydrogen is fixed for all samples (due to the 'principle of identical treatment') can be written as $\delta^2H_{hair,\,xch} = \delta^2H_{waterA}$, thus leaving $\delta^2H_{hair,\,true}$

as the only unknown:

$$\delta^2 H_{hair,\ true} = \frac{\delta^2 H_{hair,total} - (f_{Hxch} \times \delta^2 H_{waterA})}{(1 - f_{Hxch})} \quad \text{(II.12)}$$

It could be argued that for the purpose of the above calculations one should use $\delta^2 H$ values for the vapour fraction of waters A and B used for equilibration, and strictly speaking this is correct given that we have learned earlier in Part I of this book that equilibrium fractionation between liquid water and water vapour results in an isotopic fractionation of about −63 to −73‰. However, both water A and water B will be subject to the same liquid/vapour equilibrium fractionation with the numerical difference between the terms $[\delta^2 H_{waterA} - \delta^2 H_{waterB}]_{liquid}$ and $[\delta^2 H_{waterA} - \delta^2 H_{waterB}]_{vapour}$ being of the order of 3–6‰ depending on actual ^2H isotopic composition of the waters, equilibrium temperature and corresponding equilibrium fractionation factor α. In addition, the exchange between vapour hydrogen and keratin hydrogen is of course also subject to an equilibrium fractionation, which is presumed to be 1.08 (Chesson *et al.*, 2009) (i.e. very close to the liquid/vapour fraction factor for water of 1.081 or 1.079 at 23 or 25 °C, respectively). Accounting for both fractionation processes or treating the entire exchange process as having a net fraction factor of 1 results in a difference for calculated true $\delta^2 H$ values of only 2–4‰ (i.e. creates an error that is of the same order as the uncertainty associated with the measurement of hair itself). The resulting difference for calculated molar exchange fractions is therefore quite small (less than 1%).

II.6.3.3 Ionization Quench Effect

The concurrent presence of a gas other than the analyte gas in the ion source of an IRMS can impair accuracy of isotope ratio measurement and, hence, derived δ values. The effect we are referring to here is not the effect caused by isobaric interference – the interference of a gas giving rise to ions of the same mass/charge ratio as the analyte gas such as the well-known isobaric interference between N_2 and CO affecting both major ions of interest, namely *m/z* 28 and *m/z* 29.

The pitfall we are looking at here is an interference, which in the absence of a more apt description we have termed 'ionization quench' (IQ) where molecules of the interfering gas X compete for or prevent ionization of the target gas T. We originally made this observation more than 15 years ago when we prepared and analysed CO_2 gas standards comprised of 5 vol% CO_2 in argon for ^{13}C isotope abundance analysis in breath gas CO_2 on a dual-inlet IRMS. While these gas mixtures invariably yielded measured $\delta^{13}C$ values that were significantly different compared to known accepted $\delta^{13}C$ values for the CO_2 reference gas used, standards prepared by mixing CO_2 from the same source with either nitrogen or helium yielded measured $\delta^{13}C$ values in line with accepted $\delta^{13}C$ values. See Figure II.12.

Owing to this observation, one naturally wonders if and what effect the concurrent presence of nitrogen would have on measured $\delta^2 H$ values of a material with known

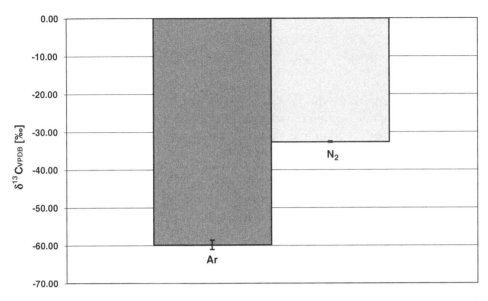

Figure II.12 Effect of argon concurrently present in the ion source on measured $\delta^{13}C$ values of same sized aliquots of CO_2 (accepted $\delta^{13}C_{VPDB}$: $-32.56‰$).

^2H isotopic composition, given that when running samples for ^2H-BSIA using typical operating conditions such as GC temperature set to 85° C and a carrier gas flow of approximately 90 ml/min (in line with manufacturer's recommendations) there is a chance of peak overlap between the hydrogen peak and the subsequently eluting nitrogen peak, especially in the case of nitrogen-rich compounds (e.g. explosives such as ammonium nitrate, cyclotrimethylenetrinitramine (RDX, a constituent of Semtex) and hexamethylentriamine (hexamine, the organic precursor of RDX); see Section III.3.1).

The example shown in Figure II.13 clearly demonstrates that approximately 50% if not more of the following N_2 peak is concurrently present in the ion source during the period when the analyser registers the latter part of the H_2 peak.

Figure II.14 illustrates the results of this peak in terms of peak height, peak shape and even retention time of the H_2 peak.

Immediately noticeable is the difference in peak height, which is reduced by approximately 2000 mV for the H_2 peak overlapped by a following N_2 peak. The peak shape of the affected H_2 peak appears slightly broadened, which is also reflected by a minute increase in retention time (here 0.6 s).

As minute as the latter two effects may seem, the 20% decrease in peak height already provides an indication of how serious the IQ effect will be when we come to examine and compare observed δ^2H values. The H_2 peak shown in the left panel of Figure II.14 represents 6.9 μmol of H_2 created by the high-temperature conversion of 96 μg of the international reference material IAEA-CH7 (polyethylene). Its δ^2H value against VSMOW is certified by the IAEA as $-100.3‰$ and on this occasion was measured as $-101.0‰$ (i.e. not significantly different from its certified δ^2H value given

II.6: GENERIC CONSIDERATIONS FOR STABLE ISOTOPE ANALYSIS

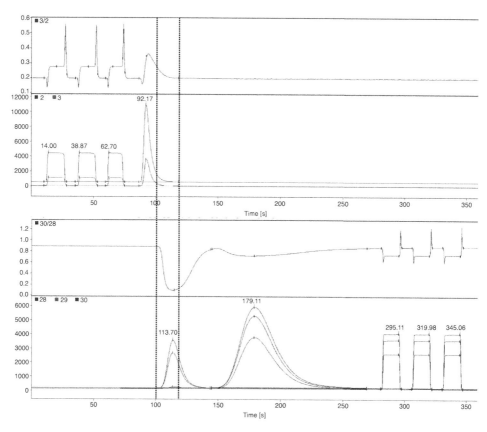

Figure II.13 Illustration of the potential interference on a H_2 peak caused by a partial overlap with a following N_2 peak.

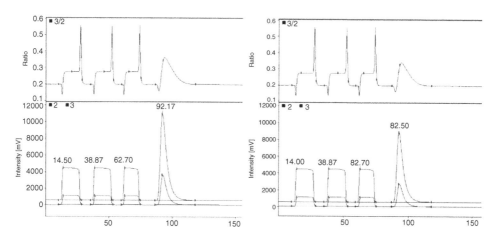

Figure II.14 Comparison of peak heights, peak shape and retention time of a H_2 peak in the absence (left) and presence (right) of N_2. © 1990 Intercept, reproduced with permission from Rautenschlein, Habfast and Brand (1990).

the measurement precision of ±1–2‰). However, the H_2 peak in the right panel of Figure II.13 is supposed to represent 7.2 μmol of H_2 from the conversion of 101 μg of IAEA-CH7, but in this case 568 μg of $AgNO_3$ has been added simultaneously yielding 1.68 μmol of N_2. This time the apparent δ^2H value of IAEA-CH7 was measured as −122.2‰ (i.e. 21.9‰ too negative when compared to its IAEA certified δ^2H value of −100.3‰).

Fortunately, this problem only manifests itself when analysing compounds with a high nitrogen content such as trinitrotoluene (N/H ratio = 1 : 1.67), ammonium nitrate (N/H ratio = 1 : 2) or cellulose nitrate (N/H ratio = 1 : 2.33). Compounds with a low N/H ratio such as MDMA or proteins with N/H ratios of 1 : 15 or approximately 1 : 10, respectively, are not affected since here conditions typically used result in a baseline separation of H_2 and N_2 due the (small) size of the N_2 peak.

However, this problem still needs to be overcome when, for example, explosives or compounds used in the manufacture of explosives have to be analysed. For the solution to this analytical challenge, the reader is referred to Question II.6.3(2) in the Set Problems in Chapter II.8.

II.6.4 Generic Considerations for CSIA

II.6.4.1 Isotopic Calibration during GC/C-IRMS

In contrast to BSIA by EA-IRMS or TC/EA-IRMS, it is not possible in GC/C-IRMS to calibrate target compounds against a standard of known isotopic composition, analysing an isotopic reference material in *exactly* the same way as the sample. There are only four feasible means of isotopic calibration: (1) introduction of a reference gas directly into the ion source, (2) introduction of a combustible reference gas into the carrier gas stream prior to the combustion interface, (3) addition of reference compounds to the sample or (4) a combination of the above.

The results of an extensive study into methods of isotopic calibration by Merritt *et al.* emphasized these demands (Merritt, Brand and Hayes, 1994). Comparing the use of internal reference compounds with the introduction of reference gas pulses directly in the ion source of the IRMS, Merritt *et al.* found an offset of more than 2‰ between the two methods in the case of incomplete combustion and other systematic errors affecting only the analytes. These systematic errors affected both the analytes and the co-injected reference compounds, but were not reflected by the external reference gas pulses. Other groups reported similar observations (Caimi and Brenna, 1996; Caimi, Houghton and Brenna, 1994). In the absence of such systematic errors, Merritt *et al.* reported that both methods of isotopic calibration gave consistent results as long as multiple reference peaks were used to permit drift correction. Only one reference peak for isotopic calibration, albeit from an internal reference compound, is not enough to compensate for the influence of GC parameters such as analyte/stationary phase interaction or column temperature on measured isotope ratios (Meier-Augenstein, Watt and Langhans, 1996).

There are several stages during GC/C-IRMS analysis where mass discrimination and, hence, isotopic fractionation can occur. Closer inspection identifies seven potential sources:

 (i) Isotopic fractionation during sample injection (which can be overcome by on-column or time programmed splitless injection).

 (ii) Chromatographic isotope effect.

 (iii) Chromatographic peak distortion (leading edge of peak and trailing peak tail).

 (iv) Combustion/reduction (or thermal conversion) process.

 (v) Peak distortion of N_2/CO_2 (H_2/CO) gas peak during passage of the combustion (thermal conversion/pyrolysis) interface.

 (vi) Influence of N_2O and NO_2 on IRMS analysis of CO_2 (isobaric interference).

 (vii) Changing flow conditions at the open split prior to the IRMS.

Obviously, external reference gas pulses introduced directly into the ion source of the IRMS do not compensate for any of the points listed above, although they do reflect the ion source conditions to which the analyte gas is subjected, whereas internal reference compounds reflect all of the aforementioned. I have designed and developed a method for isotopic calibration that, provided a combustible gas was used, could reflect the systematic errors caused by items (iv)–(vii). This method combines the convenience and practicability of external reference gas calibration with the advantage of reflecting the majority of physical influences to which analytes are subjected in a GC/C-IRMS system (Meier-Augenstein, 1997).

To this day most laboratories engaged in CSIA use method (1) since this forms an integral part of most instruments as supplied by the manufacturers. Combining this method with method (3), the addition of at least two organic reference compounds to the sample, although highly desirable, happens only on a small scale. One of the reasons is of course the lack of suitable international reference materials specifically developed for the needs of CSIA by GC/C-IRMS.

II.6.4.2 Isotope Effects in GC/C-IRMS during Sample Injection

Other phenomena to affect accurate and precise isotope abundance measurement of gaseous or volatile compounds are mass-discriminatory effects during sample injection on to the GC and the gas chromatographic process, both causing isotopic fractionation. In either case, we are looking at thermodynamic isotope effects. However, while mass discrimination during injection can be avoided, mass discrimination during chromatography cannot – it can only be mitigated.

During the process of sample injection highly volatile, volatile and semivolatile organic compounds are much more susceptible to mass discrimination caused by differences in molar volume, vapour pressure and, hence, boiling point. To avoid partial sample loss and, hence, mass discrimination during injection it is recommended to inject samples exclusively in splitless/split mode – keeping the split valve of the split/splitless injector closed for a time of 6–30 s (the optimum time depends on the particular mass transfer conditions of the individual system used) and open the split valve thereafter for the remainder of the analytical run (Meier-Augenstein, Watt and Langhans, 1996). Even compounds such as pentafluoroproprionyl-isopropylester derivatives of amino acids that are not as volatile as, for example, alkanes or alkyl esters of short-chain fatty acids are subject to mass discrimination when injected in split mode with a split ratio of 1 : 12 (Glaser and Amelung, 2002).

II.6.4.3 Chromatographic Isotope Effect in GC/C-IRMS

It would seem counterintuitive for organic compounds even with a molecular weight above 150 g/mol to be subject to a mass-discriminatory effect resulting in the isotopically heavier species to emerge faster from a chemo-physical process than its corresponding isotopically lighter counterpart. However, there is evidence to the contrary demonstrating the chromatographic process (i.e. the two-phase partitioning of the analyte or solute between mobile phase and stationary phase) results in mass discrimination already measurable during liquid/solid phase interaction as is the case in liquid chromatography (LC) and high-performance liquid chromatography (HPLC). Caimi and Brenna reported that the beginning of a HPLC peak had an isotope ratio sharply enriched relative to the parent material, while the end of the peak is mildly depleted (Caimi and Brenna, 1997). Hofmann et al. observed a shift of $\delta^{15}N$ values to more negative values of approximately 2‰ for amino acids that were isolated by ion-exchange chromatography (Hofmann et al., 1995). A similar observation was made in a study aimed at the isolation of MDMA and other active ingredients from illicit Ecstasy tablets. The method employed supported liquid extraction with cartridges containing a modified form of diatomaceous earth (de Korompay et al., 2008). No significant isotopic fractionation for ^{13}C was observed as a result of the extraction process. However, reported $\delta^{15}N$ values were suggestive of an influence of the extraction process and this observation has already prompted further studies. These observations show that quantitative peak collection of the entire LC peak is important for accurate isotope ratio analysis. Hence, LC techniques must be applied with caution when used for sample preparation or sample clean-up of complex mixtures for isotope ratio analysis. A comprehensive overview of isotopic fraction during chromatography was presented by Filer in 1999 (Filer, 1999).

In contrast to dual-inlet IRMS machines, CF-IRMS systems involving GC separation of organic compound mixtures or combustion gas mixtures produce almost Gaussian-shaped signals. Again, we are dealing with an analytical system employing a form of two-phase partitioning where slight differences in solute/stationary phase interaction leads to mass discrimination. Due to the 'inverse chromatographic isotope effect' (Filer,

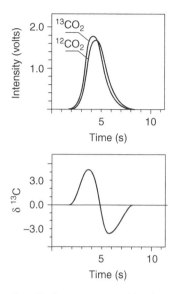

Figure II.15 Illustration of the time displacement caused by the 'inverse' chromatographic isotope effect between the $^{13}CO_2$ and $^{12}CO_2$ aspects of a compound CO_2 peak and the resulting S-shaped 45/44 ratio signal. © 1990 Intercept, reproduced with permission from Rautenschlein, Habfast and Brand (1990).

1999; Matucha *et al.*, 1991; Matucha, 1995) the *m/z* 45 signal ($^{13}CO_2$) precedes the *m/z* 44 signal ($^{12}CO_2$) by 150 ms on average (Brand, 1996; Rautenschlein, Habfast and Brand, 1990) (Figure II.15). This effect, albeit to a lesser degree, can also be observed in EA- and TC/EA-IRMS systems where combustion or pyrolysis products are separated on a packed GC column.

It is important to realize that the chromatographic isotope effect is not caused by a vapour pressure effect, which would result in the leading part of a peak being isotopically lighter and the end of a peak being isotopically heavier. A vapour pressure or mass-dependent isotope effect can be observed when looking at a hydrogen peak entering an IRMS after emerging from a TC/EA. Even though TC/EA modules comprise a GC with a packed column (packed with a 5-Å molecular sieve) the hydrogen peak is not retained by this stationary phase. However, lateral diffusion processes within the carrier gas envelope as well as differences in path or way length for individual H_2 molecules when passing through the molecular sieve result in a mass-dependent mass discrimination where $^1H_2$ molecules travel faster and, hence, emerge slightly earlier from the interface than their heavier $^1H^2H$ (HD) counterparts. Looking at the *m/z* 3 : 2 ratio trace recorded in real-time with the ion trace for the individual masses (*cf*. Figures II.13 and II.14) one can see the 3 : 2 ratio trace swinging downwards at the beginning of the H_2 peak, thus indicating a ratio dominated by *m/z* 2. However, towards the apex (i.e. the latter part of the H_2 peak) the 3 : 2 ratio trace swings upwards indicating a ratio dominated by *m/z* 3.

The so-called 'inverse' chromatographic isotope effect is rather the result of different solute/stationary phase interactions that are dominated by van der Waals dispersion forces leading to an earlier elution of the heavier isotopomer (Matucha *et al.*, 1991). This difference in chromatographic solute/stationary phase interaction is caused by lower molar volumes of the labelled and thus heavier compounds. The reason for the decrease in molar volume is the increased bond strength and thus shortened bond length between $^{13}C-H$ or $^{12}C-^{2}H$ and, to a lesser degree, $^{12}C-^{13}C$, and $^{12}C-H$ and $^{12}C-^{12}C$, respectively (*cf*. Section I.4.1).

You may have exploited this effect already in conventional GC-MS using perdeuterated internal standards for quantification of GC-MS peaks. This technique is known as stable isotope dilution assay and relies on the fact that, for example, perdeuterated benzene (C_6D_6) elutes prior to and can be baseline-resolved from benzene (C_6H_6).

In GC/C-IRMS the time displacement observed, for example, between the *m/z* 45 and the *m/z* 44 signal depends on the nature of the compound and on the chromatographic parameters, such as polarity of the stationary phase, column temperature and carrier gas flow (Meier-Augenstein, Watt and Langhans, 1996). Therefore, loss of peak data due to unsuitably set time windows for peak detection and, hence, partial peak integration will severely compromise the quality of the isotope ratio measurement by GC/C-IRMS. The same is true for traces of peak data from another sample compound due to close proximity resulting in peak overlap with the sample peak to be analysed. Owing to the fact that isotope ratios cannot be determined accurately from the partial examination of a GC peak, high-resolution cGC (HRcGC) resulting in true baseline separation for adjacent peaks is of paramount importance for high-precision CSIA.

Chapter II.7
Summary of Part II

As with any hyphenated analytical technology, CF-IRMS is sensitive to potential problems particular to this technique. These problems include isotopic fractionation during sampling, sample storage or sample preparation, and effects affecting isotopic integrity of the peak or signal representing the sample as it travels through and is analysed by the instrument. However, all these problems can either be overcome by having appropriate procedures in place and their strict observance or be compensated for through the use of appropriate standards and reference materials, which are applied following the principle of identical treatment. In short, in common with every other analytical technique applied in a forensic context, methods and techniques for the forensic application of stable isotope analysis have to be robust and fit-for-purpose.

Chapter II.8
Set Problems

Section II.3.1

(1) Calculate stretch and shift correction equations for the $\delta^{13}C$ values given in Table II.6.

(2) Calculate the consequence an inappropriate chosen end-member pair would have for the corrected δ^2H value of GISP (see Table II.4) if IAEA-CH7 ($\delta^2H_{VSMOW} = -100.3‰$) and NBS-22 ($\delta^2H_{VSMOW} = -118.5‰$) would be used instead of VSMOW/SLAP.

Section II.6.3

(1) Which of the following compounds contains exchangeable hydrogen atoms? Draw the molecular structures and circle the hydrogen atoms prone to exchange.
Diethyl ether; *n*-butanol; benzene; phenol.

(2) What options do you think are open to you to achieve a baseline separation between H_2 and N_2 peak for nitrogen rich compounds? Which is the most likely to succeed?

Appendix II.A
How to Set Up a Laboratory for Continuous Flow Isotope Ratio Mass Spectrometry

Should you find yourself in the position of acquiring an IRMS instrument for your analytical chemistry or forensic chemistry laboratory, please consider where and how this instrument will be accommodated or sited before you even prepare the tender documents for the instrument purchase. The performance of your instruments and the quality of the results they produce not only depend on the specification of the instrument or the amount of time and effort you put into sampling, sample preparation, sample separation and instrument maintenance, but also on the quality of the environment your instrument will operate in.

I have received, and still do receive, numerous requests for advice from people regarding poor performance or other problems with their IRMS equipment. It saddens me to say that a good number of these enquiries boil down to instrument siting or accommodation that is not fit-for-purpose. This is not to say the instrument operators are at fault here. More often than not the reasons for inadequate instrument siting are a combination of several factors, such as budgetary constraints, budget management (departmental equipment budgets versus institution-wide estates budget) and space allocation/space management – factors that sometimes are dictated by available space but sometimes can be the result of unrealistic institutional policies.

In the following sections, consideration is given to all the points that both concern instrument installation and affect instrument performance. The resulting advice is the distillate of some foresight seasoned with a good dose of hindsight in conjunction with experiences made or rather suffered by myself (and many a colleague) when setting up IRMS facilities, in my case at three universities and one government research institute. Since individual circumstances and precise requirements vary, a broad and general outline of bare necessities is given that can easily be adapted to and incorporated into the situation you might find yourself in.

Stable Isotope Forensics: An Introduction to the Forensic Application of Stable Isotope Analysis   Wolfram Meier-Augenstein
© 2010 John Wiley & Sons, Ltd

II.A.1 Pre-Installation Requirements

All IRMS manufacturers distribute their own pre-installation requirement guides – unfortunately mostly only once the order has been placed. Any potential IRMS user would be well advised to request this documentation together with the sales brochures and application notes beforehand. Information provided in the pre-installation requirements will make for a much more meaningful dialogue with Estates and Buildings as well as Health and Safety officers at your organization so a room suitable not only to accommodate the new instrument, but also making provision and installation of required services both physically and financially feasible is identified prior to issuing the purchase order.

One thing to bear in mind is that pre-installation requirements have to be read in a similar way to computer system requirements published by software manufacturers. In other words, a minimum requirement is just that. Meeting this means the product will be operational but not at the desired level of peak performance. As with personal computers, one should aim for the highest level of quality one can afford.

II.A.2 Laboratory Location

While planning a new or completely refurbished IRMS laboratory, the first matter to decide on is where the IRMS instrument should be sited. The choice of locale should be dictated by four considerations: (i) access, (ii) floor space, (iii) floor stability and (iv) environmental conditions (e.g. temperature and humidity).

The first three points almost always lead to a choice of ground floor or basement level accommodation, although this is not a foregone conclusion. The new laboratory should have level access or be accessed via a gently sloped ramp since the IRMS instruments are exceptionally heavy with uneven weight distribution and must therefore be transported on a forklift trolley. The doors must be wide enough to manoeuvre the instrument through with ease, if possible still within its shipping crate. The floor must be rock solid (i.e. should not transmit vibrations) and be able to support the weight of the instrument. Available floor space should also be generous enough to allow easy access to the instrument from all sides (e.g. instrument floor space plus 0.6–1 m on each side). In addition, the spatial requirements of ancillary instruments, peripherals and future equipment need to be considered as well as distance of computers from the IRMS instrument's magnet.

Another facet of the floor space consideration is temperature control. Placing an IRMS instrument in something akin to a broom closet even if compliant with access to the instrument from all sides will inevitably cause problems later with temperature control. IRMS instruments are sensitive to temperature changes in excess of 1 °C/h. Close proximity of air ducts or fans blowing hot (or cold) air directly on to the instrument will undermine any effort made to keep the laboratory's average room temperature constant. Temperature probes providing feedback information to an air-conditioning system are typically not installed at the precise spot where the instrument will be sited.

Being cognizant of the fact that some universities base laboratory space allocation on figures as low as 10 m²/person it is worth knowing that a report published in April 1998 by the National Institute of Justice and the NIST entitled *Forensic Laboratories: Handbook for Facility Planning, Design, Constructions, and Moving* (http://www.ncjrs.gov/pdffiles/168106.pdf) mentions a minimum figure of 65.03 m²/person, although this figure is a composite comprising *pro rata* space allowances for laboratory work, desk work and even communal space ('coffee room').

II.A.3 Temperature Control

Virtually all analytical instruments will benefit from an environment with constant ambient temperature, preferably around 18–24 °C and IRMS instruments are no exception. For these instruments to function reliably, constant room temperature is essential with a temperature drift of less than 1 °C/h. Air conditioning is therefore a must on the checklist for any IRMS laboratory. If money is no object, one should opt for an air-conditioning system that comes with humidity control as well. The size and capacity of the air-conditioning unit depends on the number of instruments, their respective heat output at peak as well as number of people typically present in your laboratory. In the list on the basis of which the required capacity of the air-conditioning unit will be calculated, do not forget to include additional GCs, vacuum pumps, personnel and add some spare capacity for future equipment.

Even though one can and should exploit the space volume of a laboratory as a buffer against violent temperature fluctuations (hence, the need for a spacious room rather than a broom closet), one should bear in mind in the early stages of planning where and how the heat exchangers for the air-conditioning unit will be installed since this can and often will limit the options for the location of the laboratory. So-called window-sill units will not have the capacity required to control room temperature in such a way that it meets instrument requirements.

II.A.4 Power Supply

When planning your new laboratory, make sure it will have its own dedicated power supply line from the nearest local substation. In other words, insist that your laboratory will not be the last in line of an already several times over-extended campus grid serving several other major consumers such as elevators, nuclear magnetic resonance equipment, walk-in cold storage rooms, X-ray or similar. A power supply line delivering 240 V to your IRMS instrument at 01 : 00 h in the morning but only 190 V during normal working hours is not fit for purpose.

All analytical instruments are sensitive to fluctuation in the power supply and again IRMS instruments are no exception. Apart from maintaining a constant voltage, your power supply should be free of surges, spikes and brown-outs. If that cannot be guaranteed, make sure a line conditioner or line filter is installed. Another option is to have one or more 'uninterrupted power supply' (UPS) units installed and connected between wall socket and your instrument since modern UPS units act as surge filters or line conditioners that provide a constant voltage even if your mains power supply does not. In addition, a UPS will enable you to save your data and shut down your IRMS instrument in a controlled fashion should there be a power cut. A UPS unit providing up to 2 h of backup supply is probably the best compromise in terms of expense, space requirements and additional heat output in a situation where you are not served by an emergency power generator. However, even if your mains power supply is backed up by an emergency power generator, you should still consider investing in a smaller UPS unit providing 15–30 min of power backup since there will be a short interruption of the power supply before the emergency power generator kicks in and a spike in the power when it does.

Another point worth mentioning is the number of sockets in the new laboratory in addition to the dedicated socket for the IRMS instrument served by its own circuit breaker. In a variation of Murphy's Law there never seems to be enough sockets in a laboratory and if there are, they are never where you need them. This inevitably leads to the intricate and intertwined system of tripwires across the floor otherwise known as extension leads and extension sockets. As a general rule, for an IRMS laboratory you will require at least one dedicated three-phase, 16-A five-connector power line that must be fused separately, although some IRMS manufacturers now offer the option of being run off a standard single-phase power supply. In addition, each wall should contain one dedicated, separately fused circuit feeding four blocks of four sockets each, two at floor level and two at bench-top level. In addition, consider the possibility of one or two socket cubes suspended from the ceiling in the centre of the lab.

I have always found having the three-phase power supply brought to the instrument from the ceiling to be an advantage. It removes a major trip hazard from the floor and it gives one a certain degree of latitude or flexibility to allow for minor differences between the planning stage and where the instrument is finally sited.

II.A.5 Gas Supply

In the context of an IRMS facility, gas supply means the supply of gases required to run your instruments, gases such as helium, oxygen, hydrogen, nitrogen, CO_2, CO and synthetic air (for use in a gas chromatograph's flame ionization detector) as well as a supply of clean compressed air. The over-riding rule for gas supply in instrumental analysis, in general, and IRMS, in particular, is 'the cleaner the better' – a variation on the theme of 'garbage in, garbage out'. This maxim can only be achieved if every step in the gas train is set up in such a way as to exclude leaks, atmospheric break-in and contamination. To this end, all materials used for gas supply that will come in direct contact with the gases should be made of stainless steel (i.e. all tubing, connectors, unions, valves and ferrules). This includes pressure regulators that should at least contain a stainless steel diaphragm. Regulator diaphragms made from other materials will deteriorate over time and start leaking. Note that using stainless steel tubing and then connecting it with brass unions and brass ferrules is not only a false economy, but also makes no sense from a material characteristics point of view. Brass ferrules are softer than stainless steel and will therefore not cut into the stainless steel tubing and, hence, not afford a leak tight seal. Conversely, it neither makes sense to use copper tubing and then connect that with stainless steel ferrules considering how much harder stainless steel is than copper. Stainless steel ferrules can quite easily cut through a copper tube and thus cause a massive leak. Stainless steel tubing must be clean on the inside (i.e. free of any contaminants associated with its manufacture).

Gases should be of highest purity, namely 99.998% (= 4.8 in European notation) or, even better, 99.999% (= 5.0). Ideally, the carrier gas should be of even higher quality, namely 99.9999% (= 6.0). However, helium of this quality is rather expensive and the same quality can be achieved by using 5.0 helium in conjunction with a high-capacity gas purifier. In my experience the best gas purifier on the market for the purpose of purifying helium for IRMS applications is Supelco's (Bellefonte, PA, USA) thermochemical absorption system. Apart from the demand that the carrier gas should not contain contaminants such as CO, water or hydrocarbons that would interfere with isotope ratio measurement, the other main reason for the high purity demand on the carrier gas helium are the separation columns in gas chromatographs and elemental analysers coupled to an IRMS instrument. Stationary phases of medium to high polarity used in GC are extremely sensitive to oxygen being present in the carrier gas, which leads to column deterioration and excessive column bleed. While stationary phases for packed columns used in elemental analysers are not as susceptible to breakdown as stationary phases used in gas chromatographs, they are still susceptible to contamination, which will lead to a loss of column performance. To achieve the best possible accuracy and precision of IRMS measurements, the carrier gas must therefore be free of oxygen, organic contaminants and moisture.

With the possible exception of the smaller reference gas cylinders (lecture theatre bottles), ideally all the gases should be stored outside the laboratory in well-ventilated safety cabinets or cylinder storage cages to meet Health and Safety guidelines. To safeguard against running out of helium or oxygen (for combustion) when one can afford it least, two cylinders should be connected simultaneously to a change-over regulator with a pressure dependent toggle switch. Admittedly these change-over arrays are typically five times more expensive than a standard regulator, but given the cost of a complete IRMS system are well worth the investment. The moment more than one IRMS system has to be supplied from one gas line entering the laboratory, especially carrier gas hungry systems such as EA-IRMS systems with a carrier gas throughput of 90 ml/min, a two-cylinder change-over array is a must both from a practicality point of view and from an investment protection consideration. See Figure II.A.1.

From the gas cylinder storage point, 1/2-inch stainless steel tubing should be run into the laboratory and along the walls at head height, terminating in a shut-off valve followed by a pressure regulator with a gauge. From these regulators supplied by 1/2-inch lines, 1/8-inch stainless steel tubing can be taken off and run to the instrument. Prior to its first use such a gas supply line set-up should be flushed with clean, dry nitrogen to remove air and moisture. See Figure II.A.2.

Similarly, compressed air should be supplied from an external compressor and brought into the laboratory via a dedicated line that incorporates an oil mist filter.

Figure II.A.1 Pressure triggered change-over unit for helium supply.

If your institute has such a facility, make sure it incorporates storage or buffer tanks between compressor and delivery points so in the event of a compressor failure or maintenance there will be sufficient pressure on the line to keep your system operational until such time the compressor is on-line again. In the absence of a centrally fed supply line for compressed air you will require a small compressor in the laboratory since most IRMS systems use pneumatically driven valves for one purpose or other. Running pneumatic valves that operate every couple of minutes from compressed air cylinders is not a practical or financially sustainable option. When purchasing a small compressor, the choice should be made on the basis of noise (or lack thereof), differential pressure regulation and size of the compressed air tank. For these reasons my preferred choice are Jun-Air (Benton Harbor, MI, USA) compressors (see Table II.A.1). They may not be cheap to buy, but are worth every penny in the long term.

Consider a natural gas supply line for your Bunsen burners if there is a need for minor wet chemistry work in the laboratory or for sealing off glass ampoules. Finally, an exhaust line for the vacuum pumps should be installed, either plumbed into the exhaust of an existing fume cupboard or leading directly outside. Similarly, IRMS instruments set up to analyse ^{18}O isotope abundance of CO generated from high-temperature pyrolysis of samples will require an exhaust line to extract CO from the laboratory. Should your set-up require handling of CO you will also need to install one

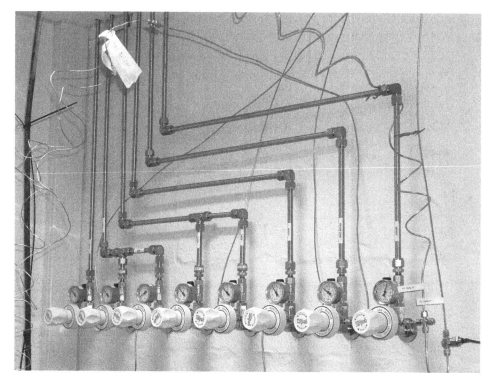

Figure II.A.2 Laboratory gas delivery manifold fed from an external gas supply.

or more CO monitors that display CO concentration (in parts per million) and sound an audible alarm when safe limits are exceeded.

As with temperature control, supplying a laboratory with gases from cylinders sited outside the building will influence the decision of where the IRMS laboratory should be located. Running stainless steel supply lines from the ground floor level to a laboratory on the third floor is an expensive exercise and, depending on total length (and volume) of the pipes, may not be feasible due to the associated drop in pressure from cylinder to delivery point.

II.A.6 Forensic Laboratory Considerations

Access to the building or building floor housing the secure suite of laboratories must be access controlled, such as by means of a swipe card system that not only controls but also monitors access and egress of a secure area. Even with such a system in place, access to the IRMS suite should be limited by a secondary access control system (e.g. swipe card plus a keypad lock).

Generally, sample reception and log-in, sample storage and sample preparation should take place in separate rooms. The installation of a safe and evidence lockers with secure locking mechanisms and anchored either in the floor or a supporting wall

Table II.A.1 List of useful tools and equipment in an IRMS laboratory.

Item	Source/supplier
High-capacity gas purifier	Supelco
OMI™ gas purifier	Supelco
Stainless steel tubing (1/8- and 1/6-inch)	Supelco
Stainless steel fittings, valves, unions, etc.	Swagelok
Tube cutter for 1/8- and 1/6-inch tubing	Alltech
Tube bender	
Heat gun	
Hand-held blow torch (crème brûlée torch)	
Retort stand	
Soap bubble flow meter	
Electronic flow meter for helium	Varian
Cable ties	
Assorted tools and tool box (one per work station)	
Torque wrench	
Needle file set	Supelco
Crimp sealer (8, 11 and 20 mm)	
Teflon tape	
Compressor	Jun-Air
Dentist's mirror	
Ferrule drill and drill bit set	
Ferrule removal tool	Valco

should be considered when deciding on the floor space for sample storage. Lockable freezers for the storage of temperature-sensitive evidence should also be included on the must-have list for this area. While sample reception and sample storage may be shared with other units within the same secure suite of forensic laboratories, the sample preparation laboratory for IRMS analysis should definitely be isolated from other wet chemistry laboratories and be fitted with secondary-level access control.

From a practical as well as a Good Laboratory Practice point of view, I cannot emphasize enough to keep sample preparation, in particular any wet chemistry, storage of chemicals and solvents strictly separate from the analytical laboratory housing the IRMS instrument to minimize the risk of cross-contaminating samples waiting for analysis as well as contamination and accidental damage of the instrument. Even weighing out samples should take place in a separate room, isolated from the instrument room, although this balance room could conceivably be a room backing on to the IRMS laboratory.

II.A.7 Finishing Touches

When choosing the floor cover, practicality should overrule appearances. A carpet-like floor cover might look nice, but is prone to induce static electricity discharges and is difficult to clean. A chemically inert surface is better suited, especially if the laboratory floor has been fitted with a drainage point at the lowest point of the floor.

Make provisions for ample work surface/bench-top area with underneath storage facilities (mobile cupboard and drawer units on lockable castors are ideal) and plenty of shelf space.

Good, meaning sufficient and evenly spaced, lighting is also important. Soft fluorescent light tubes should be suspended from the ceiling at a height of approximately 2.5 m.

From a practical point of view, a sink with running hot and cold water is desirable as long as it is not directly adjacent to the mass spectrometer. Similarly, a supply line of deionized water is a valuable thing to have. Another item on the list should be a connection to a central vacuum line should such a facility already be incorporated into the fabric of your institute.

MS laboratories are notoriously noisy chiefly due to the high vacuum systems. Accommodating the rotary pumps in ventilated yet noise-insulated cupboards will go a long way to reduce the noise level in the laboratory. However, this still leaves the high-pitched noise of the turbo-molecular pumps. Again, if money is no object a sound engineer should be consulted to suggest appropriate sound-proofing measures tailor-made to the situation at hand. Given how much the solution proposed by a sound engineer may cost, be prepared to settle for wearing ear mufflers or ear defenders.

If you want to make sure that all your planning and requests for services and infrastructure come to fruition, do not sit back and wait for the day the purchase order for the IRMS instrument will go out. Trying to address all the requirements in the manufacturer's pre-installation requirement now will be too late. If the laboratory does not meet the manufacturer's specification this may not delay delivery, but it will definitely delay installation. Most companies have a clause in their terms and conditions specifying a time frame in which the instrument has to be installed lest the warranty period is forfeit. Become engaged in the entire process from identifying the right room for the instrument to the installation of all required services. Be prepared for specifications to be watered down, the drainage point not being at the lowest point of the floor and doors not being wide enough despite crated instrument dimensions having been provided umpteen times over. By the same token, be prepared to discover that the architect, engineers, builders, plumbers, electricians and so on, will all work on the assumption that you, the forensic specialist and end-user of the IRMS laboratory, will have no idea of what you are talking about, and what you and your instrument really require. Arm yourself with a tape measure, a calliper, a volt meter (or even better a voltage monitor) and the pre-installation requirement booklet of your IRMS system, and be prepared to fight your corner every step of the way.

Table II.A.1 lists items I have found to be indispensible for running an IRMS laboratory. When mentioned, names of sources or suppliers are listed to give the reader a starting point where to obtain certain materials, tools and so on, but are not to be understood as endorsement of the best or one and only supplier. It only means I am using them and so far these products have not let me down.

Table II.A.2 List of secondary organic standards for stable isotope analysis (courtesy of Arndt Schimmelmann, University of Indiana).

	δ^2H_{VSMOW} (‰) (mean ± 1σ)	$\delta^{13}C_{VPDB}$ (‰) (mean ± 1σ)
n-Alkanes available pure and in mixtures		
C-1 methane CH$_4$ #1, CAS # 74-82-8	−160.8 ± 2.1	−38.25 ± 0.03
C-1 methane CH$_4$ #2, CAS # 74-82-8	−41.3 ± 1.3	−37.60 ± 0.03
C-12 dodecane C$_{12}$H$_{26}$, CAS # 112-40-3	−62.5 ± 2.2	−31.99 ± 0.04
C-14 tetradecane C$_{14}$H$_{30}$, CAS # 629-59-4	−71.7 ± 1.4[a]	−30.69 ± 0.03[a]
C-15 pentadecane C$_{15}$H$_{32}$, CAS # 629-62-9	−88.4 ± 1.2	−29.25 ± 0.01
C-16 hexadecane C$_{16}$H$_{34}$, CAS # 544-76-3	−76.7 ± 1.7	−30.66 ± 0.03
C-17 heptadecane C$_{17}$H$_{36}$, CAS # 629-78-7	−142.4 ± 1.7	−31.16 ± 0.03
C-18 octadecane C$_{18}$H$_{38}$, CAS # 593-45-3	−53.8 ± 2.1	−31.11 ± 0.02
C-19 nonadecane C$_{19}$H$_{40}$, CAS # 629-92-5	−118.0 ± 2.0	−33.17 ± 0.02
C-20 icosane (eicosane) C$_{20}$H$_{42}$, CAS # 112-95-8	−52.6 ± 0.8	−32.35 ± 0.04
C-21 heneicosane C$_{21}$H$_{44}$, CAS # 629-94-7	−214.7 ± 2.0[a]	−29.10 ± 0.03[a]
C-22 docosane C$_{22}$H$_{46}$, CAS # 629-97-0	−62.8 ± 1.6	−32.87 ± 0.03
C-23 tricosane C$_{23}$H$_{48}$, CAS # 638-67-5	−48.8 ± 1.4	−31.77 ± 0.01
C-24 tetracosane C$_{24}$H$_{50}$, CAS # 646-31-1	−53.0 ± 1.6	−33.34 ± 0.02
C-25 pentacosane C$_{25}$H$_{52}$ #2, CAS # 629-99-2	−125.8 ± 1.6	−30.00 ± 0.03
C-26 hexacosane C$_{26}$H$_{54}$, CAS # 630-01-3	−54.9 ± 1.5	−33.03 ± 0.03[a]
C-27 heptacosane C$_{27}$H$_{56}$ #1, CAS # 593-49-7	−227.3 ± 2.0	−28.61 ± 0.02
C-27 heptacosane C$_{27}$H$_{56}$ #2, CAS # 593-49-7	−178.2 ± 2.5	−29.56 ± 0.01
C-28 octacosane C$_{28}$H$_{58}$, CAS # 630-02-4	−49.0 ± 1.5	−32.21 ± 0.03
C-29 nonacosane C$_{29}$H$_{60}$ #1, CAS # 630-03-5	−179.3 ± 2.7[a]	−31.08 ± 0.02[a]
C-29 nonacosane C$_{29}$H$_{60}$ #2, CAS # 630-03-5	−242.6 ± 2.9	−30.07 ± 0.01
C-30 triacontane C$_{30}$H$_{62}$ #1, CAS # 638-68-6	−46.3 ± 2.1	−33.15 ± 0.02[a]
C-30 triacontane C$_{30}$H$_{62}$ #2, CAS # 638-68-6	−213.4 ± 1.2	−29.86 ± 0.01
C-32 dotriacontane C$_{32}$H$_{66}$, CAS # 544-85-4	−212.4 ± 1.0	−29.47 ± 0.02
C-34 tetratriacontane C$_{34}$H$_{70}$, CAS # 14167-59-0	−231.8 ± 1.4	−29.54 ± 0.02
C-35 pentatriacontane C$_{35}$H$_{72}$, CAS # 630-07-9	−194.8 ± 0.9	−29.84 ± 0.01
C-36 hexatriacontane C$_{36}$H$_{74}$, CAS # 630-06-8	−246.7 ± 1.4	−30.00 ± 0.03
C-37 heptatriacontane C$_{37}$H$_{76}$, CAS # 7194-84-5	−180.1 ± 1.8	−30.24 ± 0.03
C-38 octatriacontane C$_{38}$H$_{78}$, CAS # 7194-85-6	−102.6 ± 1.3	−31.49 ± 0.01
C-39 nonatriacontane C$_{39}$H$_{80}$, CAS # 7194-86-7	−218.6 ± 2.3	−28.68 ± 0.01
C-40 tetracontane C$_{40}$H$_{82}$, CAS # 4181-95-7	−106.7 ± 0.3	−32.20 ± 0.04
C-41 hentetracontane C$_{41}$H$_{84}$, CAS # 7194-87-8	−206.0 ± 1.7[a]	−28.97 ± 0.01[a]
C-44 tetratetracontane C$_{44}$H$_{90}$, CAS # 7098-22-8	−199.9 ± 2.0	−29.12 ± 0.02
C-50 pentacontane C$_{50}$H$_{102}$, CAS # 6596-40-3	−191.3 ± 1.0	−27.79 ± 0.03
Fatty acid esters		
Decanoic acid methyl ester (C10:0); C$_{11}$H$_{22}$O$_2$, methyl decanoate (solution in hexane, in sealed glass ampoule) CAS # 110-42-9	−215 ± 4	29.67 ± 0.02
Tetradecanoic acid methyl ester (C14:0), #1 C$_{15}$H$_{30}$O$_2$, ≥99%, methyl myristate, CAS # 124-10-7	−223.9 ± 1.7	−26.69 ± 0.01
Tetradecanoic acid methyl ester (C14:0), #14M C$_{17}$H$_{34}$O$_2$, ≥99%, methyl myristate, CAS # 124-10-7	−231.2 ± 1.4	−29.98 ± 0.02

Table II.A.2 *(Continued)*

	$\delta^2 H_{VSMOW}$ (‰) (mean ± 1σ)	$\delta^{13} C_{VPDB}$ (‰) (mean ± 1σ)
Tetradecanoic acid ethyl ester (C14:0), #n14E $C_{16}H_{32}O_2$, 99%, ethyl myristate, CAS # 124-06-1	−231.2 ± 2.7	−29.13 ± 0.03
Hexadecanoic acid methyl ester (C16:0) #1 $C_{17}H_{34}O_2$, ≥99%, methyl palmitate, CAS # 112-39-0	−227.9 ± 1.6	−30.74 ± 0.01
Hexadecanoic acid methyl ester (C16:0) #16M $C_{17}H_{34}O_2$, ≥99%, methyl palmitate, CAS # 112-39-0	+89.1 ± 0.8	−30.44 ± 0.02
Hexadecanoic acid methyl ester (C16:0) #n16M $C_{17}H_{34}O_2$, ≥99.5%, methyl palmitate, CAS # 112-39-0	−166.8 ± 1.7	−29.90 ± 0.03
Hexadecanoic acid ethyl ester (C16:0), #IU 16E $C_{18}H_{36}O_2$, ≥99%, ethyl palmitate, CAS # 628-97-7	−211.0 ± 1.7	−30.92 ± 0.02
Hexadecanoic acid ethyl ester (C16:0), #16E $C_{18}H_{36}O_2$, ≥99%, ethyl palmitate, CAS # 628-97-7	+275.6 ± 2.1	−27.66 ± 0.03
Hexadecanoic acid propyl ester (C16:0), #16P $C_{19}H_{38}O_2$, ≥99%, propyl palmitate, CAS # 2239-78-3	+449.3 ± 2.2	−30.03 ± 0.02
Hexadecanoic acid butyl ester (C16:0), #16B $C_{20}H_{40}O_2$, ≥99%, butyl palmitate, CAS # 111-06-8	+552.8 ± 1.7	−27.21 ± 0.01
Octadecanoic acid methyl ester (C18:0), #n18M $C_{19}H_{38}O_2$, ~99%, methyl stearate, CAS # 112-61-8	−206.2 ± 1.7	−23.24 ± 0.01
Octadecanoic acid ethyl ester (C18:0), #18E $C_{20}H_{40}O_2$, ~99%, ethyl stearate, CAS # 111-61-5	−214.2 ± 0.7	−28.22 ± 0.01
Icosanoic acid methyl ester (C20:0) #2, $C_{21}H_{42}O_2$, ≥99% (formerly named eicosanoic acid methyl ester), methyl icosanoate, CAS # 1120-28-1	−166.7 ± 0.3	−30.68 ± 0.02
Icosanoic acid methyl ester (C20:0) #20M, $C_{21}H_{42}O_2$, ≥99% (formerly named eicosanoic acid methyl ester), methyl icosanoate, CAS # 1120-28-1	+508.8 ± 1.9	−28.42 ± 0.01
Icosanoic acid methyl ester (C20:0) #X, $C_{21}H_{42}O_2$, ≥99%(formerly named eicosanoic acid methyl ester), methyl icosanoate, CAS # 1120-28-1	+75.7 ± 1.1	−6.91 ± 0.04
Icosanoic acid ethyl ester (C20:0), #20E $C_{22}H_{44}O_2$, ≥99% (formerly named eicosanoic acid ethyl ester), ethyl icosanoate, CAS # not available	+340.8 ± 1.9	−24.80 ± 0.01
Icosanoic acid ethyl ester (C20:0), #20E2 $C_{22}H_{44}O_2$, ≥99% (formerly named eicosanoic acid ethyl ester), ethyl icosanoate, CAS # not available	−195.5 ± 1.2	−26.10 ± 0.03
Icosanoic acid propyl ester (C20:0), #20P $C_{23}H_{46}O_2$, ≥99% (formerly named eicosanoic acid propyl ester), propyl icosanoate, CAS # not available	+191.9 ± 1.6	−29.00 ± 0.02
Icosanoic acid butyl ester (C20:0), #20B $C_{24}H_{48}O_2$, ≥99% (formerly named eicosanoic acid butyl ester), butyl icosanoate, CAS # 26718-91-2	+1.5 ± 1.4	−28.63 ± 0.02
Tetracosanoic acid methyl ester (C24:0), $C_{25}H_{50}O_2$, ≥99%, methyl lignocerate, CAS # 2442-49-1	−179.3 ± 1.7	−26.57 ± 0.02
Tricontanoic acid methyl ester (C30:0), $C_{31}H_{62}O_2$, CAS # 629-83-4	−189.4 ± 2.0[a]	−26.33 ± 0.02[a]

(Continued)

Table II.A.2 (Continued)

	δ^2H_{VSMOW} (‰) (mean ± 1σ)	$\delta^{13}C_{VPDB}$ (‰) (mean ± 1σ)
Other compounds for GC/C-IRMS		
n-Butylcyclohexane, ≥99%, CAS # 1678-93-9	−53.3 ± 1.4	−24.47 ± 0.01
t-Butylcyclohexane, ≥99%, CAS # 1678-98-4	−70.6 ± 1.9	−26.08 ± 0.03
5α-Androstane, $C_{19}H_{32}$ #1, CAS # 438-22-2	−256.2 ± 1.4	−28.61 ± 0.01
5α-Androstane, $C_{19}H_{32}$ #2, CAS # 438-22-2	−297.4 ± 2.2	−31.64 ± 0.01
Coronene, $C_{24}H_{12}$, 99%, CAS # 191-07-1	−48.3 ± 0.9	−26.81 ± 0.04
Coumarin, $C_9H_6O_2$, ≥99.5%, CAS # 91-64-5	+82.3 ± 1.2	−35.60 ± 0.01
Dibenzothiophene, $C_{12}H_8S$, 99.4%, CAS # 132-65-0	+84.9 ± 1.8	−27.68 ± 0.01
Glyceryl tripalmitate, $C_{51}H_{98}O_5$, ≥99.0%, CAS # 555-44-2	−215.1 ± 0.9	−30.12 ± 0.01
Phenanthrene, $C_{14}H_{10}$, ≥99.5%, CAS # 85-01-8	−84.1 ± 1.3	−25.39 ± 0.03
Squalane, $C_{30}H_{62}$ (2,6,10,15,19,23-hexamethyltetracosane), CAS # 111-01-3	−168.9 ± 1.9	−20.49 ± 0.02
Derivatizing agents for GC/C-IRMS		
Acetic acid anhydride, $(C_2H_3O)_2O$, CAS # 108-24-7; 99.5%, aliquots of ∼1 ml sealed in 2005 under argon in brown glass ampoules, stored in freezer	−133.2 ± 2.1	−20.98 ± 0.03
Phthalic acid, $C_8H_6O_4$, CAS # 88-99-3; δ^2H measured in Na-phthalate to exclude exchangeable carboxyl H; $\delta^{13}C$ measured in free acid	−95.5 ± 2.2	−27.21 ± 0.02
Methanol, CH_3OH, 99.8%, anhydrous, CAS # 67-56-1 (the δ^2H values characterize bulk hydrogen and methyl hydrogen (calculated after subtracting the OH hydrogen that was liberated in reactions between MeOH and sodium metal); $\delta^{13}C$ was determined in bulk methanol; 5 ml sealed in glass ampoule)	bulk methanol: −112.6 ± 0.8; 3 methyl hydrogen: −141 ± 3	−46.77 ± 0.04
Compounds for EA- or TC/EA-IRMS		
Hexatriacontane, $C_{36}H_{74}$, CAS # 630-06-8	−246.7 ± 1.4	−30.00 ± 0.03
Acetanilide, $C_6H_5NHCOCH_3$, CAS # 103-84-4; $\delta^{15}N_{AIR}$ = +1.15 ± 0.05‰		−29.52 ± 0.02
Benzoic acid #A, C_6H_5COOH, CAS # 65-85-0; $\delta^{18}O_{VSMOW}$ = +23.2‰ (IAEA)		−28.81 (Coplen et al., 2006)
Benzoic acid #B, C_6H_5COOH, CAS # 65-85-0; enriched in ^{18}O; $\delta^{18}O_{VSMOW}$ = +71.4‰ (IAEA)		−28.85 (Coplen et al., 2006)
Urea, CH_4N_2O, ≥99.5%, CAS # 58069-82-2; $\delta^{15}N_{AIR}$ = +017 ± 0.02‰		−34.21 ± 0.05
Corn starch, $(CH_2O)n$, ≥99.5%, CAS # not available		−11.01 ± 0.02
Coumarin, $C_9H_6O_2$, ≥99.5%, CAS # 91-64-5	+82.3 ± 1.2	−35.60 ± 0.01
Phthalic acid, $C_8H_6O_4$, CAS # 88-99-3 δD is not stable due to exchangeable carboxyl hydrogen		−27.21 ± 0.02

[a]Additional repeat measurements are pending.
CAS #, Chemical Abstracts Service number.

Last, but by no means least, apart from ordering international calibration and reference materials from the IAEA in Vienna (http://www.iaea.org/programmes/aqcs/ordering_rms.htm), start to think about which organic secondary standards you may require. One source of such material has already been mentioned briefly in Chapter II.6.1 and a full listing of all the compounds available from Arndt Schimmelman's laboratory at the University of Indiana is given in Table II.A.2.

References Part II

Aitken, C.G.G. and Lucy, D. (2002) Estimation of the quantity of a drug in a consignment from measurements on a sample. *Journal of Forensic Sciences*, **47**, 968–975.

Aitken, C.G.G. and Lucy, D. (2004) Evaluation of trace evidence in the form of multivariate data. *Journal of the Royal Statistical Society Series C Applied Statistics*, **53**, 109–122.

Aitken, C.G.G., Lucy, D., Zadora, G. and Curran, J.M. (2006) Evaluation of transfer evidence for three-level multivariate data with the use of graphical models. *Computational Statistics and Data Analysis*, **50**, 2571–2588.

Aitken, C.G.G., Zadora, G. and Lucy, D. (2007) A two-level model for evidence evaluation. *Journal of Forensic Sciences*, **52**, 412–419.

Barrie, A., Bricout, J. and Koziet, J. (1984) Gas chromatography – stable isotope ratio analysis at natural abundance levels. *Biomedical Mass Spectrometry*, **11**, 583–588.

Begley, I.S. and Scrimgeour, C.M. (1996) On-line reduction of H_2O for delta H-2 and delta O-18 measurement by continuous-flow isotope ratio mass spectrometry. *Rapid Communications in Mass Spectrometry*, **10**, 969–973.

Begley, I.S. and Scrimgeour, C.M. (1997) High-precision delta H-2 and delta O-18 measurement for water and volatile organic compounds by continuous-flow pyrolysis isotope ratio mass spectrometry. *Analytical Chemistry*, **69**, 1530–1535.

Bell, J.G., Preston, T., Henderson, R.J., Strachan, F., Bron, J.E., Cooper, K. and Morrison, D.J. (2007) Discrimination of wild and cultured European sea bass (*Dicentrarchus labrax*) using chemical and isotopic analyses. *Journal of Agricultural and Food Chemistry*, **55**, 5934–5941.

Benson, S.J. (2009) Introduction of isotope ratio mass spectrometry (IRMS) for the forensic analysis of explosives. PhD Thesis. University of Technology, Sydney.

Birchall, J., O'Connell, T.C., Heaton, T.H.E. and Hedges, R.E.M. (2005) Hydrogen isotope ratios in animal body protein reflect trophic level. *Journal of Animal Ecology*, **74**, 877–881.

Bolck, A., Weyermann, C., Dujourdy, L., Esseiva, P. and van den Berg, J. (2009) Different likelihood ratio approaches to evaluate the strength of evidence of MDMA tablet comparisons. *Forensic Science International*, **191**, 42–51.

Bowen, G.J., Chesson, L., Nielson, K., Cerling, T.E. and Ehleringer, J.R. (2005) Treatment methods for the determination of delta H-2 and delta O-18 of hair keratin by continuous-flow isotope-ratio mass spectrometry. *Rapid Communications in Mass Spectrometry*, **19**, 2371–2378.

Bowen, G.J., Wassenaar, L.I. and Hobson, K.A. (2005) Global application of stable hydrogen and oxygen isotopes to wildlife forensics. *Oecologia*, **143**, 337–348.

Boyd, T.J., Osburn, C.L., Johnson, K.J., Birgl, K.B. and Coffin, R.B. (2006) Compound-specific isotope analysis coupled with multivariate statistics to source-apportion hydrocarbon mixtures. *Environmental Science and Technology*, **40**, 1916–1924.

Brand, W.A. (1996) High precision isotope ratio monitoring techniques in mass spectrometry. *Journal of Mass Spectrometry*, **31**, 225–235.

Brand, W.A. and Coplen, T.B. (2001) An interlaboratory study to test instrument performance of hydrogen dual-inlet isotope-ratio mass spectrometers. *Fresenius' Journal of Analytical Chemistry*, **370**, 358–362.

Brand, W.A., Tegtmeyer, A.R. and Hilkert, A. (1994) Compound-specific isotope analysis – extending toward N-15 N-14 and O-18 O-16. *Organic Geochemistry*, **21**, 585–594.

Brand, W.A., Coplen, T.B., Aerts-Bijma, A.T., Boehlke, J.K., Gehre, M., Geilmann, H., Groning, M., Jansen, H.G., Meijer, H.A.J., Mroczkowski, S.J., Qi, H.P., Soergel, K., Stuart-Williams, H., Weise, S.M. and Werner, R.A. (2009a) Comprehensive inter-laboratory calibration of reference materials for $d^{18}O$ versus VSMOW using various on-line high-temperature conversion techniques. *Rapid Communications in Mass Spectrometry*, **23**, 999–1019.

Brand, W.A., Huang, L., Mukai, H., Chivulescu, A., Richter, J. and Rothe, M. (2009b) How well do we know VPDB? Variability of $d^{13}C$ and $d^{18}O$ in CO_2 generated from NBS19-calcite. *Rapid Communications in Mass Spectrometry*, **23**, 915–926.

Brenna, J.T. (1994) High-precision gas isotope ratio mass-spectrometry – recent advances in instrumentation and biomedical applications. *Accounts of Chemical Research*, **27**, 340–346.

Brenna, J.T., Corso, T.N., Tobias, H.J. and Caimi, R.J. (1997) High-precision continuous-flow isotope ratio mass spectrometry. *Mass Spectrometry Reviews*, **16**, 227–258.

Brereton, R.G. (2006) *Chemometrics – Data Analysis for the Laboratory and Chemical Plant*, John Wiley & Sons, Ltd, Chichester.

Caimi, R.J. and Brenna, J.T. (1993) High-precision liquid chromatography-combustion isotope ratio mass spectrometry. *Analytical Chemistry*, **65**, 3497–3500.

Caimi, R.J. and Brenna, J.T. (1995a) High-sensitivity liquid-chromatography combustion isotope ratio mass spectrometry of fat-soluble vitamins. *Journal of Mass Spectrometry*, **30**, 466–472.

Caimi, R.J. and Brenna, J.T. (1995b) High-sensitivity, high-precision liquid-source IRMS of proteins and nonvolatile lipids. *Abstracts of Papers of the American Chemical Society*, **210**, 41-GEOC.

Caimi, R.J. and Brenna, J.T. (1996) Direct analysis of carbon isotope variability in albumins by liquid flow-injection isotope ratio mass spectrometry. *Journal of the American Society for Mass Spectrometry*, **7**, 605–610.

Caimi, R.J. and Brenna, J.T. (1997) Quantitative evaluation of carbon isotopic fractionation during reversed-phase high-performance liquid chromatography. *Journal of Chromatography A*, **757**, 307–310.

Caimi, R.J., Houghton, L.A. and Brenna, J.T. (1994) Condensed-phase carbon isotopic standards for compound-specific isotope analysis. *Analytical Chemistry*, **66**, 2989–2991.

Carter, J.F., Hill, J.C., Doyle, S. and Lock, C. (2009) Results of four inter-laboratory comparisons provided by the Forensic Isotope Ratio Mass Spectrometry (FIRMS) network. *Science & Justice*, **49**, 127–137.

Champod, C., Evett, I.W. and Kuchler, B. (2001) Earmarks as evidence: a critical review. *Journal of Forensic Sciences*, **46**, 1275–1284.

Chesson, L.A., Podlesak, D.W., Cerling, T.E. and Ehleringer, J.R. (2009) Evaluating uncertainty in the calculation of non-exchangeable hydrogen fractions within organic materials. *Rapid Communications in Mass Spectrometry*, **23**, 1275–1280.

Coplen, T.B. (1988) Normalization of oxygen and hydrogen isotope data. *Chemical Geology*, **72**, 293–297.
Coplen, T.B. (1994) Reporting of stable hydrogen, carbon, and oxygen isotopic abundances. *Pure and Applied Chemistry*, **66**, 273–276.
Coplen, T.B. (1995) Reporting of stable hydrogen, carbon, and oxygen isotopic abundances [Technical report]. *Geothermics*, **24**, 708–712.
Coplen, T.B. (1996) More uncertainty than necessary. *Paleoceanography*, **11**, 369–370.
Coplen, T.B. and Ramendik, G.I. (1998) Summary of the IUPAC recommendations for the publication of 'delta' values for H, C, and O stable isotope ratios. *Geokhimiya*, **3**, 334–336.
Coplen, T.B., Brand, W.A., Gehre, M., Groning, M., Meijer, H.A.J., Toman, B. and Verkouteren, R.M. (2006a) After two decades a second anchor for the VPDB delta C-13 scale. *Rapid Communications in Mass Spectrometry*, **20**, 3165–3166.
Coplen, T.B., Brand, W.A., Gehre, M., Groning, M., Meijer, H.A.J., Toman, B. and Verkouteren, R.M. (2006b) New guidelines for delta C-13 measurements. *Analytical Chemistry*, **78**, 2439–2441.
Corso, T.N. and Brenna, J.T. (1997) High-precision position-specific isotope analysis. *Proceedings of the National Academy of Sciences of the United States of America*, **94**, 1049–1053.
Cox, D.R. and Solomon, P.J. (2002) *Components of Variance*, Chapman & Hall/CRC Press, Boca Raton, FL.
de Groot, P.A. (2004) *Handbook of Stable Isotope Analytical Techniques – Volume 1*, Elsevier, Amsterdam.
de Korompay, A., Hill, J.C., Carter, J.F., NicDaéid, N. and Sleeman, R. (2008) Supported liquid–liquid extraction of the active ingredient (3,4-methylenedioxymethylamphetamine) from ecstasy tablets for isotopic analysis. *Journal of Chromatography A*, **1178**, 1–8.
Eakin, P.A., Fallick, A.E. and Gerc, J. (1992) Some instrumental effects in the determination of stable carbon isotope ratios by gas chromatography-isotope ratio mass spectrometry. *Chemical Geology*, **101**, 71–79.
Ehleringer, J.R., Bowen, G.J., Chesson, L.A., West, A.G., Podlesak, D.W. and Cerling, T.E. (2008) Hydrogen and oxygen isotope ratios in human hair are related to geography. *Proceedings of the National Academy of Sciences of the United States of America*, **105**, 2788–2793.
Ellis, L. and Fincannon, A.L. (1998) Analytical improvements in irm-GC/MS analyses: advanced techniques in tube furnace design and sample preparation. *Organic Geochemistry*, **29**, 1101–1117.
Evershed, R.P. (1996) High-resolution triacylglycerol mixture analysis using high-temperature gas chromatography mass/spectrometry with a polarizable stationary phase, negative ion chemical ionization, and mass-resolved chromatography. *Journal of the American Society for Mass Spectrometry*, **7**, 350–361.
Evett, I., Jackson, G., Lambert, J.A. and McCrossan, S. (2000) The impact of the principles of evidence interpretation on the structure and content of statements. *Science & Justice*, **40**, 233–239.
Farmer, N., Meier-Augenstein, W. and Lucy, D. (2009) Stable isotope analysis of white paints and likelihood ratios. *Science & Justice*, **49**, 114–119.
Filer, C.N. (1999) Isotopic fractionation of organic compounds in chromatography. *Journal of Labelled Compounds and Radiopharmaceuticals*, **42**, 169–197.
Fraser, I. and Meier-Augenstein, W. (2007) Stable ^{2}H isotope analysis of human hair and nails can aid forensic human identification. *Rapid Communications in Mass Spectrometry*, **21**, 3279–3285.

Fraser, I., Meier-Augenstein, W. and Kalin, R.M. (2006) The role of stable isotopes in human identification: a longitudinal study into the variability of isotopic signals in human hair and nails. *Rapid Communications in Mass Spectrometry*, **20**, 1109–1116.

Glaser, B. and Amelung, W. (2002) Determination of C-13 natural abundance of amino acid enantiomers in soil: methodological considerations and first results. *Rapid Communications in Mass Spectrometry*, **16**, 891–898.

Hobson, K.A., Atwell, L. and Wassenaar, L.I. (1999) Influence of drinking water and diet on the stable-hydrogen isotope ratios of animal tissues. *Proceedings of the National Academy of Sciences of the United States of America*, **96**, 8003–8006.

Hobson, K.A., Bowen, G.J., Wassenaar, L.I., Ferrand, Y. and Lormee, H. (2004) Using stable hydrogen and oxygen isotope measurements of feathers to infer geographical origins of migrating European birds. *Oecologia*, **141**, 477–488.

Hofmann, D. and Brand, W.A. (1996) Microcombustion of ng amounts of carbon in nonvolatile materials for isotope ratio evaluation. *Isotopes in Environmental and Health Studies*, **32**, 255–262.

Hofmann, D., Jung, K., Segschneider, H.-J., Gehre, M. and Schüürmann, G. (1995) $^{15}N/^{14}N$ analysis of amino acids with GC-C-IRMS – methodical investigations and ecotoxicological applications. *Isotopes in Environmental and Health Studies*, **31**, 367–375.

Juchelka, D., Beck, T., Hener, U., Dettmar, F. and Mosandl, A. (1998) Multidimensional gas chromatography coupled on-line with isotope ratio mass spectrometry (MDGC-IRMS): progress in the analytical authentication of genuine flavor components. *Journal of High Resolution Chromatography*, **21**, 145–151.

Kelly, J.F., Bridge, E.S., Fudickar, A.M. and Wassenaar, L.I. (2009) A test of comparative equilibration for determining non-exchangeable stable hydrogen isotope values in complex organic materials. *Rapid Communications in Mass Spectrometry*, **23**, 2316–2320.

Kelly, J.F., Ruegg, K.C. and Smith, T.B. (2005) Combining isotopic and genetic markers to identify breeding origins of migrant birds. *Ecological Applications*, **15**, 1487–1494.

Krummen, M., Hilkert, A.W., Juchelka, D., Duhr, A., Schluter, H.J. and Pesch, R. (2004) A new concept for isotope ratio monitoring liquid chromatography/mass spectrometry. *Rapid Communications in Mass Spectrometry*, **18**, 2260–2266.

Leckrone, K.J. and Hayes, J.M. (1998) Water-induced errors in continuous-flow carbon isotope ratio mass spectrometry. *Analytical Chemistry*, **70**, 2737–2744.

Matthews, D.E. and Hayes, J.M. (1978) Isotope-ratio-monitoring gas chromatography-mass spectrometry. *Analytical Chemistry*, **50**, 1465–1473.

Matucha, M. (1995) Isotope effects (IEs) in gas chromatography (GC) of labelled compounds (LCs), in *Synthesis and Applications of Isotopically Labelled Compounds* (ed. J. Allen), John Wiley & Sons, Ltd, Chichester, pp. 489–494.

Matucha, M., Jockisch, W., Verner, P. and Anders, G. (1991) Isotope effect in gas-liquid-chromatography of labeled compounds. *Journal of Chromatography*, **588**, 251–258.

McCullagh, J.S.O., Juchelka, D. and Hedges, R.E.M. (2006) Analysis of amino acid C-13 abundance from human and faunal bone collagen using liquid chromatography/isotope ratio mass spectrometry. *Rapid Communications in Mass Spectrometry*, **20**, 2761–2768.

Meehan, T.D., Giermakowski, J.T. and Cryan, P.M. (2004) GIS-based model of stable hydrogen isotope ratios in North American growing-season precipitation for use in animal movement studies. *Isotopes in Environmental and Health Studies*, **40**, 291–300.

Meier-Augenstein, W. (1995) Online recording of C-13 C-12 ratios and mass-spectra in one gas-chromatographic analysis. *Journal of High Resolution Chromatography*, **18**, 28–32.

Meier-Augenstein, W. (1997) A reference gas inlet module for internal isotopic calibration in high precision gas chromatography combustion-isotope ratio mass spectrometry. *Rapid Communications in Mass Spectrometry*, **11**, 1775–1780.

Meier-Augenstein, W. (1999a) Applied gas chromatography coupled to isotope ratio mass spectrometry. *Journal of Chromatography A*, **842**, 351–371.

Meier-Augenstein, W. (1999b) Use of gas chromatography-combustion-isotope ratio mass spectrometry in nutrition and metabolic research. *Current Opinion in Clinical Nutrition and Metabolic Care*, **2**, 465–470.

Meier-Augenstein, W. (2002) Stable isotope analysis of fatty acids by gas chromatography-isotope ratio mass spectrometry. *Analytica Chimica Acta*, **465**, 63–79.

Meier-Augenstein, W. (2004) GC and IRMS technology for ^{13}C and ^{15}N analysis of organic compounds and related gases, in *Handbook of Stable Isotope Analytical Techniques* (ed. P.A. de Groot), Elsevier, Amsterdam, pp. 153–176.

Meier-Augenstein, W. and Fraser, I. (2008) Forensic isotope analysis leads to identification of a mutilated murder victim. *Science & Justice*, **48**, 153–159.

Meier-Augenstein, W., Hoffmann, G.F., Holmes, B., Jones, J.L., Nyhan, W.L. and Sweetman, L. (1993) Use of a thick-film capillary column for the analysis of organic acids in body fluids. *Journal of Chromatography B*, **615**, 127–135.

Meier-Augenstein, W., Brand, W., Hoffmann, G.F. and Rating, D. (1994) Bridging the information gap between isotope ratio mass spectrometry and conventional mass spectrometry. *Biological Mass Spectrometry*, **23**, 376–378.

Meier-Augenstein, W., Rating, D., Hoffmann, G.F., Wendel, U., Matthiesen, U. and Schadewaldt, P. (1995) Determination of ^{13}C enrichment by conventional GC-MS and GC-(MS)-C-IRMS. *Isotopes in Environmental and Health Studies*, **31**, 261–266.

Meier-Augenstein, W., Watt, P.W. and Langhans, C.D. (1996) Influence of gas-chromatographic parameters on measurement of ^{13}C/^{12}C isotope ratios by gas-liquid chromatography combustion isotope ratio mass spectrometry. 1. *Journal of Chromatography A*, **752**, 233–241.

Merritt, D.A., Brand, W.A. and Hayes, J.M. (1994) Isotope-ratio-monitoring gas chromatography-mass spectrometry – methods for isotopic calibration. *Organic Geochemistry*, **21**, 573–583.

Merritt, D.A., Freeman, K.H., Ricci, M.P., Studley, S.A. and Hayes, J.M. (1995) Performance and optimization of a combustion interface for isotope ratio monitoring gas chromatography/mass spectrometry. *Analytical Chemistry*, **67**, 2461–2473.

Metges, C.C. and Petzke, K.J. (1999) The use of GC-C-IRMS for the analysis of stable isotope enrichment in nitrogenous compounds, in *Methods of Investigation of Amino Acid and Protein Metabolism* (ed. A.E. El-Khoury), CRC Press, Boca Raton, FL, pp. 121–134.

Mudge, S.M., Belanger, S.E. and Nielsen, A.M. (2008) *Fatty Alcohols: Anthropogenic and Natural Occurrence in the Environment*, RSC Publishing, Cambridge.

Nelson, S.T. (2000) A simple, practical methodology for routine VSMOW/SLAP normalization of water samples analyzed by continuous flow methods. *Rapid Communications in Mass Spectrometry*, **14**, 1044–1046.

Nitz, S., Weinreich, B. and Drawert, F. (1992) Multidimensional gas chromatography-isotope ratio mass spectrometry (MDGC-IRMS). A. System description and technical requirements. *Journal of High Resolution Chromatography*, **15**, 387–391.

Pierrini, G., Doyle, S., Champod, C., Taroni, F., Wakelin, D. and Lock, C. (2007) Evaluation of preliminary isotopic analysis (C-13 and N-15) of explosives: a likelihood ratio approach to assess the links between Semtex samples. *Forensic Science International*, **167**, 43–48.

Pietzsch, J., Nitzsche, S., Wiedemann, B., Julius, U., Leonhardt, W. and Hanefeld, M. (1995) Stable-isotope ratio analysis of amino acids – the use of N(O)-ethoxycarbonyl ethyl-ester derivatives and gas chromatography-mass spectrometry. *Journal of Mass Spectrometry*, **30**, S129–S135.

Preston, T. and Owens, N.J.P. (1983) Interfacing an automatic elemental analyser with an isotope ratio mass spectrometer: the potential for fully automated total nitrogen and nitrogen-15 analysis. *Analyst*, **108**, 971–977.

Preston, T. and Owens, N.J.P. (1985) Preliminary ^{13}C measurements using a gas chromatograph interfaced to an isotope ratio mass spectrometer. *Biomedical Mass Spectrometry*, **12**, 510–513.

Preston, T. and Slater, C. (1994) Mass-spectrometric analysis of stable-isotope-labeled amino-acid tracers. *Proceedings of the Nutrition Society*, **53**, 363–372.

Prosser, S.J. and Scrimgeour, C.M. (1995) High-precision determination of H-2/H-1 in H_2 and H_2O by continuous-flow isotope ratio mass spectrometry. *Analytical Chemistry*, **67**, 1992–1997.

Qi, H.P., Coplen, T.B., Geilmann, H., Brand, W.A. and Bohlke, J.K. (2003) Two new organic reference materials for delta C-13 and delta N-15 measurements and a new value for the delta C-13 of NBS 22 oil. *Rapid Communications in Mass Spectrometry*, **17**, 2483–2487.

Rautenschlein, M., Habfast, K. and Brand, W.A. (1990) High-precision measurement of ^{13}C/^{12}C ratios by on-line combustion of GC eluates and isotope ratio mass spectrometry, in *Stable Isotopes in Paediatric, Nutritional and Metabolic Research* (eds T.E. Chapman, R. Berger, D. Reijngourd and A. Okken), Intercept, Andover, pp. 133–148.

Rennie, M.J., MeierAugenstein, W., Watt, P.W., Patel, A., Begley, I.S. and Scrimgeour, C.M. (1996) Use of continuous-flow combustion MS in studies of human metabolism. *Biochemical Society Transactions*, **24**, 927–932.

Reynard, L.M. and Hedges, R.E.M. (2008) Stable hydrogen isotopes of bone collagen in palaeodietary and palaeoenvironmental reconstruction. *Journal of Archaeological Science*, **35**, 1934–1942.

Ricci, M.P., Merritt, D.A., Freeman, K.H. and Hayes, J.M. (1994) Acquisition and processing of data for isotope-ratio-monitoring mass spectrometry. *Organic Geochemistry*, **21**, 561–571.

Rieley, G. (1994) Derivatization of organic compounds prior to gas chromatographic combustion-isotope ratio mass-spectrometric analysis – identification of isotope fractionation processes. *Analyst*, **119**, 915–919.

Schimmelmann, A. (1991) Determination of the concentration and stable isotopic composition of nonexchangeable hydrogen in organic matter. *Analytical Chemistry*, **63**, 2456–2459.

Sharp, Z.D., Atudorei, V. and Durakiewicz, T. (2001) A rapid method for determination of hydrogen and oxygen isotope ratios from water and hydrous minerals. *Chemical Geology*, **178**, 197–210.

Sharp, Z.D., Atudorei, V., Panarello, H.O., Fernandez, J. and Douthitt, C. (2003) Hydrogen isotope systematics of hair: archeological and forensic applications. *Journal of Archaeological Science*, **30**, 1709–1716.

Smith, A.D. and Dufty, A.M. (2005) Variation in the stable-hydrogen isotope composition of Northern Goshawk feathers: relevance to the study of migratory origins. *Condor*, **107**, 547–558.

Teffera, Y., Kusmierz, J.J. and Abramson, F.P. (1996) Continuous-flow isotope ratio mass spectrometry using the chemical reaction interface with either gas or liquid chromatographic introduction. *Analytical Chemistry*, **68**, 1888–1894.

Tegtmeyer, A.R., Lenz, A., Ricci, M. and Brand, W.A. (1992) Isodat – a complex data system for online isotope ratio determination. *Isotopenpraxis*, **28**, 113.

Wassenaar, L.I. and Hobson, K.A. (2000) Stable-carbon and hydrogen isotope ratios reveal breeding origins of red-winged blackbirds. *Ecological Applications*, **10**, 911–916.

Wassenaar, L.I. and Hobson, K.A. (2003) Comparative equilibration and online technique for determination of non-exchangeable hydrogen of keratins for use in animal migration studies. *Isotopes in Environmental and Health Studies*, **39**, 211–217.

Wassenaar, L.I. and Hobson, K.A. (2006) Stable-hydrogen isotope heterogeneity in keratinous materials: mass spectrometry and migratory wildlife tissue subsampling strategies. *Rapid Communications in Mass Spectrometry*, **20**, 2505–2510.

Weber, D., Gensler, M. and Schmidt, H.L. (1997) Metabolic and isotopic correlations between D-glucose, L-ascorbic acid and L-tartaric acid. *Isotopes in Environmental and Health Studies*, **33**, 151–155.

Werner, R.A. and Brand, W.A. (2001) Referencing strategies and techniques in stable isotope ratio analysis. *Rapid Communications in Mass Spectrometry*, **15**, 501–519.

Woodbury, S.E., Evershed, R.P. and Rossell, J.B. (1998) Delta C-13 analyses of vegetable oil fatty acid components, determined by gas chromatography combustion isotope ratio mass spectrometry, after saponification or regiospecific hydrolysis. *Journal of Chromatography A*, **805**, 249–257.

Part III
Stable Isotope Forensics: Case Studies and Current Research

Chapter III.1
Forensic Context

Chemical analyses carried out in forensic science laboratories primarily aim for the *identification* and *comparison* of various samples with the intention of linking the samples of concern to a specific person or event. *Identification* (i.e. establishing chemical and structural identity by traditional methods of chemical analysis) is often achieved through the characterization of one or more specific sample constituents, while *comparison* often involves the identification and quantification of multiple components in the samples of interest. With automated instrumentation widely available, highly specific spectroscopic and mass spectrometry (MS)-based technologies are now, in most modern forensic science laboratories, the most valuable tool used to achieve these analytical goals.

That being said, it cannot be emphasized enough that state-of-the-art technology and seamless documentation are not a substitute for proficiency, neither do they guarantee a particular procedure or an analytical method to be fit-for-purpose or appropriately applied in a given context. Stating the obvious, a method for extracting and analysing the composition of organic stains in fabric samples by gas chromatography (GC)-MS is not fit-for-purpose for analysing the composition of inclusions in a glass fragment. Similarly, a GC-flame ionization detection (FID) method for determining the presence of, for example, aromatic hydrocarbons using relative retention is no longer fit-for-purpose if key conditions of the originally developed GC-MS method on which the GC-FID method is based are no longer met. Such key conditions could be the polarity and, hence, selectivity of the stationary phase of the GC column or the time window in which the internal standard and retention time anchor elutes from the GC column to be free of any other peaks since a flame ionization detector only detects a peak, but does not identify the compound causing it. Even though these examples may seem glaringly obvious mistakes and, thus, easy to avoid, one should never assume this sort of thing does not happen or that peer review would provide a cast-iron guarantee for such flawed methods not to be published. Like GC-FID, compound-specific isotope analysis (CSIA) by continuous flow isotope ratio MS (CF-IRMS) systems coupled to a gas chromatograph via an on-line combustion interface (GC/C-IRMS) only detects a peak without being able to ascertain the chemical identity of the compound causing it. First, the peak detected by the IRMS is comprised solely of CO_2 because, secondly, the compound was combusted after eluting from the GC since as we now know IRMS systems measure isotope ratios of permanent gas ions such as CO_2^+. So, peak and, hence, compound identification is either achieved through use of a GC(-MS)/C-IRMS hybrid system capable of simultaneous and unambiguous compound peak identification (as described in Section II.2.4.1) or rests on the comparison of chromatographic parameters such as relative retention established by the GC-MS method on which the GC/C-IRMS should be based. 'Should' is the operative word here, since there are laboratories out there who still base compound identification on comparing retention parameters established using an apolar column such as a DB-5 with retention parameters obtained from a polar column such as a DB-1701 (Saudan *et al.*, 2006) without giving any thought to the fact that different polarity and, hence, different selectivity of a stationary phase can result in a reversal of elution order of two or more compound peaks. In other words, while on the apolar column compound A would elute before compound B, on a polar column this order becomes reversed, with compound B now eluting before compound A.

It was a long-established legal requirement that scientific methods used to arrive at any conclusion being given in evidence in a court of law had to be recognized and accepted by the scientific authorities, governing bodies or communities (peers) of the science area in question (*Frye versus United States*, 1923; 54 App. D.C. 46, 47, 293 F. 1013, 1014). However, the seemingly sound Frye principle did not give judges much discretion to probe the word of so-called expert witnesses. This situation changed with the case of *Daubert versus Merrell Dow Pharmaceuticals, Inc.* (1993; 509 U.S. 579, 125L. Ed. 2d 469, 113 S. Ct. 2786) that put emphasis on the 'sound methodology' test giving judges the discretion, if not the duty, to examine the evidence themselves to determine whether it is fit-for-purpose rather than to rely merely on a potentially flawed token of scientific acceptance such as publication in an obscure yet peer-reviewed journal. The latter is a particularly contentious point given the generally accepted distinction between high- and low-impact-factor journals, although even this metric is subject to a certain bias since it is not adjusted for the relative size of a particular area of scientific research. However, the issue of high- and low-impact-factor journals aside, it is an undisputable fact that articles get published without being properly scrutinised because the article was technically outside the scope of the journal it was submitted to or because neither the review nor the editorial process spotted a problem such as the aforementioned peak matching of peaks from GC columns of different selectivity (Saudan *et al.*, 2006) or flawed conclusions drawn on the basis of observations made using analytical instrumentation lacking the resolution required for the task at hand (Sharma and Lahiri, 2005; Sharma, Purkait and Lahiri, 2005; Nic Daéid and Meier-Augenstein, 2008).

The US Supreme Court decision in *Daubert versus Merrell Dow Pharmaceuticals, Inc.* set out four principal questions judges should consider in determining whether an area or field of science is reliable enough to be given in evidence in a court of law. It should be noted this decision explicitly stated these four questions should not be regarded as 'a definitive checklist or test', thus providing judges with the latitude to employ discretion and criteria of their own. The four so-called Daubert's principles are:

(i) Is the evidence based on a testable theory or technique?

(ii) Has the theory or technique been peer reviewed?

(iii) Does the technique have a known error rate and standards controlling its operation?

(iv) Is the underlying science generally accepted?

The above questions may not capture everything that as a scientist one would intuitively like to see captured. For example, question (iii) could or should be expanded to include whether data generated by the technique were reported and interpreted in the light of known error rates and population statistics. Another question that to my mind is missing would ask if the laboratory, scientist or technician applying the technique is proficient to do so. In this regard one should not mistake accreditation with proficiency. That being said, these four questions provide a certain framework and a framework at that, which is used in court to assess the evidence based on our work and the expert testimony we are about to give. It therefore behoves us to do the very best job we possibly can when developing and validating our methods, subjecting them and ourselves to true peer review, taking part in proficiency testing, and to be honest when it comes to pointing out the limitations of our chosen analytical technique.

Last, but not least, we must never forget to exercise our jobs with integrity, without bias and with professional detachment, although the latter is a fine balancing act between not getting too close, but at the same time not forgetting the personal dimension. If we get it wrong, people will suffer. One of the most important lessons I ever learned was taught to me by Dave Barclay, the former Head of Physical Evidence of the UK National Crime and Operations Faculty as then was: 'Forensic science is the interpretation of results in context, not the results *per se*'. The definition given in Box III.1 is largely based on a slide prepared and shown by Dave whenever he was asked to give a presentation on forensic science and crime scene investigation. In other words, relying on the results of your chosen analytical technique and excluding all other findings is an exercise in futility that does not accomplish anything and will not help anybody. Any discrepancy in results and their interpretation has to be addressed in context. Tests may have to be repeated and additional, different tests may have to be carried out until a consistent picture emerges where all pieces of the jigsaw puzzle fit. This timeless piece of wisdom was nicely captured by Sir Arthur Conan Doyle who let his creation Sherlock Holmes speak the immortal words: 'When you have eliminated the impossible, whatever remains, however improbable, must be the truth'.

Box III.1 Forensic Science

- Forensic science is the application and interpretation of scientific tests and observations in the context of individual legal case circumstance.

- Forensic science is the interpretation of results in context, not the results *per se*.

- Forensic science is absolutely context dependent.

- Forensic science's primary function is to provide intelligence[1]:

    ○ to clarify circumstance

    ○ to direct resource

    ○ to set elimination criteria

    ○ to point to a suspect

    ○ to assist at interview

    ○ ... and to provide clues!

- Intelligence detects – evidence convicts.

- Intelligence is open – evidence is closed.

---

[1] Intelligence will become or turn into evidence the moment it is relied upon to make a case against a suspect for prosecution in a court of law.

Chapter III.2
Distinguishing Drugs

Differentiation or, better, examination of drug samples based on variation in natural stable isotope abundance can be based either on determination of isotope composition of the bulk sample, of specific compounds or both. Early approaches have more commonly attempted to differentiate natural and semisynthetic drugs based on different geographical origin. However, there is also a great deal of interest in determining if it would be possible to differentiate synthetic drug products based on different synthetic procedures and, if possible, even from different manufacturing batches but the same synthetic route.

III.2.1 Natural and Semisynthetic Drugs

III.2.1.1 Marijuana

In 1997, evidence was presented in Australia to the Supreme Court of the Northern Territory in the case of *Queen versus Thomas Ivan Brettschneider* that forensic comparisons of cannabis samples seized from a truck with those seized from a plantation showed no difference in ^{13}C and ^{15}N isotopic abundance at 95% confidence level. Since then systematic scientific investigations have been carried out to determine to which level of confidence stable isotope signatures of cannabis or marijuana samples can be used to infer at which geographic location cannabis plants have been grown and as to whether the plants were cultivated indoors or outdoors.

Scientists in Brazil developed a methodology to assist with tracking marijuana trafficking routes into São Paulo based on ^{13}C and ^{15}N isotope analysis of seized marijuana samples, and their relation to climate and growing conditions. In the first instance they found while typical δ^{13}C and δ^{15}N values for marijuana samples from dry regions such as Bahia and Pernambuco were -26.28 ± 1.55 and $1.51 \pm 3.11‰$, respectively, corresponding figures for marijuana samples from wet regions such as Mato Grosso do Sul and Para were -29.77 ± 1.05 and $5.89 \pm 1.73‰$, respectively (Shibuya *et al.*, 2006).

Stable Isotope Forensics: An Introduction to the Forensic Application of Stable Isotope Analysis Wolfram Meier-Augenstein
© 2010 John Wiley & Sons, Ltd

Based on this research, the same group built a model to classify the origin of unknown samples using linear discriminant analysis based on about 150 samples seized in the main producing regions of the country. Results for 76 samples seized in the city of São Paulo showed that most of them were cultivated in a humid region with the same origin as those from Mato Grosso do Sul. Even when taking into account the small number of samples relative to the size of the area under investigation, their work demonstrated the potential of the use of $\delta^{13}C$ and $\delta^{15}N$ values in conjunction with linear discriminant analysis to determine the provenance of marijuana samples seized in different regions of Brazil. The authors of these studies pointed out that $\delta^{13}C$ and $\delta^{15}N$ data alone could not pinpoint the origin of the drugs; however, when used together with existing intelligence such as probable trafficking routes, stable isotope data appeared to be a powerful tool for monitoring drug trade and might even be able to detect the appearance of new trafficking routes in the country. The authors concluded that most of the samples seized in the São Paulo area seemed to share the same origin as those seized in Mato Grosso do Sul and were probably coming through the Paraguay–Mato Grosso do Sul route. In spite of the status of Maranhão as the main cannabis-producing state in Brazil, their work showed no evidence of trafficking routes to exist between this region and the city of São Paulo (Shibuya et al., 2007).

Scientists at the University of Utah under the lead of Jim Ehleringer took a different approach when studying the same problem – determination of marijuana provenance and marijuana trafficking routes in the United States. While they agreed with their Brazilian counterparts that $\delta^{13}C$ and $\delta^{15}N$ values of marijuana are good descriptors of growing conditions (West, Hurley and Ehleringer, 2009), they felt another discriminator was needed for a more accurate determination of geographic origin. Work by the same group focused on developing predictive models based on $\delta^{2}H$ values of marijuana samples of known origin to aid provenance determination of specimens with a sample history unknown to the scientists. When testing their model using ^{2}H isotope data of unknown samples in a blind study, the proportion of incorrect region-of-origin assignments ranged from 3 to 10%, with the latter figure representing incorrect assignment of samples as being of Eastern United States origin when in fact they had been of Eastern Canada origin. These findings are rather encouraging and it is hoped that with further refinement future predictive models in conjunction with isotopic landscapes or 'isoscapes' (www.isoscapes.org) will exhibit improved levels of specificity and sensitivity.

III.2.1.2 Morphine and Heroin

Since ^{13}C isotopic composition of heroin depends on the ^{13}C isotopic composition of raw opium which in turn depends to a degree on the climatic conditions at the geographical origin of the *Papaver somniferum* from which the morphine was extracted and on the ^{13}C isotopic composition of the acetic anhydride used to convert morphine to heroin (i.e. diacetylmorphine), Desage et al. hypothesized this method may be effective for differentiating heroin batches, but may not be used directly for assigning geographic origins of samples based on specific $\delta^{13}C$ values alone. Even though the results confirmed this hypothesis, the study was not as successful as anticipated, with

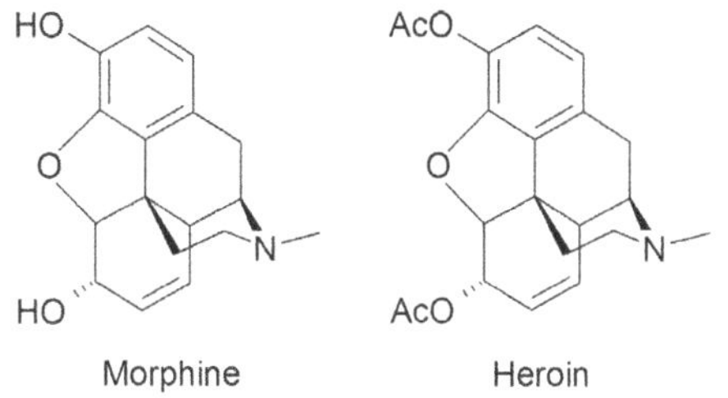

Figure III.1 Morphine and heroin.

δ^{13}C values falling into a narrow range of −31.5 to −33.5‰ when considering that sample origins covered a wide geographic area ranging from India to Turkey (Desage et al., 1991).

There is of course a kinetic isotope fractionation affecting ^{13}C abundance during the acetylation of morphine to heroin and as a result heroin will almost always exhibit more negative δ^{13}C values than the parent morphine (Figure III.1). There is also an isotopic fractionation against ^{15}N during the acetylation of morphine base to heroin base, but the net ^{15}N isotopic fractionation of this reaction is zero since all of the heroin base is precipitated during the manufacturing process. When heroin is deacetylated to morphine, the morphine produced yields δ^{13}C values that are indistinguishable from the original morphine (Casale et al., 2005).

Since the conversion of morphine into heroin did not result in a measurable ^{15}N isotopic fractionation, other studies set out to explore the possibility of using δ^{15}N values as the determining parameter for direct geographic origin assignments. Unfortunately, these studies found δ^{15}N values to scatter in narrow bands of −3.6 to +1.7, −1.6 to +1.3 and −4.30 to +0.40‰ (Besacier et al., 1997). On the other hand, for heroin coming from Korea and Columbia, much wider δ^{15}N ranges of −2.9 to +7.2 and −6.7 to +1.3‰, respectively, have been reported (Galimov et al., 2005).

In 2006, the examination by stable isotope analysis of a type of heroin hydrochloride hitherto not encountered was reported. This new type of heroin had been seized in Australia from the merchant vessel *Pong Su* in 2003. The major alkaloids profiles were deemed to be consistent with heroin of Southeast Asian origin. However, the acid/neutrals signature profiles (SIG II) were inconsistent with the kind of heroin typical for coming out of Southeast Asia. It was therefore concluded that these samples were sufficiently different from typical Southeast Asian heroin and they were therefore classified as from an unknown origin/process. Various drug enforcement authorities speculated this heroin might be from a new region or new illicit process due to the unusual chromatographic impurity profiles that were present. Samples from 20 different kilogram packages were examined for isotopic composition to determine if the samples fit isotopic patterns of heroin samples from known origins or if they were unique to any known origins. Authentic specimens from Southeast Asian ($n=59$), Southwest Asia ($n=37$), South America ($n=104$) and Mexico ($n=21$) were concomitantly examined for comparison purposes using both elemental analysis (EA)-IRMS and GC/C-IRMS techniques. Heroin samples were also converted to morphine, without apparent isotopic fractionation, utilizing methanolic HCl for hydrolysis, followed by CSIA using GC/C-IRMS.

Average δ^{13}C and δ^{15}N values for the 20 *Pong Su* samples as analysed by EA-IRMS were -32.8 ± 0.17 and -1.5 ± 0.51‰, respectively. Observed δ^{15}N values varied depending on heroin purity and ranged from -0.8 to -2.5‰. ^{13}C-CSIA of the 20 heroin samples and their parent material morphine after acid hydrolysis yielded δ^{13}C values of -32.3 ± 0.10 and -31.5 ± 0.16‰, respectively, with heroin being more depleted in ^{13}C than its parent morphine, as mentioned before (Casale *et al.*, 2005). By direct comparison of these stable isotope data with data obtained from heroin and morphine of known provenance, the *Pong Su* samples were found to be isotopically and isotopically/alkaloidally distinct from known origin/process classifications of authentic Southwest Asian, Southeast Asian, South American and Mexican specimens (Casale *et al.*, 2006).

Given how well δ^2H and, to a degree, δ^{18}O values of other plant tissues and plant products are correlated with geographic origin, our laboratory has recently started analysing seized heroine samples for their ^{2}H and ^{18}O isotopic composition. Obviously, heroin contains two additional oxygen atoms from the two acetyl groups as well as six additional hydrogen atoms also introduced into the molecule during acetylation. However, the aim of our pilot study was to determine what if any differences in ^{2}H and ^{18}O isotopic composition between the different seizures existed, and if they would be of any diagnostic value. Initial results were rather promising showing ranges for δ^2H and δ^{18}O values from -179 to -80 and 11.5 to 32.1‰, respectively (see Table III.1), again demonstrating the added dimension δ^2H values can provide to isotopic profiling of physical evidence.

III.2.1.3 Cocaine

Research carried out on coca leaves and cocaine (Figure III.2) base has come to the same conclusion reached for marijuana and heroin – the potential probative power of stable isotope signatures to yield information on drug provenance provided they are interpreted in context and, ideally, used in conjunction with complementary data such as impurity or minor constituent profiling. In one study this was achieved by analysing the isotopic composition of whole coca leaves as well as that of the minor alkaloids trimethoxycocaine and truxilline extracted from coca leaves sampled at different geographic locations in South America. Coca leaves from South America were found to vary in their δ^{13}C (-32.4 to -25.3‰) and δ^{15}N (0.1–13.0‰) values (Ehleringer *et al.*, 2000). Humidity levels and the length of the rainy season and differences in soils were thought to affect the fixation processes and cause the observed subtle variations in ^{13}C and ^{15}N isotopic abundance, respectively. In conjunction with the variations of trace alkaloids content (truxilline and trimethoxycocaine) found in cocaine, researchers were able to correctly identify 96% of 200 cocaine samples originating from the regions studied.

In a different study, ^{13}C and ^{15}N isotope signatures of cocaine from different regions in Colombia were investigated and results from bulk isotope analysis by EA-IRMS were compared to results obtained from CSIA using GC/C-IRMS. With the exception of a

Table III.1 Observed ranges for δ^2H, δ^{13}C, δ^{15}N and δ^{18}O values of natural and hemisynthetic drugs.

Drug	Plant source	Country of origin	Observed δ^2H$_{VSMOW}$ range (‰)	Observed δ^{18}O$_{VSMOW}$ range (‰)	Observed δ^{13}C$_{VPDB}$ range (‰)	Observed δ^{15}N$_{AIR}$ range (‰)
Marijuana[a]	*Cannabis sativa* L.	Brazil			−32.5 to −23.5	−3.4 to +8.5
Marijuana[b]	*Cannabis sativa* L.	USA	−160 to −125		−51.8 to −20.3	−7.9 to +29.5
Marijuana[c]	*Cannabis sativa* L.	Russia, Ukraine			−28.4 to −26.4	−3.2 to +9.7
Heroin[d]	(*Papaver somniferum*)	unknown; seized in Glasgow (UK)	−179 to −80	+11.5 to +23.1		
Heroin[c]	(*Papaver somniferum*)	Colombia			−35.9 to −32.7	−6.7 to +1.3
Heroin[c]	(*Papaver somniferum*)	South Korea			−38.3 to −35.1	−2.9 to +7.2
Heroin[e]	(*Papaver somniferum*)	Mexico, Southwest Asia, Southeast Asia, South America			−34.2 to −32.0	
Morphine[e]	(*Papaver somniferum*)	Mexico, Southwest Asia, Southeast Asia, South America			−31.0 to −29.8	
Cocaine[b]	*Erythroxylum coca*	Bolivia, Colombia, Ecuador, Peru			−34.5 to −25.3	−11.8 to +13.0
Cocaine[c]	*Erythroxylum coca*	Colombia			−39.9 to −30.9	−10.4 to −2.2

[a]Data from Shibuya et al. (2006).
[b]Data from Ehleringer et al. (2000); West, Hurley and Ehleringer (2009).
[c]Data from Galimov et al. (2005).
[d]Author's own data.
[e]Data published by Casale et al. (2006).

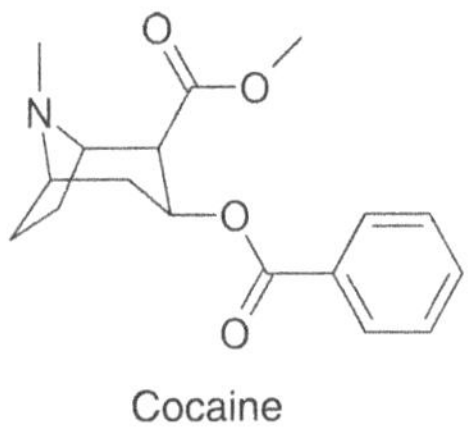

Figure III.2 Cocaine.

cocaine samples shown to contain impurities, δ^{13}C and δ^{15}N values for pure cocaine samples from either instrumental method were not significantly different. Analysis by EA-IRMS and GC/C-IRMS yielded δ^{13}C values of -35.54 ± 0.32 and -36.03 ± 0.46‰, respectively, while the corresponding δ^{15}N values were -4.8 ± 1.2 and -4.9 ± 1.3‰, respectively (Galimov et al., 2005).

As part of a criminal investigation, the stable isotope laboratory of the Bundeskriminalamt (BKA) in Wiesbaden (Germany) analysed 132 cocaine samples from a seizure of 1.2 tons of cocaine by German police in 2002. Measured δ^{15}N values ranged from -17 to -2 ‰ – a finding that in the context of observations described above indicated that this seizure was comprised of cocaine from various different locations (Sewenig et al., 2007).

III.2.2 Synthetic Drugs

III.2.2.1 Amphetamines

The first preliminary study to trace the origin of different batches of confiscated MDMA (Ecstasy) tablets by ^{13}C-CSIA allowed the discrimination of four different groups of MDMA tablets based on variations in their δ^{13}C values. The same study showed that further discrimination might be obtained when using δ^{15}N values of MDMA (Mas et al., 1995), and in later studies δ^{15}N values of illicit MDMA have been shown to vary between -17.4 and $+18.8$ ‰ (i.e. covering a range of 36.2‰) (Palhol, Lamoureux and Naulet, 2003; Palhol et al., 2004).

Combining ^{2}H isotope analysis with ^{13}C and ^{15}N isotopic analysis of MDA (3,4-methylenedioxyamphetamine) and MDMA extracted from seized Ecstasy tablets provided an isotopic 'fingerprint' of the active ingredient, and it was suggested that such a fingerprint could yield sufficient information for individually seized tablets to be linked to their original common batch (Carter et al., 2002b). Indeed, in 2003, this suggestion was put to the test in an actual criminal investigation concerning a person who when apprehended was found to be in possession of 21 Ecstasy tablets. A subsequent search of the premises where the person had been arrested found a concealed larger quantity of visually similar tablets. The suspect denied all knowledge of the second seizure or link between the second seizure and the 21 tablets found on their person. GC-MS analysis of tablet extracts showed all to contain MDMA with a purity of greater than 98%. Since no other compounds or impurities were identified in the extracts, comparison by GC-MS profiling provided very limited discrimination. As the police required forensic evidence to link the two batches of tablets, stable isotope profiling was chosen to investigate if a relationship existed between the two seizures. Mass Spec Analytical Ltd (MSA), a member of the FIRMS Network (see Section II.5.1) was engaged by Nottinghamshire Police Constabulary in the United Kingdom to carry out stable isotope analyses of tablets from the two seizures of Ecstasy tablets to determine if they were related. No significant difference for both ^{13}C and ^{15}N isotopic composition at 95% confidence between tablets from the two seizures was found (Carter et al., 2005). MSA stressed that if these data were to be used in evidence, further analyses would be necessary such as excipient profiling of the existing tablet extracts or additional isotopic analysis of further tablets to increase the power of the statistical data analysis. However, when presented with the initial stable isotope findings in conjunction with other evidence, the defendant chose to enter a guilty plea at Nottingham Crown Court on 12 November 2003.

One of the first, if not the first, law enforcement agency in Europe to embrace stable isotope profiling as a new forensic tool and to set up a dedicated stable isotope laboratory was the BKA in Wiesbaden, and the initial focus of studies and investigations carried out there was on synthetic drugs. One reason was and still is the increasing rate of synthetic drug abuse and other drug-related offences in Germany; the other reason is the role Germany plays as a final destination as well as a transit country for the illegal traffic in precursor compounds such as 1-phenyl-2-propanone (P2P) or piperonyl-methyl-ketone (PMK) from sources in the East to Central European countries. Employing a combined approach of bulk-specific isotope analysis (BSIA) and CSIA, scientists at the BKA determined multivariate stable isotope profiles for seven samples of illicit P2P from four different seizures as well as for 27 samples of nine batches of legally produced P2P (three samples per batch). While δ^{13}C values for both seized and control samples fell within the same range of −32.0 to −27.3‰, the difference in range for δ^{2}H values was quite remarkable. Whereas mean δ^{2}H values of illicit P2P samples ranged from −195 to −86‰, average δ^{2}H values of P2P samples from the nine control batches only ranged from −71.3 to −45.3‰ (Schneiders, Holdermann and Dahlenburg, 2009). Not one of the P2P control samples exhibited a δ^{2}H value below −77.5‰. One could surmise the comparatively small δ^{2}H range of 26‰ of the legally produced P2P samples reflected well-controlled production conditions at the facilities of legitimate manufacturers, whereas the δ^{2}H range of 109‰ of the illicit P2P samples seemed to reflect the not so tightly controlled reaction conditions of a clandestine production facility. Bivariate plots of δ^{2}H versus δ^{13}C values clearly distinguished illicit P2P sample provenances according to seizure. Three of the seven samples could be associated with case 1, two with case 4, leaving one sample each for cases 2 and 3. Measured δ^{2}H, δ^{13}C and δ^{18}O values for all 27 control samples of P2P were statistically analysed and pair-wise compared for T^2 distances by Hotelling's T^2 test (Pierrini et al., 2007; Farmer et al., 2009). Despite the fact that observed ranges for δ^{2}H, δ^{13}C and δ^{18}O values were 26, 3.6 and 11.2‰, respectively, and thus relatively narrow, only 2.28% of all 351 possible sample pairs were incorrectly associated with each other as being from the same batch.

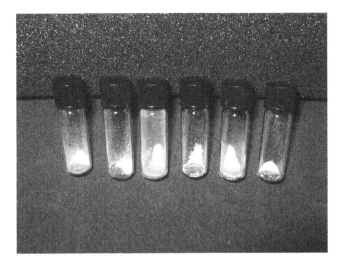

Figure III.3 Six amphetamine powders from the 18 seizures isotopically profiled in Figure III.4.

Results such as the aforementioned suggest that multidimensional bulk isotope analysis of aliquots from entire tablets and tablet extracts can yield valuable information on drug provenance and, hence, drug supply chains within turnaround times of as little as 9 h, but typically less than 48 h. For example, subsamples of 18 amphetamine seizures seized from six different sources had been analysed in our laboratory for their ^{13}C and ^{15}N isotopic composition (Figure III.3). The measured δ^{13}C/δ^{15}N values ranged from -25.5 ± 0.1 to -17.6 ± 0.4 and -14.1 ± 0.2 to -5.7 ± 0.5‰, respectively, and could be grouped in eight statistically significant different clusters, indicating that two of the 18 seizures had different histories compared to the other sources (Figure III.4). Differences like these and how they may relate to entirely different origins, or merely to different production conditions or 'recipes' used in their synthesis, are the subject of ongoing research.

I feel, however, that I have to stress the point that whilst the use of authentic street drugs is a first step into determining the width and breadth of stable isotope abundance per element per drug by generating and populating a drug isotope database, due to the mostly unknown provenance of these samples the type of systematic research that can be undertaken on such samples is limited. Without knowing how synthetic routes ('recipes') and variation of reaction conditions within a given synthetic route influence stable isotopic composition of the final product, only limited conclusions or inferences can be drawn from stable isotope profiles or signatures of seized drugs. In other words, inferences made regarding provenance and sample history from stable isotope analysis of seized samples are tentative even if blind analytical results as presented in Figure III.4 are consistent with known intelligence. Although seized tablets can be 'grouped' according to IRMS and/or other spectroscopic data, groupings with regards to similarity and from that inferred shared sample history and sample provenance may not necessarily be conclusive.

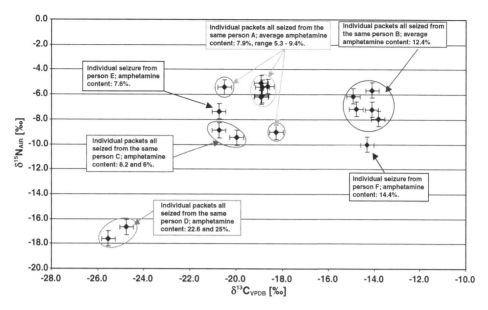

Figure III.4 Bivariate plot of $\delta^{15}N$ versus $\delta^{13}C$ values for 18 amphetamine seizures.

III.2.2.2 MDMA: Synthesis and Isotopic Signature

III.2.2.2.1 Three Different Synthetic Routes – Controlled Conditions

In probably the first study of this kind and extent, MDMA was chosen as a model compound for investigating the influence of synthetic route and changes in reaction conditions on stable isotope signature of drugs. Niamh Nic Daéid's (Centre for Forensic Science, University of Strathclyde, Glasgow, United Kingdom) and her research group's expertise in organic synthesis of drugs as well as traditional forensic analysis was combined with our expertise in stable isotopes and stable isotope analytical techniques. For this study, six batches of MDMA per synthetic route were synthesized via three commonly used reductive amination routes that are all accessible to the clandestine chemist; in total 18 batches of MDMA were prepared (Buchanan *et al.*, 2008). The three reductive aminations of PMK make use of different reducing agents and reaction conditions: Al/Hg amalgam, $NaBH_4$ and Pt/H_2. Since access to the key ketone PMK is technically impossible without holding the appropriate licenses and thus can neither be easily diverted from commercial sources, clandestine synthesis of MDMA must start from pre-precursors such as safrole or isosafrole. It is not unheard of that these pre-precursors are converted into PMK in Asia and then shipped to separate locations (e.g. in Europe) to be refined into the final product MDMA in a way similar to that known to exist for the precursor P2P and its final products amphetamine and methamphetamine (Schneiders, Holdermann and Dahlenburg, 2009). To mimic conditions of clandestine synthesis as closely as possible, a large batch (approximately 200 g) of PMK was synthesized from a single batch of safrole. From this single precursor pool

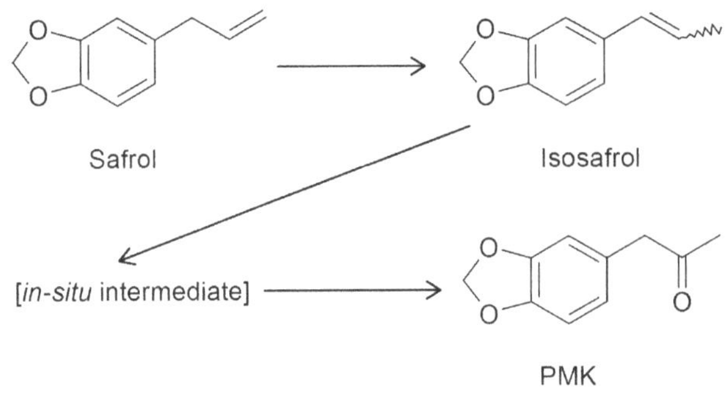

Figure III.5 Schematic synthetic route for PMK from safrole.

18 individual batches of MDMA hydrochloride were synthesized using three different synthetic routes, with the resultant products being subsequently analysed for ^{2}H, ^{13}C and ^{15}N isotopic composition. See Figures III.5 and III.6.

Upon receipt by our laboratory, sample ID and weight of synthesized MDMA samples were entered into a chain of custody sheet as well as into the laboratory's drug book. Aliquots sufficient for stable isotope analysis were weighed out and dried in a desiccator for 7 days to remove any traces of moisture (*in vacuo* over P_4O_{10}). Weights of the remaining drug samples were recorded in the laboratory's drug book and the samples locked into the drug safe.

To prepare samples for isotope analysis, drug aliquots were removed from the desiccator and approximately 0.2 and 0.4 mg weighed out in triplicate into silver and tin capsules for analysis by thermal conversion (TC)/EA-IRMS and EA-IRMS, respectively. Capsules were subsequently crimped and placed into 96-well plates already preloaded with blank and crimped capsules filled with appropriate reference materials. Batch-run-ready well-plates were placed into another desiccator, where they were kept *in vacuo* over P_4O_{10} until analysis.

Results from the isotope analyses of the 18 MDMA batches are graphically represented in Figures III.7 and III.8. Each data point represents the mean of triplicate analyses

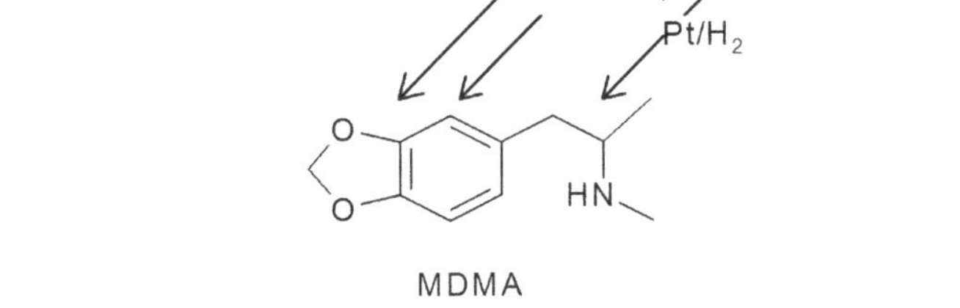

Figure III.6 Schematic synthetic routes for MDMA from PMK.

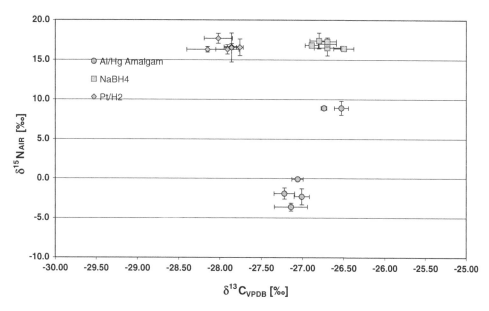

Figure III.7 Bivariate plot of $\delta^{15}N$ versus $\delta^{13}C$ values of 18 MDMA batches from three different synthetic routes.

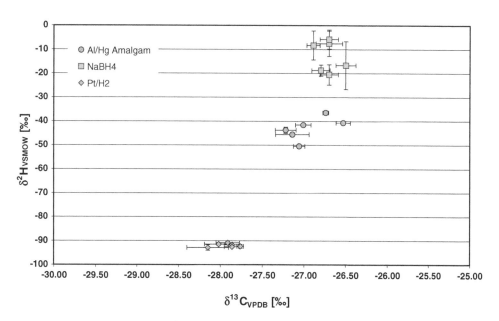

Figure III.8 Bivariate plot of $\delta^{2}H$ versus $\delta^{13}C$ values of 18 MDMA batches from three different synthetic routes.

and the error bars indicate 1σ around the mean for each individual sample. The observed range for the $\delta^{13}C$, $\delta^{15}N$ and $\delta^{2}H$ values of the 18 batches were in line with published δ values for MDMA (Billault et al., 2007; Carter et al., 2002a, 2002b; Palhol et al., 2004; Mas et al., 1995).

A cursory examination of Figures III.7 and III.8 shows the least variation in δ values is associated with changes in ^{13}C isotopic composition (range: 1.6‰). The small variation in $\delta^{13}C$ values observed for all the samples in this study should not come as a great surprise given what we have learned in previous parts of this book. If one considers that 10 of the 11 carbon atoms on the final MDMA molecule are contributed by the precursor safrole and that all of these samples were synthesized using the same batch of safrole, only the 11th carbon – the N-methyl carbon atom in the MDMA molecule – does not originate from safrole, but is contributed by methylamine hydrochloride (Al/Hg amalgam and $NaBH_4$ routes) or aqueous methylamine (Pt/H_2 route). In general, $\delta^{13}C$ data points *within* a synthetic route cluster together visually, indicating the fractionation induced within the synthetic process is reproducible under tightly controlled conditions (i.e. if the synthesis is carried out by the same chemist using the same method and materials). Interestingly enough, all six batches from the Pt/H_2 route can be discriminated from the batches of the other two synthetic routes on the basis of $\delta^{13}C$ values alone, while the range of $\delta^{13}C$ values observed for the Al/Hg amalgam and $NaBH_4$ routes overlaps.

Examination of the $\delta^{15}N$ values reveals variation over a range of 21.3‰ and the Al/Hg amalgam batches exhibit the widest variation within any synthetic route, with two of the batches showing considerable ^{15}N enrichment relative to the remaining four (Figure III.7), yet the Al/Hg amalgam route can be discriminated from the other two sets by virtue of $\delta^{15}N$ values alone despite the large variation in $\delta^{15}N$ values within that set. Notably, MDMA synthesis with the Al/Hg amalgam was the hardest to reproduce as the temperature of the exothermic reaction is difficult to control and a series of reagents must be added by hand in quick succession, making these two parameters likely to vary from batch to batch. Based on these observations, $\delta^{15}N$ values appear to be the most sensitive to these inadvertent inconsistencies in preparative method, confirming the observations by Carter et al. (2002a) during the synthesis of methamphetamine. It should be noted that a paper was published in 2007 reporting the synthesis of MDMA by a different Al/Hg amalgam method, which yielded consistent $\delta^{15}N$ values for MDMA (Billault et al., 2007).

Looking at the $\delta^{15}N$ values of these 18 MDMA batches in the context of the nitrogen-contributing precursor yields a further interesting insight. The nitrogen atom on the MDMA molecule is contributed, along with the N-methyl carbon, by methylamine hydrochloride (both Al/Hg amalgam and $NaBH_4$ route) or aqueous methylamine (Pt/H_2). Should the ^{15}N isotopic composition of the nitrogen-contributing reagent be the sole contributing factor to the final MDMA $\delta^{15}N$ value, one would expect the $\delta^{15}N$ values to fall into two groups, with Al/Hg amalgam and $NaBH_4$ data forming one group and Pt/H_2 data forming the other. Instead, observed $\delta^{15}N$ values for both $NaBH_4$ and Pt/H_2 fall within a narrow range of 1.4‰, thus indicating the synthetic process itself is responsible for nitrogen isotopic fractionation. This observation confirms previous studies suggesting that observed changes or differences in $\delta^{15}N$ values are largely the

Table III.2 Summary δ^2H, $\delta^{13}C$ and $\delta^{15}N$ values of MDMA hydrochloride samples synthesized from aliquots of the same precursor PMK, but by three different synthetic routes of reductive amination.

Route	Sample (case)	Sample ID	δ^2H (‰)	1σ	$\delta^{15}N$ (‰)	1σ	$\delta^{13}C$ (‰)	1σ
Al/Hg amalgam	1	HSB75B	−40.7	0.2	8.9	0.9	−26.53	0.08
	2	HSB76B	−50.5	0.3	−0.1	0.1	−27.06	0.07
	3	HSB78B	−36.5	0.6	8.9	0.2	−26.74	0.03
	4	HSB79B	−45.6	0.9	−3.6	0.5	−27.14	0.20
	5	HSB80B	−43.8	1.2	−1.9	0.7	−27.22	0.12
	6	HSB81B	−41.6	0.7	−2.3	1.0	−27.01	0.09
NaBH$_4$	7	HSB101	−18.8	2.3	17.4	1.0	−26.80	0.11
	8	HSB102	−7.6	5.3	16.5	1.0	−26.70	0.16
	9	HSB104	−16.7	10.0	16.4	0.1	−26.50	0.12
	10	HSB105	−20.6	4.2	17.0	0.8	−26.70	0.11
	11	HSB106	−5.8	4.0	17.3	0.0	−26.70	0.11
	12	HSB108	−8.3	6.1	16.8	0.2	−26.89	0.08
Pt/H$_2$	13	HSB112	−92.8	1.2	16.3	0.3	−28.15	0.25
	14	HSB113	−91.7	0.1	16.7	0.4	−27.86	0.09
	15	HSB114	−90.8	0.4	16.3	0.6	−27.91	0.14
	16	HSB115	−91.4	0.5	17.7	0.6	−28.02	0.17
	17	HSB117	−92.3	0.8	16.6	1.1	−27.76	0.04
	18	HSB121	−92.5	0.1	16.5	1.8	−27.86	0.09

Based on data from Buchanan *et al.* (2008).

result of fractionation due to the synthetic process rather than being a reflection of ^{15}N isotopic composition of the precursor. See Table III.2.

The results of ^{2}H-BSIA from the 18 MDMA batches show the highest variation in δ^2H values (range: 98.6‰) overall. However, δ^2H values fall into three distinct clusters with ranges of 14, 14.8 and 2‰ for the Al/Hg amalgam, NaBH$_4$ and Pt/H$_2$ route, respectively (see Figure III.9). Equally noteworthy is the observation that as far as δ^2H values are concerned, Pt/H$_2$ and Al/Hg amalgam routes yield the most reproducible and homogenous synthetic product with 95% confidence levels of 0.8 and 5.0‰, respectively. The corresponding figure for the NaBH$_4$ route is 6.8‰. Lastly, average δ^2H values per six batches per route are −91.9, −43.1 and −12.9‰ for Pt/H$_2$, Al/Hg amalgam and NaBH$_4$, respectively. One could argue the observed variation in δ^2H values between the three different routes might be expected due to the large number of possible hydrogen contributors, but closer examination permits us to reduce the number of possible suspects.

First, in all three routes the same HCl and 25% NaOH have been used to precipitate MDMA as MDMA hydrochloride. Second, all other chemicals have been used consistently with the exception of methylamine hydrochloride (Al/Hg amalgam and NaBH$_4$) and methylamine (Pt/H$_2$) as well as the reducing agents themselves. In other words, all things being equal, the potential net contributors to δ^2H values of MDMA by the Pt/H$_2$ route are methylamine and hydrogen together with kinetic isotope effects during chemical reaction, with differences in δ^2H values between this route and the other two routes being directly attributable to these two reagents and they way they

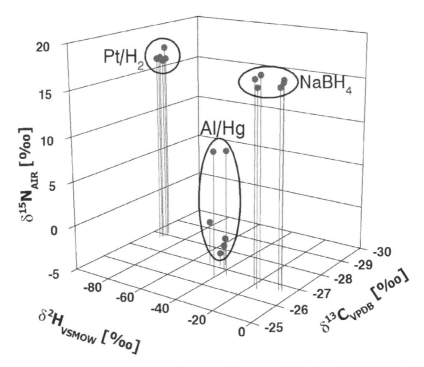

Figure III.9 Three-dimensional plot of δ^2H versus δ^{15}N versus δ^{13}C values of MDMA hydrochloride samples synthesized from aliquots of the same precursor PMK but by three different routes of reductive amination. Datasets from each synthetic route encircled and labelled accordingly.

react. Similarly, the main difference in δ^2H values between the Al/Hg amalgam and NaBH$_4$ MDMA batches of approximately 30‰ would be attributable to the reducing agent and associated kinetic isotope effects during its reaction.

Since visual discrimination based on two- or three-dimensional data plots of observed δ values may be influenced by some subjectivity (wishful thinking), multivariate datasets should be analysed by multivariate chemometric techniques such as Hotelling's T^2 test, PCA and HCA or multivariate likelihood ratio approaches (Bolck *et al.*, 2009). Cluster analysis is one method to find an algorithm that could accurately link the data according to synthetic route. HCA has been used in similar published works (Palhol *et al.*, 2004), but in the study described in this chapter the validity of the resulting clusters can be assessed because the samples are of known provenance. HCA of the 18 batches based on their observed δ^2H, δ^{13}C and δ^{15}N values grouped all samples according to synthetic route. HCA was performed using the statistical software package SYSTAT 11 (www.systat.com), Euclidean distance measure and nearest-neighbour (single-linkage) clustering method (Figure III.10). In the resulting dendrogram HCA grouped data into three clusters corresponding to a particular synthetic route each.

To ascertain the discriminating power of each of the isotopic abundance, HCA was also performed on all combinations of δ^2H, δ^{13}C and δ^{15}N values taken one or two at a time. Accurate discrimination of the synthetic routes was only possible when δ^2H

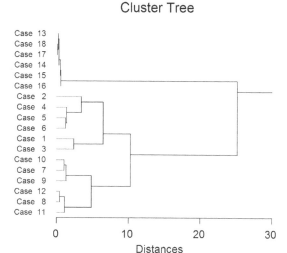

Figure III.10 HCA of 18 batches of MDMA; three variables, Euclidean distance, single linkage. Cases 1–6, 7–12 and 13–18 refer to synthetic routes Al/Hg, NaBH$_4$ and Pt/H$_2$, respectively.

values were included (i.e. when δ^2H values were subjected to HCA on their own or in conjunction with δ^{13}C and/or δ^{15}N values). This work clearly indicated that ^{2}H isotope analysis of MDMA is a vital factor for the discrimination of samples by synthetic route, while δ^{13}C or δ^{15}N values on their own could not distinguish between all three routes at 95% confidence level.

The 18 synthesized MDMA samples were impurity profiled by the Collaborative Harmonisation of Methods for Profiling of Amphetamine Type Stimulants (CHAMP) GC-MS method to see if full discrimination between all three synthetic routes could be achieved. (The CHAMP method is the outcome of an European research project funded under Framework Programme 6.) To compare the impurity profiles as the CHAMP method recommends, Pearson correlation coefficients were calculated for every pair of samples in the six datasets using the van Deursen impurities normalized to the sum of the targets and pretreated with the fourth root. Since the 'true' relationships of the batches are known, a threshold value for the calculated coefficients was sought such that values above the threshold indicate related samples and values below the threshold indicate unrelated samples. The lowest coefficient calculated for a pair of samples from within a synthetic route was calculated to set a maximum threshold that would allow the six samples within each route to be deemed similar. If the CHAMP method provided complete discrimination of the MDMA batches by synthetic route, then all pairings should have been below the calculated maximum threshold. However, only 68% of these pairings yielded coefficients below the maximum threshold, meaning 32% of the samples were incorrectly linked as belonging to the same batch. Statistical analysis of the impurity profiles using different selections of key impurities and data pretreatments did not improve matters, but resulted in even worse rates of false linkages of 75 and even 99%. In this particular study, the CHAMP profiling method could not accurately discriminate between the synthesized MDMA batches on the basis of all three synthetic routes when utilizing Pearson correlation coefficients calculated using either of the target impurity lists suggested in the literature (van Deursen, Lock and Poortman-van der Meer, 2006; Weyermann *et al.*, 2008). However, one could argue that this might have been a consequence of the high level of experimental control and skill exercised during these syntheses – something that is unlikely to be encountered when dealing with samples produced in clandestine laboratories.

III.2.2.2.2 One Synthetic Route – Variable Conditions

To obtain answers to the obvious question as to how variations in reaction conditions would influence stable isotopic composition of MDMA (and GC-MS profiles), and thus mimicking 'real-life' circumstances and conditions encountered in clandestine laboratories, Niamh Nic Daéid and her PhD student decided on the Pt/H_2 synthesis route as the model system (Buchanan, 2009). The Pt/H_2 method is one of the most popular synthesis routes with clandestine laboratories, and it also produced the six batches with the most consistent δ^2H, $\delta^{13}C$ and $\delta^{15}N$ values during the study into different synthetic routes when reaction conditions were kept as constant as possible. To keep the number of syntheses under different reaction conditions to a manageable size, a factorial design was chosen based on five factors. However, for reasons of confidentiality it is not possible to impart any more details about the study or its outcomes in this chapter. Information about this work and relevant publications will be made available on the companion website of this book as they become available (www.wiley.com/go/meieraugenstein).

III.2.2.3 Methamphetamine: Synthesis and Isotopic Signature

An extensive study into methamphetamine and its precursors L-ephedrine and D-pseudoephedrine was carried out in Japan where abuse of methamphetamine is the most serious of all amphetamine-type stimulants. Originally focusing on ^{13}C and ^{15}N isotopic composition of ephedrine in its relation to source – natural (extracted from *Ephedra sinica*, also known as *ma huang*), semisynthetic or synthetic – scientists at the Japanese Central Customs Laboratory and the University of Tokyo have extended their studies to now include ^2H isotopic composition as well (Kurashima *et al.*, 2004, 2009).

In the study looking at δ^2H, $\delta^{13}C$ and $\delta^{15}N$ values of ephedrine and methamphetamine (Kurashima *et al.*, 2009), of the 27 control ephedrine hydrochloride samples analysed, 10 were of natural origin (Group 1), five were synthetic products (Group 2), seven were semisynthetic, molasses-based products imported from India (Group 3) while five were semisynthetic, pyruvic acid-based products imported from Germany (Group 4). The results of their analyses of these samples are summarized in Table III.3.

While not correcting their measured δ^2H values and reporting true δ^2H values, hydrogen exchange experiments were carried out (i) to determine influence on measured

Table III.3 Reported δ^2H, $\delta^{13}C$ and $\delta^{15}N$ values of ephedrine hydrochloride and pseudo-ephedrine from various sources.

Source	Country of origin	Observed δ^2H_{VSMOW} values (‰)	Observed $\delta^{13}C_{VPDB}$ values (‰)	Observed $\delta^{15}N_{AIR}$ values (‰)
Ephedra sinica[a]	China	−193 to −151	−31.1 to −26.0	−2.2 to +10.6
Ephedra sinica[b]	Unknown; purchased in the USA	−170	−28.6	+4.3
Synthetic[a]	China, Taiwan, Japan	−70 to +30	−29.2 to −28.0	−11.0 to −6.4
Synthetic[b]	Germany, supplied by Fluka	−135	−26.2	+3.6
Semisynthetic – molasses[a]	India	−74 to +243	−23.7 to −22.0	+2.4 to +7.3
Semisynthetic – pyruvic acid[a]	Germany	+75 to +148	−27.5 to −25.3	−4.1 to −2.8
Semisynthetic[b]	USA, India	+159 to +171	−26.2 to −23.3	−0.1 to +5.1

[a] Data from Kurashima *et al.* (2009).
[b] Data from Collins *et al.* (2009).

δ^2H values and (ii) to establish a defined reference point for the synthesis of methamphetamine from ephedrine, although the authors remarked that observed shifts in δ^2H values due to hydrogen exchange were deemed to be too small to have a great influence on data interpretation. However, as already pointed out in Section II.6.3.2, this approach is only acceptable for self-contained studies aiming to look at differences between samples while observing the 'principle of identical treatment' (Bowen *et al.*, 2005; Farmer, Meier-Augenstein and Kalin, 2005; Fraser and Meier-Augenstein, 2007; Werner and Brand, 2001). For the purpose of establishing a database, and for comparison and exchange of information between forensic laboratories and law enforcement agencies on both national and international levels, measured and true δ^2H values should be reported, with the latter determined following a well-controlled protocol that has proven to yield reproducible and repeatable results. Incidentally, based on the δ^2H values published by the Japanese scientists, molar exchange fractions f_{ex} for ephedrine hydrochloride and methamphetamine hydrochloride were calculated as 0.09454 and 0.09106, respectively. The similarity of these exchange fractions is remarkable considering three out of 16 hydrogen atoms are theoretically free to exchange in ephedrine hydrochloride, while the number is reduced to two out of 16 for methamphetamine hydrochloride.

That being said, six repeat syntheses of methamphetamine from ephedrine samples of different ^2H isotopic composition using red phosphorus and hydroiodic acid ('Nagai' route) yielded methamphetamine with δ^2H values that were consistently and reproducibly lower by −38.8‰ compared to the δ^2H values of the corresponding ephedrine precursor material. Since substituting the hydroxyl group with hydrogen via chlorination with $SOCl_2$ and subsequent catalytic hydration ('Emde' route) yielded a similar ^2H isotopic shift of −41‰, it was concluded that the aromatic hydrogen atoms were more depleted in ^2H than the hydrogen of the hydroxyl group. Based on the data reported by the Japanese group, I calculated the mean fractionation factor α for the conversion of

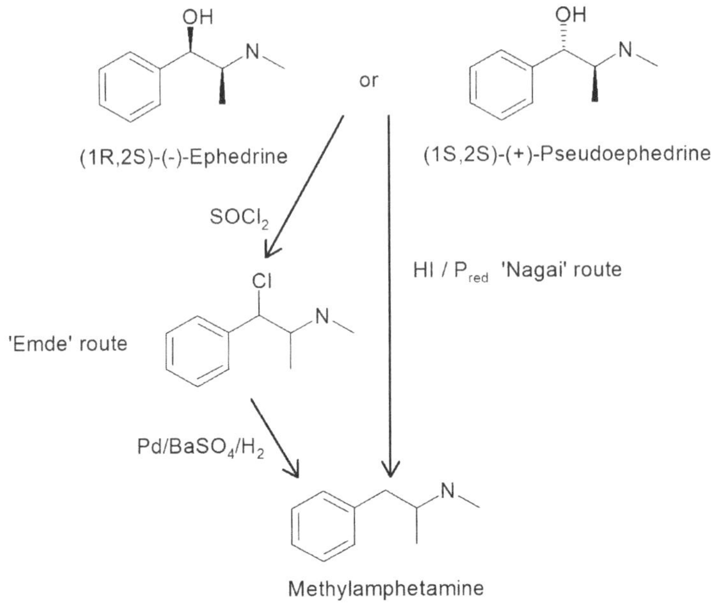

Figure III.11 Schematic synthetic routes 'Emde' and 'Nagai' for methamphetamine from ephedrine or pseudoephedrine.

ephedrine into methamphetamine by the red phosphorus/hydroiodoc acid route to be 0.9613 ± 0.00464, whereas the corresponding enrichment factor ε for this reaction was calculated as $-38.6 \pm 4.6‰$.

These findings were corroborated by results from a study carried out in Australia (Collins et al., 2009). Stable isotope signatures for semisynthetic ephedrine (and pseudoephedrine) as well as for ephedrine extracted from *Ephedra* sp. fell into the same ranges reported by the Japanese research group (*cf*. Table III.3). Similarly, the Australian research group observed a shift in $δ^2H$ values between precursor ephedrine (pseudoephedrine) and product methamphetamine by the 'Nagai' method (Figure III.11) that was in good agreement with that reported by the Japanese research group. Based on the data reported by each group, I calculated enrichment factors ε of -34.0 ± 1.8 and $-38.6 \pm 4.6‰$ for the 'Nagai' route synthesis as carried out by the Australian and Japanese research groups, respectively. However, there was a difference in ^2H isotopic fractionation observed by the two research groups between precursor ephedrine (pseudoephedrine) and product methamphetamine when synthesized by the 'Emde' route (Figure III.11). While the data from one synthesis reported by the Japanese research group suggested an enrichment factor ε of $-34.7‰$ (i.e. similar to that observed for the 'Nagai' route), $δ^2H$ values reported by the Australian research group for seven syntheses by the 'Emde' route yield an enrichment factor ε of $-2.4 \pm 1.7‰$. Since the Australian group made no mention about the matter of hydrogen exchange and how this was addressed, one possible reason for this discrepancy could be hydrogen exchange. However, given the good agreement of the ^2H enrichment factors for the 'Nagai' route between the two research groups, it is conceivable that differences in δ^2H values for methamphetamine synthesized by the 'Emde' route may reflect the ^2H isotopic composition of the hydrogen gas used during the catalytic hydration or may even be linked to the activity of the palladium/barium sulfate catalyst.

III.2.3 Conclusions

Despite the careful control exercised during each individual synthetic procedure, product $\delta^{15}N$ values showed no significant difference ($P = 0.2$) between MDMA from the NaBH$_4$ and the Pt/H$_2$ route (Table III.2). However, $\delta^{15}N$ values were significantly different between these two routes and the Al/Hg amalgam route ($P < 0.01$) even though $\delta^{15}N$ values varied considerably between the six batches *within* the Al/Hg synthetic route (Table III.2). For this reason, $\delta^{15}N$ values may be of lesser diagnostic use than previously thought when it comes to 'blind' grouping batches of unknown provenance presumed to be synthesized by the same preparative route or even the same clandestine laboratory; rather, $\delta^{15}N$ values may 'only' be of diagnostic use when applied in conjunction with other analytical data such as GC-MS impurity profiles providing information as to the synthetic route. If MDMA hydrochloride samples from the controlled synthesis studies had been seized on the street and subjected to IRMS analysis alone, only δ^2H values would have allowed tentative visual discrimination into three groups corresponding to the synthetic route used for manufacture. However, the scenario that emerged from the study based on one synthetic route, but variable reaction conditions, is much more likely to be encountered when tasked with the forensic examination of seized drugs. It was therefore gratifying to see that in real case work stable isotope signatures can provide added value to the forensic investigative process of a synthetic product whose manufacturing process or conditions are inherently and notoriously poorly controlled. In summary, one can say stable isotope signatures yield information complementary to data obtained from traditional techniques such as GC-MS impurity profiling and, in combination with multivariate forensic statistical techniques, offer the potential to discriminate between synthetic drug seizures by synthetic route and by laboratory product.

The study into the effect of synthetic route and reaction conditions on stable isotopic profiles of MDMA formed part of a wider study aiming to determine in a systematic and scientifically sound approach what contribution stable isotope techniques could make to the forensic examination of illegal drugs, and to combating serious and organized drug crime. This was only made possible through the financial support from project grants and studentship awards (EPSRC EP/D0403451/1 and RSC07/G68). In return for this investment not only were methods developed to extract information on sample history and provenance from one of the most prolific illegal synthetic drugs in Europe and the United Kingdom, but the foundations were also laid to a database that at present comprises IRMS, inductive coupled plasma (ICP)-MS and GC-MS data of more than 140 individual seizures of three illegal drugs from four European countries.

Based on the subset of seized Ecstasy tablets within this dataset it was possible for the first time to approximate the probative power of multivariate stable isotope signatures of seized drugs based on the stable isotope abundances of ^{2}H, ^{13}C, ^{15}N and ^{18}O presumed to represent four independent variables. If one defines the Dynamic Range (DR, unitless) of the δ value for a particular stable isotope as the observed isotopic range for a given suite of samples divided by the typical standard deviation of a replicate measurement of the same sample (e.g. 14‰/0.2‰ = 70), one can calculate the probability for a random match by simple multiplying DR figures for each isotope. This approach of course assumes that all isotope abundance figures are independent variables. The range of 'natural' variability for the stable isotopic composition of Ecstasy tablets of unknown provenance was 148, 12.5, 14.0 and 43.4‰ for ^{2}H, ^{18}O, ^{13}C and ^{15}N, respectively. Given the influence of sample (in)homogeneity on the error of measurement we used conservative figures for the standard deviation of a replicate measurement of the same sample of 2, 0.5, 0.5 and 0.5‰ for measured $δ^2$H, $δ^{18}$O, $δ^{13}$C and $δ^{15}$N values.

Multiplication of the resulting DRs for each of the isotope abundances in the Ecstasy sample suite of 74, 25, 28 and 86 for ^{2}H, ^{18}O, ^{13}C and ^{15}N, respectively, yielded a random match probability of 1 : 4 454 800. In other words, to a first approximation, there is approximately only a 1 : 4.4 million chance that the observed isotopic signature for a given Ecstasy tablet would be encountered for two wholly unrelated Ecstasy tablets. Using a more conservative approach still, by multiplying DRs based on standard deviations of 2‰ for $δ^2$H values and 1‰ for all others would result in a random match probability of 1 : 534 576, which still makes multivariate stable isotope analysis a very powerful forensic tool. Clandestine synthesis and movement of drugs are a global problem run by many international criminal networks. Increasing information on drug linkage by developing robust datasets and mathematical models for interpretation will lead to an increase in objective scientific intelligence about international networks, will improve opportunities to disrupt such networks and will secure "safer" convictions (i.e. a conviction less likely to be challenged successfully).

Chapter III.3
Elucidating Explosives

Probably the first forensic stable isotope laboratory that was created to deal with stable isotope profiling of explosives and related substances is the stable isotope facility at the Forensic Explosive Laboratory (FEL), which is part of the UK's Defence, Science and Technology Laboratories. This is chiefly thanks to the vision and foresight of FEL's Principal Scientist, Sean Doyle, who was also a key person in founding the FIRMS Network (see Section II.5.1). However, this is not the only stable isotope forensics laboratory actively engaged in characterizing explosives isotopically. Government laboratories in the United States conducting research in this area are the Federal Bureau of Investigation (FBI)'s Stable Isotope Forensics Laboratory at the Counterterrorism and Forensic Science Research Unit of the FBI and the Chemical Science Division at Oak Ridge National Laboratory, which is managed by UT-Battelle for the US Department of Energy. Last, but not least, the Australian Federal Police's Forensic Operations Laboratory is also in the process of establishing a stable isotopes capability for characterizing explosives and other materials of forensic significance.

For obvious reasons there are limits to what can be disclosed in this book and in this chapter. However, it must be acknowledged that huge efforts are being made by scientists engaged in stable isotope research to support law enforcement agencies and to protect the public (Benson et al., 2009b; Quirk et al., 2009; Widory, 2009). For example, stable isotope analyses of explosives and their precursors have already been carried out in certain high-profile cases of terrorism involving British citizens such as Richard Reid (the 'shoe bomber') and his convicted accomplice Sajid Badat or the failed bomb attacks on London of 21 July 2005. However, despite these initial successes, similar to the issues surrounding stable isotope analysis of synthetic drugs, there is the potential that at some point in the future inter-batch variability (i.e. 'natural' or manufacturing process inherent scatter) of isotopic abundance may reduce the discriminatory power of stable isotope signatures, potentially leading to a situation in which two completely unrelated samples of explosives might become indistinguishable. In the following I touch upon some examples of ongoing as well as published research seeking to gain insights into natural variability observed in samples of explosives or precursors obtained

---

Stable Isotope Forensics: An Introduction to the Forensic Application of Stable Isotope Analysis Wolfram Meier-Augenstein
© 2010 John Wiley & Sons, Ltd

from controlled sources as well as into the underlying reaction mechanisms so as to develop protocols enabling forensic scientists and law enforcement to maximize the information locked into stable isotope signatures of explosives with the greatest possible confidence.

III.3.1 Bulk Isotope Analysis of Explosives and Precursors

Taking a simplified approach one could divide explosives and their precursors that become part of a criminal investigation into three groups:

(i) Explosives manufactured for military or commercial use.

(ii) Improvised explosive devices (IEDs) using commercial or military explosives.

(iii) Improvised or home-made explosives.

Since military/industrial explosives are manufactured under controlled conditions, have known compositions, and have been used by paramilitary and terrorist organizations, efforts made in studying the potential of stable isotope signatures of explosives as corroborative evidence have focused on compounds and materials such as ammonium nitrate, cyclotrimethylenetrinitramine (RDX, a constituent of Semtex), hexamethylentetramine (hexamine, the organic precursor of RDX), pentaerythritoltetranitrate (PETN, a constituent of Semtex, detonating cord and detonators), Semtex (a mixture of RDX, PETN, dye, antioxidant, plasticizer and binder), trinitrocellulose or trinitrotoluene. Research articles published thus far concerning stable isotope analysis of explosives come exclusively from four sources: BRGM – Centre Scientifique et Technique (Orléans, France), the Forensic Operations Laboratory of the Australian Federal Police (Canberra, Australia), the Forensic Explosives Laboratory (FEL, Dstl, Fort Halstead, United Kingdom) and the author's research laboratory (Benson *et al.*, 2009a, 2009b; Lock and Meier-Augenstein, 2008; Meier-Augenstein, Kemp and Lock, 2009; Pierrini *et al.*, 2007; Widory, 2006, 2009), although this should not be taken to mean there are no other research laboratories engaged in the field of organic explosives, in general, or stable isotopic composition of explosives, in particular (Phillips *et al.*, 2003). I should also mention work is and has been carried out studying stable isotopic composition of inorganic oxidizers such as chlorates and perchlorates used in pyrotechnics and home-made explosive devices (Ader *et al.*, 2001; Begley *et al.*, 2003; Bohlke *et al.*, 2005; Sturchio *et al.*, 2007) as well as black powder (Gentile, Siegwolf and Delemont, 2009). The results of the latter study emphasized a need already highlighted in this book on several occasions – the need to gain comprehensive knowledge of intra- and inter-batch variability as well as factors such as storage (Fraser, Meier-Augenstein and Kalin, 2008) and chemical ageing that might affect a compound's or material's isotopic composition so that the real value or benefit of stable isotope evidence can be assessed.

III.3.1.1 Ammonium Nitrate

Using stable isotope signatures of ammonium nitrate has to contend with several challenges (see also Section III.3.3). The natural abundance range of ^{15}N in ammonium nitrate is rather narrow, with δ^{15}N values from –4 to +2‰ (Benson *et al.*, 2009a; Widory, 2009). This is a reflection of the fact that ammonia is made from atmospheric nitrogen and nitric acid is made by oxidation of ammonia. Studies carried out on ammonium nitrate by my PhD student, Claire Lock, showed a range of 13‰ for their δ^{18}O values (Lock, 2009). Combining δ^{15}N and δ^{18}O values into a bivariate plot (*cf.* Figure III.12) showed suggestive grouping for samples of unknown origin, yet all but identical in physical appearance (samples AN28 and AN29), but also for samples known to be of different origin and whose physical appearance was distinctly different (AN2, small prill, opaque white; AN3, medium-sized prill, opaque peach coloured). A 'prill' is a small aggregate of one or more materials, most often in the shape of a dry ball or sphere, formed from a molten liquid, which apart from ammonium nitrate contains a mineral desiccant or anti-caking agent such as calcium silicate or sodium aluminosilicate (Figure III.13). In a forensic context, ammonium nitrate is often encountered in the form of crushed ammonium nitrate prills mixed with fuel oil – an explosive mixture called ANFO (ammonium nitrate fuel oil) that has been widely used by criminals and terrorists.

Comparing δ^2H values for ammonium nitrate observed in our laboratory with published data has shown a wide range of 277‰ (minimum: –257‰; maximum +20‰), suggesting a higher degree of discrimination ought to be achievable through

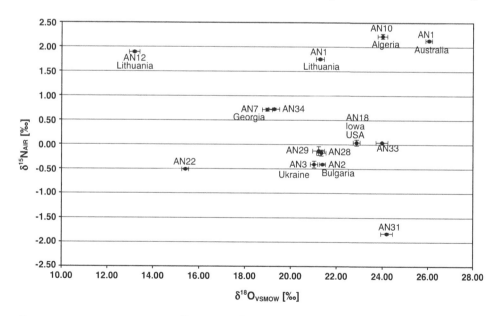

Figure III.12 Bivariate plot of δ^{15}N versus δ^{18}O values for ammonium nitrate prills from various sources (country of origin shown where known). Based on data generated by or for my PhD student Claire Lock (Lock, 2009).

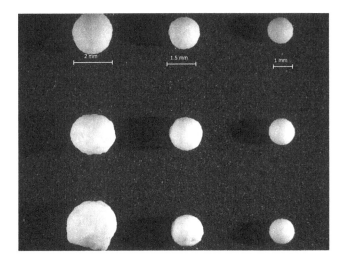

Figure III.13 Ammonium nitrate prills from various sources. Image provided courtesy of Sarah Benson (Forensic Operations Laboratory, Australian Federal Police, Canberra) (Benson, 2009).

the inclusion of δ^2H values to create a multivariate stable isotope profile (Benson et al., 2009a; Widory, 2009). It should be noted that the observed range of about 60‰ for sample δ^2H values in the Australian study was not sufficient to resolve all of the samples although they came from different sources and countries of origin (Benson et al., 2009a). It should also be noted that hydrogen data from the Australian study were neither scale nor exchange corrected, therefore reported δ^2H values are not 'true' δ^2H values (although this has little or no impact on the reported δ^2H ranges). In fact, the authors state in their article:

> As reference materials were not measured in the sequences (due to a lack of available [isotope relevantly] calibrated reference materials bracketing the isotope values of the QC standards), the data from these experiments are not comparable with data generated outside of these experiments.

Despite this shortfall caused by a lack of matrix-matched reference materials calibrated on the relevant isotope scales, these results provide a springboard for future work. It is already quite obvious that considerable research has still to be undertaken before the processes driving ^2H isotopic signatures of ammonium nitrate are fully understood so they can be exploited with confidence in a forensic context (see Section III.4.3).

III.3.1.2 Hexamine, RDX and Semtex

One of the first, if not the first, systematic study to determine the probative value of bivariate stable isotope signatures of an industrial/military explosive looked into the statistical interpretation of ^{13}C and ^{15}N isotope abundance data of 26 randomly selected samples from a sample bank of 500 Semtex H samples (Pierrini *et al.*, 2007). For the 26 Semtex samples selected, a range from −24.76 to −36.53 and −2.72 to −25.68‰ for δ^{13}C and δ^{15}N values, respectively, was observed with standard deviations for six measurements per sample being of the order of ±1‰ for either isotope. Assuming that all 26 Semtex samples came from different sources, this study concluded that likelihood ratios provide a more balanced way to test the prosecution hypothesis that two samples have come from the same source versus the defence hypothesis that the same two samples have come from different sources than Hotelling's T^2 test. A detailed description of an application of Hotelling's T^2 test to multivariate stable isotope data of physical evidence has been reported elsewhere (Farmer *et al.*, 2009).

As described in Section II.4.2, the likelihood ratio approach aims to enable the forensic scientist to calculate the ratio of the probability of the outcome of the comparison between the control and the recovered samples by testing both the prosecution hypothesis and the defence hypothesis. The prosecution hypothesis is supported if the LR > 1. In the aforementioned Semtex study, an LR > 1 was obtained for all pair-wise comparisons when the Semtex samples came from the same source irrespective of which multivariate likelihood ratio model was being used. However, when testing the defence hypothesis, LR > 1 was found in 5.5–11% of all comparisons. In other words, depending on the chosen likelihood ratio model, between 5.5 and 11% 'false positives' occurred, thus mistakenly supporting the case for the prosecution.

The authors themselves noted that each member of the population of Semtex analysed was assumed to have come from different sources and although all samples originated from real cases, no background information was available. In addition one has to consider that Semtex H is a mixture of two chemically different explosives (RDX and PETN), plasticizers, binder, antioxidant and a dye – all of which make a contribution to its bulk isotopic signature. Furthermore, percentage content of RDX and PETN varies between different types of Semtex, such as Semtex A, Semtex H or Semtex 10, and may vary between different batches of the same type.

In order to investigate to what degree the application of stable isotope signatures can provide a level of discrimination not achievable utilizing traditional forensic techniques and thus help to solve complex criminal cases, a 3-year PhD project was set up under Sean Doyle's (FEL) stewardship and my direction/supervision. One aim of this project was to investigate if stable isotope studies of single compounds such as the RDX precursor hexamine or RDX itself would give us an indication how likely or unlikely we were to achieve the ambitious ultimate goal of successfully giving stable isotope signatures of explosives in evidence in a court of law.

Preliminary work carried out on a small number of pure, crystalline RDX samples from two known different sources included their analysis for ^{13}C and ^{15}N isotope abundance. To determine intra-batch variability, two RDX samples were analysed 18 times each and the results of these analyses are illustrated in Figure III.14.

For all RDX samples, standard deviations for δ^{13}C and δ^{15}N values of 18 measurements per sample per stable isotope were 0.11 and 0.15‰ or less, respectively.

Based on the promising preliminary observation, a set of 14 hexamine samples was put together from a variety of suppliers and sources (Lock, 2009). Two of these samples, HEX3 and HEX4, were two bottles from the same batch or lot by the same supplier,

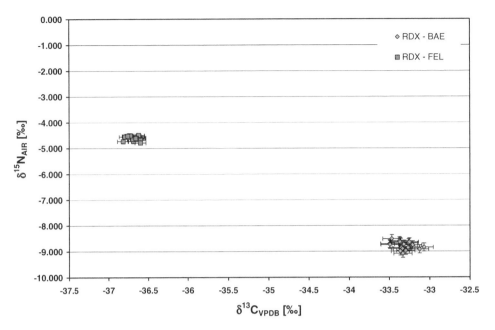

Figure III.14 Detailed bivariate plot of $\delta^{15}N$ versus $\delta^{13}C$ values for the explosive RDX from two different sources demonstrating homogeneity of the samples and reproducibility of isotope analysis ($n = 18$). Based on data generated by or for my PhD student Claire Lock (Lock, 2009).

whereas another two sample pairs, HEX1/HEX11 and HEX5/HEX6, came from two different lots of two different suppliers each. Since hexamine is the organic starting material for RDX (Figure III.15), the long-term objective was, results permitting, to carry out syntheses with hexamine of different isotopic composition to investigate the nature of the isotopic relationship between product and precursor (cf. Section III.3.2).

All samples were analysed in replicates of $n = 6$ for ^2H, ^{13}C and ^{15}N isotopic composition, and results are portrayed by the three-dimensional plot of δ^2H versus $\delta^{15}N$ versus $\delta^{13}C$ values in Figure III.16.

Using HCA of the trivariate isotope signature dataset, three pairs of hexamine samples could not be distinguished at 99% similarity level: HEX3/HEX4, HEX5/HEX6 and HEX2/HEX7 (Figure III.17). Samples HEX3 and HEX4 were from two different bottles of the same batch from the same supplier and therefore correctly grouped. Samples HEX5 and HEX6 were from the same supplier with the same country of origin, but from two different lots. Whilst this does not preclude the samples being related, the

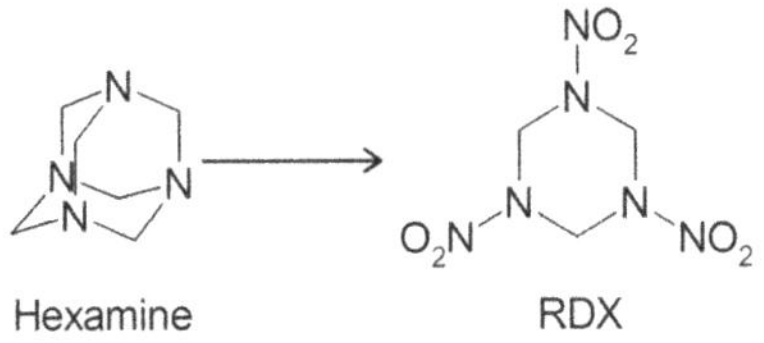

Figure III.15 Hexamine to RDX.

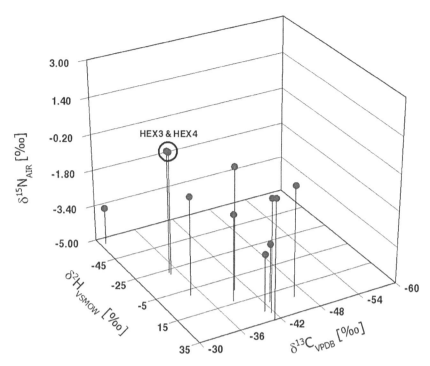

Figure III.16 Three-dimensional plot of δ^2H versus $\delta^{15}N$ versus $\delta^{13}C$ values for the RDX precursor hexamine. The encircled two data points are samples HEX3 and HEX4 from two different bottles of the same batch from the same supplier. Based on data generated by or for my PhD student Claire Lock (Lock, 2009).

difference between $\delta^{15}N$ values for either sample was above 2‰, although differences between δ^2H and $\delta^{13}C$ values for either sample were not statistically significant. Analysis of variance (ANOVA) and a Tukey's comparison at 95% confidence level both grouped HEX5 and HEX6 for ^2H and ^{13}C, but not for ^{15}N isotopic composition. Samples HEX2 and HEX7 came from two different suppliers located in two different countries. Again, this does not necessarily rule out a common source, but for this sample pair grouped by HCA only the difference between δ^2H values was not statistically significant. A Tukey's comparison and ANOVA at 95% confidence level both grouped HEX2 and HEX7 for ^2H, but neither for ^{13}C nor ^{15}N isotopic composition (Lock, 2009).

These results clearly demonstrate the potential probative power of multivariate stable isotope signatures while at the same time highlighting the need not to rely on one type of statistical analysis alone and to exercise caution when applying statistical methods that are based on a fixed cut-off point. In fact, the observations made during the statistical analysis of the hexamine data using HCA, ANOVA and Tukey comparison suggest the continuous likelihood ratio approach for multivariate datasets should be applied to single-component evidence in the same way it has already been to complex matrix evidence such as Semtex or white paint (Farmer, Meier-Augenstein and Lucy, 2009; Pierrini *et al.*, 2007).

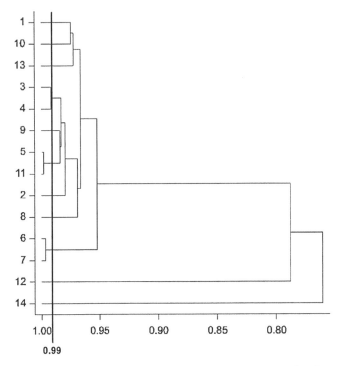

Figure III.17 Dendrogram of a HCA (single linkage, Euclidean distance) for the trivariate stable isotope dataset of 14 hexamine samples. Based on data generated by or for my PhD student Claire Lock (Lock, 2009).

III.3.1.3 Hydrogen Peroxide and Peroxides

Peroxide explosives and their use in improvised or home-made explosives have become an increasing threat to the general public because precursor materials can be obtained relatively easily and refined without the need for complex laboratory equipment. The danger inherent to both manufacturing and transporting of organic peroxides does not seem to deter perpetrators from doing either. Triacetone triperoxide (TATP), for example, is highly susceptible to shock or friction of as little as 0.3 Nm or 0.1 N, respectively. In order to determine if it was possible to distinguish between TATP produced under different reaction conditions using stable isotope signatures, four batches of TATP were synthesized using different mole ratios for reagents, different reaction temperatures and different grades of hydrogen peroxide. Products from these four synthetic routes could be fully distinguished on the basis of their δ^2H, δ^{13}C and δ^{18}O values (Benson *et al.*, 2009b).

While work on the peroxide product TATP was carried out in Australia, work in the United Kingdom focused on the peroxide precursor, hydrogen peroxide. Given FEL's natural involvement in the investigation of the 7 July 2005 London bombings and the failed 21 July 2005 London bomb attacks, research was undertaken jointly between FEL and our laboratory as part of a PhD project mentioned above

(Lock, 2009) to determine what contribution stable isotope signatures of peroxides could make to the forensic investigation of improvised explosives and their illegal manufacture.

Ten samples of hydrogen peroxide from seven different suppliers and of various concentrations were acquired to gain a preliminary insight into the natural variability of ^{2}H and ^{18}O isotope abundance signatures. Two samples, HP2 and HP8, came from two different bottles of 30% hydrogen peroxide, but from the same batch and the same supplier, to test intra-batch variability. The observed range for δ^2H and δ^{18}O values for all measured samples was 114.1 (minimum: −132.9; maximum: −18.8) and 15.33‰ (minimum: −1.90; maximum: 13.43), respectively. Statistical analysis of the data by HCA (furthest-neighbour, Euclidean distance, 95% similarity level) correctly grouped only samples HP2 and HP8 as highly similar (Lock, 2009).

Most recipes for the clandestine manufacture of organic peroxides are based on a hydrogen peroxide concentration of 30% in water. However, most consumer products contain 10% or less, while industrial products typically contain more than 30% of hydrogen peroxide, thus necessitating either concentration or dilution to the required concentration, respectively. In a preliminary study to determine the feasibility of establishing links between a diluted hydrogen peroxide solution and its parent material as well as the water used to dilute it, we carried out a series of dilution experiments in the first instance since experimental conditions for creating a defined dilution from two fixed precursor pools, hydrogen peroxide of 60% and pure deionized water, could be controlled more easily.

Since hydrogen peroxide and water contain the same mole equivalent of hydrogen our hypothesis was that the term b of the solution $y = ax + b$ of a linear regression fit applied to measured δ^2H values versus decreasing concentration in hydrogen peroxide (or increasing concentration of water) should yield a value close or identical to the δ^2H value for the water used during the dilution process. The value for b obtained from the solution for the linear regression was −46‰ and in good agreement with the δ^2H value of −47.5‰ for the water used to dilute the parent hydrogen peroxide solution (cf. Figure III.18).

Hydrogen peroxide and water do not contain oxygen in equivalent molar amounts. Hydrogen peroxide comprises 1 mol O_2 per mole of hydrogen peroxide, while water comprises 0.5 mol O_2 per mole of water. For this reason a polynomial regression fit of the form $y = a_1 x^2 + a_2 x + b$ was applied to measured δ^{18}O values versus decreasing concentration in hydrogen peroxide, which indeed provided a better curve fit to the data as well as a closer match for b, the extrapolated δ^{18}O value for pure water with the actual δ^{18}O value of the water used for dilution. The water used for dilution had a δ^{18}O value of −6.58‰, while polynomial and linear regression fits yielded values for b of −8.54 and −1.85‰, respectively (cf. Figure III.19).

Given these encouraging results it would seem entirely feasible to use hydrogen peroxide samples seized as part of a criminal investigation after initial characterization of their hydrogen peroxide content for establishing links between, for example, caches of hydrogen peroxide suspected to have been acquired and stored for illegal purposes, and hydrogen peroxide seized in a clandestine laboratory even after dilution or concentration.

178   III: STABLE ISOTOPE FORENSICS: CASE STUDIES AND CURRENT RESEARCH

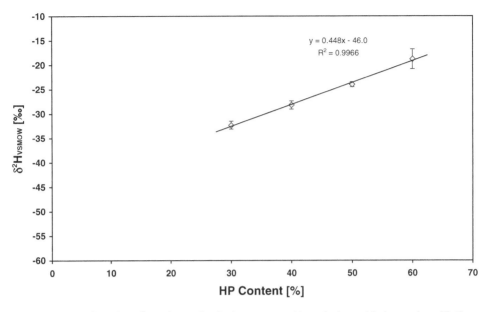

Figure III.18 Changing δ^2H values of a hydrogen peroxide solution with increasing dilution. Based on data generated by or for my PhD student Claire Lock (Lock, 2009).

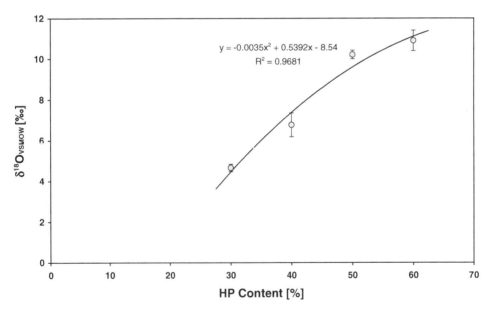

Figure III.19 Changing $\delta^{18}O$ values of a hydrogen peroxide solution with increasing dilution. Based on data generated by or for my PhD student Claire Lock (Lock, 2009).

III.3.2 Isotopic Product/Precursor Relationship

The FEL and our stable isotope forensic laboratory belong to a small circle of research laboratories engaged in the systematic investigation of explosives' precursors and the relationship between isotopic composition of precursor and eventual product. FEL investigates about 250 criminal cases per year concerning offences under the Explosives Act 1875, Explosives Substances Act 1883 and The Manufacture and Storage of Explosives Regulations 2005 (www.opsi.gov.uk). In the last 9 years approximately 5% of these cases related to the illegal manufacture of explosives. In addition, many cases were investigated with regard to intent to manufacture explosives, either from basic starting materials or commercially available explosives and pyrotechnic mixtures as evidenced from recovered recipes, chemicals or apparatus (Lock and Meier-Augenstein, 2008).

In criminal prosecutions regarding the illegal manufacture of narcotic drugs or explosives, evidence demonstrating the manufacture forms an essential plank in the case for the prosecution. Manufacturing equipment, precursors and 'recipes' are all regarded as a clear indicator of a clandestine laboratory. Being able to establish links between precursors and products would not only provide additional evidence in such circumstances, but would also help to establish links between several persons and locations involved in the manufacture of illegal substances that might otherwise be hard to substantiate.

Most chemical syntheses do not proceed in a quantitative fashion and generate product yields that, based on the amount of precursor, can range from as little as 30% to as high as 90%. As we have learned in Section I.4.2, any chemical reaction in which not all precursor material is consumed to generate a product yield of 100% is subject to isotopic fractionation. The most significant isotope effect here is the kinetic or primary isotope effect, whereby a bond containing the chemical elements under consideration is broken or formed in the rate-determining step of the reaction (Rieley, 1994). In the same way, parameters such as reaction enthalpy ΔH^0 or reaction rate constant k assume reaction-specific values, fraction factor α and enrichment factor ε are also reaction-specific constants. Thus, if a given reaction is always performed under the same reaction conditions, values for α and ε linking precursor and product will remain the same even if precursors are used from different sources and of different isotopic composition from one reaction to the next.

As part of the FEL sponsored PhD project, we set out to have a series of syntheses carried out synthesizing RDX from hexamine using the Woolwich process (Wilson, Forster and Roberts, 1950) to investigate what isotopic relationship or linkage exists between the product RDX and its precursor hexamine (Lock, 2009). From a total of 11 stock items of hexamine manufactured in five different countries and obtained from six different suppliers, five hexamine samples were selected at random and used as starting material for the synthesis of RDX. Of the five hexamine samples, Sample 2 was selected for a repeat synthesis using the same starting material, whereas the remaining four hexamine samples were only used in one synthetic process each. Measured δ^{13}C and δ^{15}N values for all hexamine samples ranged from −34.51 to −46.18 and +0.41 to −2.48‰, respectively (Lock and Meier-Augenstein, 2008).

Synthesized RDX precipitate was collected by suction filtration, washed with deionized water and left to dry on the filter for approximately 1 h. The air-dried product was subsequently transferred to a vacuum oven and dried at 75 °C overnight. Typical yields of RDX were 60%. The crystalline product isolated from the synthesis was characterized using nuclear magnetic resonance (NMR) spectrometry, Fourier transform infrared (FTIR) spectroscopy and thin-layer chromatography (TLC) (Lock, 2009). None of these analytical techniques indicated the presence of any substance other than RDX. Stable isotope analysis of RDX for ^{13}C and ^{15}N isotopic composition showed δ^{13}C values with a range of -38.34 to $-26.23‰$, while δ^{15}N values ranged from -15.62 to $-13.79‰$ with standard deviations of six measurements per sample per isotope for δ^{13}C and δ^{15}N being 0.09 and 0.22‰ or less, respectively. From the difference in δ values between corresponding precursor/product pairs the average isotope shift from hexamine to RDX was initially calculated to be $+8.35‰$ and $-14.01‰$ for $\Delta\delta^{13}$C and $\Delta\delta^{15}$N, respectively. Subsequently, individual fractionation factors α were calculated for each synthesis based on their corresponding pairs of measured δ values, which in turn were used to determine enrichment factors ε for each reaction using equations we have already encountered in Section I.4.2:

$$\alpha = \frac{\delta_{Product} + 1000}{\delta_{Source} + 1000} \tag{I.8}$$

$$\varepsilon = (\alpha - 1) \times 1000 \, (‰) \tag{I.7}$$

Fractionation factors α and enrichment factors ε as calculated for each of the syntheses are summarized in Table III.4. Average enrichment factors $^{13}\varepsilon$ and $^{15}\varepsilon$ as calculated from fractionation factors α were $+8.69 \pm 0.30$ and -14.01 ± 0.47, respectively, which compares well with the initial estimates based on differences in measured δ values (Lock and Meier-Augenstein, 2008).

The above results obtained from a series of syntheses carried out on a laboratory scale under controlled, as reproducible as possible, reaction conditions clearly demonstrated constant or reaction-specific isotopic fractionations for carbon and nitrogen between the starting material hexamine and the reaction product RDX. These data illustrate the potential of forensic stable isotope analysis at near-natural abundance levels beyond the point of direct comparative isotope analysis of chemically identical materials, enabling investigators to determine links between starting material and explosives product, which may be useful in scenarios where intelligence is required linking a clandestine laboratory or a suspect to an explosive device or a cache of its precursor.

Given the assertion made in a paper published in 2005 that no differences in stable isotopic composition existed between explosives from different sources because no differences in $(M + 1)^+/M^+$ ratios could be observed using a conventional quadrupole mass spectrometer in selected ion monitoring (SIM) mode (Sharma and Lahiri, 2005), it cannot be emphasized enough that a change in ^{13}C isotope abundance of $+8.69‰$ is equivalent to a difference at the natural abundance level of 0.0096 atom%, while a change in ^{15}N isotope abundance of $-14.01‰$ is equivalent to abundance difference of 0.0051 atom%. Neither change can be detected by a conventional mass spectrometer, not even in SIM mode since even under optimized conditions the achievable precision

Table III.4 Summary of fraction factors α and enrichment factors ε for individual hexamine/RDX precursor/product pairs.

Synthesis no.	Brand	Grade	Origin	$^{13}\alpha$	$^{13}\varepsilon$	$^{15}\alpha$	$^{15}\varepsilon$
1	Sigma-Aldrich	ultra	USA	1.0082	8.22	0.9857	−14.2684
2a	Fisher	laboratory	unknown	1.0087	8.70	0.9858	−14.1890
2b	Fisher	laboratory	unknown	1.0089	8.87	0.9858	−14.2335
3	Riedel-De Häen	extra pure	Germany	1.0084	8.44	0.9858	−14.2024
4	Riedel-De Häen	reagent	Germany	1.0089	8.92	0.9868	−13.1727
5	Fluka	puriss	South Korea	1.0090	8.99	0.9860	−14.0144
Average value				1.0087 ± 0.0003	8.69 ± 0.30	0.9860 ± 0.0005	−14.01 ± 0.47

Based on data published in Lock and Meier-Augenstein (2008).

of isotope abundance measurement using a molecular mass spectrometer in SIM mode can only be as good as 0.05 atom%, although realistically this figure is closer to approximately 0.1 or 0.5 atom% (Meier-Augenstein, 1999; Nic Daéid and Meier-Augenstein, 2008). It is therefore not surprising that the attempt reported in 2005 to identify the possibility of any isotopic substitution at natural abundance level in explosive samples using MS in SIM mode was unsuccessful. However, the system inherent detection limits of quadrupole MS instruments notwithstanding, it is difficult to understand how in the twenty-first century any scientist could conclude a phenomenon does not exist just because it could not be observed with the instrument at hand and how it could happen for such flawed a conclusion not to be picked up by the peer-review process.

III.3.3 Potential Pitfalls

Essentially, we have already touched upon the two major pitfalls in preceding parts and chapters. One pitfall concerns challenges arising during isotope analysis of nitrogen-rich compounds such as isobaric interference of nitrogen or nitrous oxides with the isotope analysis of ^{18}O (as CO) or ^{13}C (as CO_2; see Section II.6.2.1), or the 'ionization quench' (IQ) effect of nitrogen on the isotope analysis of ^{2}H (see Section II.6.3.3). The best remedial action for these problems is to avoid them by making sure:

(i) The ultimate compound conversion products are N_2 and CO or N_2 and CO_2 only.

(ii) The carrier gas entering the ion source of the IRMS is free of oxygen and water.

(iii) Gas chromatographic parameters are chosen to ensure widest possible base-line separation of N_2 from H_2, N_2 from CO and N_2 from CO_2 (Brand *et al.*, 2009a, 2009b; Meier-Augenstein, Kemp and Lock, 2009).

The second pitfall concerns the determination of true δ^2H values for compounds comprising exchangeable hydrogen atoms such as ammonium nitrate, hydrazine nitrate, nitroguanidine, pentaerythritol trinitrate, picramic acid, picric acid or urea nitrate.

Preliminary studies carried by the Forensic Operations Laboratory of the Australian Federal Police on ammonium nitrate samples from nine different manufactures in five different countries showed $\delta^{15}N$ values ranging from −4 to +2‰, which is not that surprising given how its precursor chemicals ammonia and nitric acid are made. We have already encountered this phenomenon in Section I.5.4 when looking at anthropogenic sources of nitrogen in the environment. Clearly, more information is required to increase the discriminatory power of stable isotope signatures of ammonium nitrate, and the only other two chemical elements available here are oxygen and hydrogen. Measured $\delta^{18}O$ values from the same sample pool ranged from 13 to 21‰, while measured δ^2H values extended from −20 to +45‰. On the basis of these observations the random match probability of δ value combinations would decrease from 1 : 432 to 1 : 28 080, and yet it was not possible to discriminate all samples on the basis of manufacturer and country of origin even though the inclusion of δ^2H values clearly improved the discriminatory power of the stable isotope signatures (Benson *et al.*, 2009a). However, observations made in a French laboratory and our own suggest the measured δ^2H values in ammonium nitrate may extend down to −240‰ (Widory, 2009), thus extending the range in δ^2H values that could be exploited for forensic examination from 65 to 285‰, which would decrease the random match probability to 1 : 123 120 when treating all three isotopes as independent variables.

However, neither of the two published studies addressed the issue of hydrogen exchange of NH_4^+ with ambient moisture or humidity. Given the hygroscopic nature of ammonium nitrate it is not inconceivable even for ammonium nitrate prills to adsorb sufficient quantities of water to promote hydrogen exchange between ammonium nitrate and water. Preliminary studies carried out in our laboratory yielded a mole fraction of exchange of $f_{ex} \geq 1.2$ for ammonium nitrate that was kept in a sealed container next to a beaker filled with water for 4 days. A mole fraction of 1.2 overall could be interpreted as one out of the four hydrogen atoms on NH_4^+ exchanging completely (i.e. to 100%), while a second hydrogen exchanges to 20%. Conversely, one could conclude each of the four hydrogen atoms on NH_4^+ exchanges to 30%, which seems the more likely scenario given the symmetry of the NH_4^+ molecule.

Clearly this observation has serious ramifications for ^2H isotope analysis of ammonium nitrate samples of unknown sample history such as unknown or undeterminable age, length of time of storage and/conditions during storage, thus necessitating research efforts to understand kinetics and fractionation of this exchange process, which in turn will be translated into analytical protocols that will permit us to exploit the information locked into ^2H isotope signatures of ammonium nitrate.

III.3.4 Conclusions

Information gained from research as well as actual case work has clearly demonstrated the added discriminatory power multivariate stable isotope signatures offer to the forensic investigation of cases involving explosives, improvised or home-made explosives and IEDs. That being said, from research carried out in this area thus far, it is equally clear that work in this field needs to be expanded, and that countries with a vested interest must develop mechanisms to overcome current obstacles hampering systematic research in stable isotope forensics of explosives so methods can be harmonized and validated, databases can be compiled, and international collaboration is enabled to support likelihood ratio approaches for the evaluation of stable isotope data.

As seen already with stable isotope signatures of drugs, ^2H isotopic composition provides powerful information without which discrimination between different sample histories or provenance would be very difficult. Given the importance of δ^2H values to successful discrimination and provenancing of explosives and their precursor materials, a great deal of effort by scientists as well as financial and logistical support by governments and government agencies is still required to address the challenges posed to forensic ^2H isotope analysis of, for example, ammonium nitrate by hydrogen exchange and the lack of relevant and chemically related reference materials calibrated on the VSMOW scale.

Chapter III.4
Matching Matchsticks

As part of a murder enquiry, a particular individual was suspected of perverting the cause of justice by attempting to burn evidence potentially relating to that murder. However, damp conditions on the evening of the murder prevented easy ignition of the items to be destroyed; consequently, a number of spent and unspent matches were left behind at the scene of the fire. These were recovered by police and submitted for forensic examination. A police search at the suspect's house revealed a number of matchboxes, all of the same brand, containing matches looking similar to those recovered at the fire scene. Due to the visual similarity between the recovered and seized matches it was thought stable isotope fingerprinting of the wooden part of the matchsticks could be used to substantiate the presumed link between the two sample sets.

Since nothing like this had ever been attempted before and, hence, not been described in the literature, isotope analysis of the case matchsticks was accompanied by two studies – one looking into the intra- and inter-brand variability of isotope signatures of wooden matchsticks, the other directly comparing recovered and seized matchsticks using established forensic analytical techniques, namely X-ray diffraction (XRD) analysis of match heads and microscopic analysis of the wood. These studies were carried out by my then PhD student Dr Nicola Farmer (now a forensic researcher at FEL) as part of her PhD thesis.

Owing to import restrictions imposed by heightened security measures on air travel and postal services, eventually only nine different matchstick brands from four different regions world-wide could be collected to form the basis of the inter-brand background study into natural variability of stable isotope signatures from wooden matchsticks. In addition, matches were purchased from a variety of outlets in the United Kingdom and Ireland. It was found many matchboxes do not provide batch information on the label, therefore only Swan and BoPeep matches were compared for inter-batch variability (Farmer, Meier-Augenstein and Kalin, 2005; Farmer *et al.*, 2007).

Prior to this study in support of the forensic investigation, isotope analysis of wood had almost exclusively focused on samples from tree rings in studies aiming to reconstruct past climates or investigating CO_2 uptake (Saurer *et al.*, 2003). The majority of published work focused upon the study of carbon and the factors affecting its isotopic fractionation and how this relates to growing conditions (e.g. the occurrence of droughts). Therefore, factors affecting carbon isotopic fractionation are well understood. In recent years different research groups have determined that whole wood contains the same isotopic record as cellulose (Loader, Robertson and McCarroll, 2003) and most studies therefore analyse wood samples as a whole as compared to separating it into its three main constituents of lignin, cellulose and hemicellulose (Barbour, Andrews and Farquhar, 2001; Boettger *et al.*, 2007).

Stem sections of 2–3 mm in length were cut from individual matchsticks and were subsequently prepared for IRMS analysis by grinding in a SPEX CertiPrep 6850 Freezer Mill (Glen Creston Ltd, Stanmore, Middlesex, United Kingdom), programmed to carry out three 10-min cycles and an initial 10-min period of precooling. After grinding, samples were placed in labelled glass vials and stored in a desiccator containing phosphorus pentoxide to remove residual water traces from the sample. Note, measured total δ^2H values reported below in Section III.4.3 do not reflect the 'true' ^2H isotopic composition of the wood from which the matchsticks were made since hydroxyl hydrogen atoms in cellulose have the potential to exchange with hydrogen atoms in ambient moisture. Given an international cellulose standard with an accepted true δ^2H value is not available, in order to compare ^2H data all samples were exposed to the same ambient moisture after collection so labile hydrogen atoms prone to exchange would all reflect the same ^2H level. Samples were analysed purely for the purpose of comparison on a like-for-like basis in accordance with the 'principal of identical treatment' (Bowen *et al.*, 2005; Werner and Brand, 2001).

III.4.1 ^{13}C-Bulk Isotope Analysis

To establish whether matches collected from different locations as part of a criminal investigation could be associated to each other on the basis of ^{13}C isotope analysis alone, initially the variability in ^{13}C abundance within one box of matches and between batches of the same brand was determined by removing three matches from each box and analysing each in triplicate. The maximum intra-box or intra-batch variability in ^{13}C abundance of 6.0‰ was observed within one box of Swan matches. A variation this large was not found within any of the other boxes of matches analysed (Farmer, Meier-Augenstein and Kalin, 2005).

Large-scale fractionation of ^{13}C is known to occur within individual trees. CO_2 diffusing through stomata and enzymatic processes dependent on growing conditions such as temperature, humidity and nutrient supply are all factors that can cause the fractionation of carbon within trees. Loader *et al.* performed a study to compare trees on the same site for $\delta^{13}C$ differences across tree rings over 55 years and observed an average variation in $\delta^{13}C$ values of 2.5‰ (Loader, Robertson and McCarroll, 2003). The intra-box variability of $\delta^{13}C$ values observed in our study for matches collected from around the world was on average 1.5‰ with a maximum of 2.5‰, which was consistent with the natural variation of $\delta^{13}C$ values reported by Loader *et al.* as well as observations made by other research groups (Barbour, Andrews and Farquhar, 2001). However, it was also reported that the average difference in $\delta^{13}C$ values between trees on the same site could be as high as 5.5‰ (Loader, Robertson and McCarroll, 2003). This would imply that matches produced from two different trees grown in the same plantation could produce a difference in $\delta^{13}C$ values of 5.5‰, which meant the intra-batch variation of 6.0‰ seen for Swan matches could be a result of the matches in that box having been produced from two different trees from the same plantation. However, even without this compounding factor, a variability of 2.5‰ for $\delta^{13}C$ values within the same tree will make it difficult to exploit the limited range of –20 to –30‰ for natural abundance of ^{13}C found in wood (Boettger *et al.*, 2007; Loader, Robertson and McCarroll, 2003; McCarroll and Loader, 2006). This of course meant considering $\delta^{13}C$ values of wood in isolation would be of very limited forensic value.

III.4.2 ^{18}O-Bulk Isotope Analysis

Oxygen atoms incorporated within plants can originate from three distinct sources: atmospheric CO_2, atmospheric oxygen and water taken up through the roots. Since 1974 it has been hypothesized that the oxygen value of tree rings can be correlated to the $\delta^{18}O$ value of leaf water and, hence, ultimately source water, and further studies seemed to confirm this. Recently, however, two opposing arguments have developed where the source of oxygen incorporated into synthesized molecules is under debate.

Many agree with the original studies by DeNiro and Epstein (1979), which showed oxygen incorporated into plant cellulose originates from soil water, although it was agreed that fractionation did occur in this process. It is generally believed the good correlation between $\delta^{18}O$ value of cellulose and source water is a reflection of oxygen isotope equilibrium between leaf water and carbonyl groups with exchangeable oxygen along the carbohydrate chain. However, Schmidt *et al.* hypothesize this enrichment is accidental and coincidental, as studies have been published that do not follow this enrichment trend (Schmidt, Werner and Rossmann, 2001). They suggest oxygen initially incorporated into plants originates from CO_2 and isotopic fractionation associated with oxygen exchange from carbonyl groups removes the primary enrichment. They argue that individual enrichments and depletions will finally lead to the reported global mean ^{18}O enrichment of 27 ± 4‰ of carbohydrate oxygen relative to oxygen of source water. The same group points out that $\delta^{18}O$ values of lignin and other aromatic compounds from the shikimate pathway do not follow this trend and the oxygen incorporated into such structures originates from atmospheric oxygen, introduced by mono-oxygenase reactions (similar to hydroxylation reactions in the human body that are mediated by monoamine oxidases or cytochrome P450 enzymes). Findings by Saurer *et al.*, who observed a dependence between cellulose $\delta^{18}O$ value and plant species as well as metabolic status, would seem to support Schmidt's hypothesis (Saurer, 2003; Saurer *et al.*, 2003).

To assess the discriminating power of ^{18}O signatures, the natural variability within one box of matches and between batches of the same brand was determined. Matches were sampled as mentioned above for ^{13}C isotope analysis. Maximum variability in ^{18}O abundance for all matchsticks analysed was 4.0‰ and was observed for matchstick samples from the same box of BoPeep safety matches (Farmer, Meier-Augenstein and Kalin, 2005).

Combining $\delta^{13}C$ and $\delta^{18}O$ values to produce a two-dimensional stable isotope profile enabled us to discriminate between scene and seized matches. However, it should be noted that in this instance $\delta^{13}C$ and $\delta^{18}O$ values are not truly independent variables since ^{13}C can only be sourced by trees from CO_2, yet CO_2 is also one of the possible sources of ^{18}O (Schmidt, Werner and Rossmann, 2001). For this reason, we also analysed samples for their ^2H isotopic composition to generate a bivariate stable isotope profile of two independent variables since ^{13}C and ^2H are sourced from two different precursor pools.

III.4.3 ^2H-Bulk Isotope Analysis

Isotope abundance of ^2H in wood is related to source water, and thus the climate and geographical location where a tree was grown. Although fractionation within the plant occurs, ^2H can be attributed directly and exclusively to that of the meteoric water in the area of tree growth (Keppler et al., 2007). Therefore, if ^2H isotopic composition is used in conjunction with ^{13}C isotopic composition this should increase the chances of being able to come to some conclusion about matchstick provenance as compared to using a combination of $\delta^{13}C$ and $\delta^{18}O$ values or $\delta^{13}C$ values on their own.

Matchstick samples from Swan batch A produced a narrow intra-batch variation of 8‰ for hydrogen isotope composition, indicating the trees came from the same site and thus coincidentally confirming the initial assumption for the variation in $\delta^{13}C$ values seen in one box of Swan batch A matches indicating that the wood had been harvested from different trees originating from the same plantation. In contrast, δ^2H values for the two different batches A and B of Swan matches differed by 48‰ (six times the intra-batch variation of 8‰). This indicated that wood used in the production of matches from batch B had come from a tree or trees grown and harvested at a location different to that from which wood used in batch A originated from. Typical within-box variation for matches collected from around the world covered a range of 10.4‰ with a maximum value of 17‰ (Farmer et al., 2007).

Constructing a bivariate plot of δ^2H versus $\delta^{13}C$ values, it was possible to separate and differentiate crime scene evidence from exhibits seized at the suspect's house. Control matches reported to be manufactured in Scandinavia were found to have δ^2H values of −110 to −130‰. This range is in agreement with known δ^2H precipitation values for this region. Matches collected at the crime scene fell within this range of δ^2H values. However, with an average δ^2H value of −65.0‰, matches seized from the suspect's house were clearly distinguishable from both the crime scene samples and the control group (Figure III.20). A δ^2H value of −65‰ is indicative of a growth site located at lower latitude compared to sites in Scandinavia or Eastern Europe. It should be noted, however, if matches are found to be indistinguishable on the basis of ^2H and ^{13}C isotope signatures it cannot be concluded they share any commonality other than being made from wood grown in one or more geographic regions meeting certain characteristics with regard to latitude and growing conditions, such as climate, temperature and precipitation (Farmer et al., 2007).

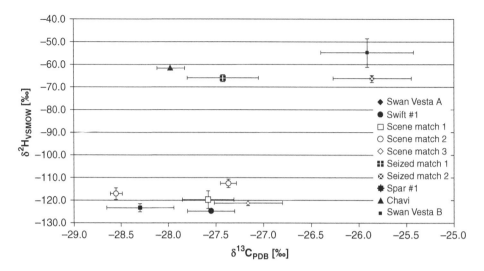

Figure III.20 Bivariate plot of δ^2H versus $\delta^{13}C$ values from matchsticks recovered at the crime scene and seized from the suspect's house as well as from the controls. © 2007 Elsevier, reproduced with permission from Farmer et al. (2007).

Comparative analysis of spent and intact strike-heads of matchsticks secured at the crime and matchsticks seized at the suspect's house by XRD confirmed the conclusions drawn from the stable isotope data (Farmer et al., 2007), demonstrating the benefit of a complementary analytical approach based on two or more independent analytical techniques.

III.4.4 Matching Matches from Fire Scenes

Given the encouraging results mentioned in the preceding chapters, Nicola Farmer set out to investigate what contribution these newly developed methods could make to the forensic investigation of arson. She soon found out that, based on official statistics, 2213 arson attacks take place per week in the United Kingdom, which on average kill two people and injure 53 per week. Arson attacks cost the economy £53.8 million pounds a week in England and Wales. However, from 1998 to 2003, fewer than one in 10 arson incidents resulted in an individual being charged, summoned or cautioned. In 2001/2002, this rate dropped from 10 to 8% (http://www.communities.gov.uk/documents/fire/pdf/130709.pdf). This is lower than the rates for 'other crimes' at 23% and 'criminal damage' at 13%. In the United States, arson and suspected arson is the largest single cause of property damage due to fire. Of the 72 000 suspicious fires recorded in 1999, 17% resulted in an individual being charged, summoned or cautioned. For the same year, the number of cases in the United Kingdom was 60 472, with the number of individuals being charged, summoned or cautioned recorded as half of that in the United States.

An initial study investigating the influence of petrol as accelerant and different types of fire extinguishers on stable isotope signatures of wooden matchsticks found no statistically significant difference when stable isotope data of unburnt and burnt matches were compared using two tailed t-tests. The range of the data was found to increase, most likely due to sample inhomogeneity (see above) as samples were not ground prior to analysis to avoid contamination (Farmer et al., 2009).

In collaboration with Dr Niamh Nic Daéid, matchsticks were placed in and retrieved from fire scenes set up for forensic training and research. Matches were thus exposed to conditions expected to occur in real fire incidents. Burnt matches collected from a fire scene were sectioned and stored *in vacuo* prior to analysis to ensure no residual accelerant traces were adhering to the match surface so as to eliminate potential artefacts. When burnt matches were compared to unburnt controls of matches retained from the same box using two-tailed t-tests as well as a multivariate analogue of the paired t-test such as Hotelling's T^2 test, no statistically significant differences were found between the samples (Farmer et al., 2009). These results provide the first indication that analysing matchsticks for geographical origin of the wood they are made of may be possible even for burnt matches, thus yielding additional information especially in cases of matchsticks made from the most commonly used wood – aspen.

III.4.5 Conclusions

In the case where incriminating evidence was set alight to pervert the cause of justice, a two-dimensional stable isotope fingerprint based on ^{2}H and ^{13}C isotopic composition could conclusively demonstrate that matches seized at the crime scene shared no commonality with matches recovered from the suspect's house. This result was confirmed by two independent techniques: thin-section microscopy and XRD (Farmer et al., 2007). During the course of this case work and associated studies it was noted that scenarios could be encountered where matches from known different sources might be indistinguishable by one or even two of these techniques, but not all three. For example, matchsticks from controls Swan A and Swan B were indistinguishable by thin-section microscopy, but could be distinguished by multidimensional isotope analysis. BSIA should therefore be added to the analytical techniques for forensic discrimination of wooden safety matches alongside established techniques such as microscopy, electron ionization-MS (EI-MS), ICP-optical emission spectrometry (ICP-OES) and XRD.

Prior to our research group and, in particular, Nicola Farmer's involvement no method for the forensic analysis of burnt wooden matches had been published. If microscopy had been carried out on samples of matchsticks secured from a fire scene these samples would have been sectioned as reported by Farmer *et al.* to reveal unburnt portions of the matchstick, thus identifying merely the type of wood (Farmer et al., 2007, 2009). Stable isotope analysis of unburnt portions of a wooden matchstick maximized the information available, producing a three-dimensional stable isotope signature based upon ^{13}C, ^{18}O and ^{2}H isotopic composition. This analytical approach was successfully demonstrated in both laboratory-based controlled experiments and a variety of real fire training scenes.

Chapter III.5
Provenancing People

From what we have learned thus far it is quite clear that stable isotope profiling of human tissue cannot be used for the identification of human remains or living people in the same way DNA fingerprinting can, either in cases where ante-mortem DNA or DNA from presumed next of kin is available for comparison or where a match on a particular DNA database can be found. However, stable isotope profiles or signatures gained from human tissue can be used to reconstruct the life history of an individual, alive or deceased (Meier-Augenstein and Fraser, 2008; Rauch *et al.*, 2007; Schwarcz and Walker, 2006). In this respect, stable isotope profiles gained from soft and hard tissues can compleiment the biological profile provided by the forensic anthropologist through examination of skeletal remains and interpretation of skeletal markers with regard to age, sex, height and race of the deceased. Information on geographic origin and geographic life history is particularly useful in cases where a body has been compromised too much or mutilated for identification to be possible by distinguishing features (scars, tattoos, fingerprints, dental records), or by methods such as cranio-facial reconstruction (Black and Thompson, 2007; Wilkinson, 2007). In addition to murder enquiries where bodies have been deliberately dismembered and mutilated to prevent identification, stable isotope profiling can also be used in cases of extremely young victims of serious crime such as infants and neonates where the lack of distinguishing features hampers identification. Probably the case best known to the general public involving a child where light element and trace element isotope profiles were used to aid and direct the investigation was the 'Adam' case of the mutilated body found in the river Thames in 2001 (O'Reilly, 2007). Other potential areas of application are disaster victim identification, people trafficking (people smuggling) and counter-terrorism.

A crucial ingredient for the success of inferring geographic life histories of unidentified human remains is the availability of global databases, contour plots or maps, and predictive models for spatial distribution of, for example, ^{2}H and ^{18}O isotopes in precipitation and source water, and how these correlate with the corresponding isotopic composition of human tissue. Thanks to the efforts of a few individuals and individual research groups, such maps and models have started to emerge. Initial efforts to develop algorithms using data collated by the Global Network of Isotopes in Precipitation

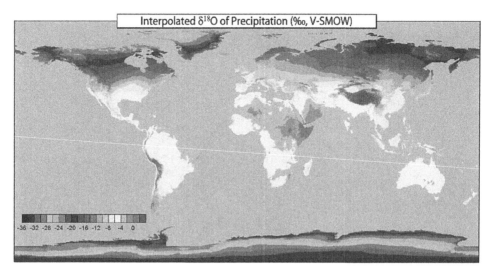

Figure III.21 Global map for $\delta^{18}O$ values in precipitation. © 2002 Geological Society of America, Inc., reproduced with permission from Bowen and Wilkinson (2002).

(GNIP; http://www-naweb.iaea.org/napc/ih/GNIP/IHS_GNIP.html) first generated an interpolated, smoothed surface map of $\delta^{18}O$ values (Bowen and Wilkinson, 2002) (cf. Figure III.21) and soon produced a web-based tool to calculate isotopic composition of precipitation for any given (terrestrial) combination of GPS coordinates and elevation, which was dubbed the On-line Isotopes in Precipitation Calculator (OIPC; www.waterisotopes.org).

Since then a lot of progress has been made in developing isotope maps (or 'isoscapes'; www.isoscapes.org) and predictive models for isotopes in precipitation or source water, human tissue and food stuffs (Ehleringer et al., 2008; van der Veer et al., 2009), and their contribution to human provenancing based on stable isotope signatures has been invaluable.

III.5.1 Stable Isotope Abundance Variation in Human Tissue

Considering what we learned about isotope effects and isotopic fractionation in Part I of this book, we can now appreciate that variations in the isotopic abundance of ^2H, ^{13}C, ^{15}N, ^{18}O and ^{34}S in various human tissue reflect the isotopic make-up of food and water consumed by a person, and as such reflect lifestyle and geographic origin/history of this person (Bol, Marsh and Heaton, 2007; Bol and Pflieger, 2002; Dickson et al., 2000; Fraser, Meier-Augenstein and Kalin, 2006; Hoogewerff et al., 2001; Macko et al., 1999; Meier-Augenstein, 2007; O'Connell and Hedges, 1999a, 1999b; O'Connell et al., 2001; Richards, Fuller and Hedges, 2001; Richards et al., 2003). In other words, diet and geo-location influence the isotopic signature of body tissues such as hair, nail,

Figure III.22 You are what and where you eat and drink – a few stable isotopes in the human body. Drawing based on a figure in Wada and Hattori (1990).

teeth and bone, and can thus be used to aid human identification in cases where no material is available for DNA comparison or where no DNA match can be found. The basic principle behind establishing lifestyle and geographic life history using stable isotope profiling is the fact that the body's only source of carbon and nitrogen is a person's staple diet (Figure III.22). Similarly, the body's major source of hydrogen is water (H_2O), be it from directly consumed water as liquid intake, from water indirectly consumed in foods such as fruit and vegetables or from water used to prepare meals and beverages such as soups, tea and coffee. Based on a tracer experiment, there is evidence suggesting that approximately 31% of hydrogen preserved in human hair is directly derived from ingested water, with the remainder derived from hydrogen present in diet be it as water or be it organically bound (Sharp *et al.*, 2003). For this reason, the ^2H and, to a degree, ^{18}O isotopic composition of human hair is highly correlated to the isotopic composition of drinking water, and virtually all the water we drink is ultimately derived from rain and snow fall (Ehleringer *et al.*, 2008; Fraser, Meier-Augenstein and Kalin, 2006; Fraser and Meier-Augenstein, 2007). In much the same way that water is the major source of hydrogen for bio-organic compounds in the human body, drinking water is the major source of oxygen for the formation of dental enamel and bone bio-apatite (Daux *et al.*, 2008; Dupras and Schwarcz, 2001; D'Angela and Longinelli, 1990; Lee-Thorp and Sponheimer, 2003a, 2003b; Longinelli, 1984).

The only source of carbon as a building block for human tissue is diet-derived carbon (Fogel and Tuross, 2003), which can result in significant differences in ^{13}C abundance body tissue for people living, for example, in Europe as compared to North America.

The main reason for this is the pervasion of the North American diet with sugar and sugar syrup derived from sugar cane and corn, with the latter being used also as animal feedstock. Corn syrup serves also as a cheap source of sugar in processed foods, including beverages such as wine and beer (Brooks *et al.*, 2002). It should be noted that even though ^{15}N isotope abundance in body tissues is more indicative of a person's dietary habit (i.e. carnivore versus omnivore versus vegan), it must be treated with caution since factors such as nutritional stress, health status or pregnancy influence the body's overall nitrogen balance and thus tissue δ^{15}N values (Fuller *et al.*, 2004, 2005). Furthermore, results from a controlled diet study suggest it may take more than 4 weeks before small changes in level and source of daily protein intake by 1 g protein/kg body weight manifest themselves in significantly changing δ^{15}N values in body compartments such as hair, plasma or urine (Petzke and Lemke, 2009).

While hair and nail can provide a record of both diet and water through ^{2}H, ^{13}C, ^{15}N and ^{34}S ranging from 14 days up to 20 months (Bol and Pflieger, 2002; Bol, Marsh and Heaton, 2007; Cerling *et al.*, 2003; O'Connell *et al.*, 2001; Fraser, Meier-Augenstein and Kalin, 2006; Sharp *et al.*, 2003), mineral material in bones and teeth reflect geographic environment by recording isotope signatures of elements present in drinking water, namely ^{18}O (D'Angela and Longinelli, 1990; Daux *et al.*, 2008; Dupras and Schwarcz, 2001; Lee-Thorp and Sponheimer, 2003a, 2003b; Longinelli, 1984). In much the same way that the body's major source of hydrogen (and hence ^{2}H) is water, be it directly or indirectly consumed, water is also the major source of oxygen for bone mineral components carbonate and phosphate, particularly for the phosphate fraction of bio-apatite. The mineral salts are primarily present in a crystallized form of tricalcium apatite called hydroxyapatite (Grynpas, 1993; Holden, Clement and Phakey, 1995; Simmons, Pritzker and Grynpas, 1991). The main component of tooth enamel is bio-apatite, often incorrectly referred to as hydroxylapatite, and the same inorganic material also gives bones their rigidity. A chemically more correct name for bio-apatite would be carbonate-rich, hydroxyl-deficient apatite. Compared to the mineral calcium hydroxylapatite $Ca_{10}(PO_4)_6(OH)_2$, in bio-apatite carbonate ions replace some of the hydroxyl ions, thus changing the crystal structure of apatite (Wopenka and Pasteris, 2005). Permanent teeth (in particular, tooth enamel) provide an ideal window into an unidentified body's geographic past. Enamel from adult teeth can provide an ^{18}O record of the region and its drinking water where a person has lived during childhood and adolescence because tooth enamel does not remodel once formed. Of particular use for this purpose are late-erupting teeth such as the second and third molars. Other teeth such as premolars are (pre)formed at such an early stage in life that they retain a record of changes in diet due to the weaning process (Fuller, Richards and Mays, 2003; Mays, Richards and Fuller, 2002; Richards, Mays and Fuller, 2002).

Isotope signatures from bone apatite can yield information on a person's origin and life history because different bones and different types of bone grow and remodel at different rates. Broadly speaking, cortical bone has a slower turnover rate than trabecular bone (Teitelbaum, 2000). Accordingly, load-bearing long bones such as the femur remodel over a period of approximately 25 years (Hedges *et al.*, 2007), while predominantly trabecular bones such as ribs remodel approximately every 5–10 years (Hill, 1998; Simmons, Pritzker and Grynpas, 1991).

Actual cases, isotope analyses carried out, conclusions drawn and outcomes of the investigation (where known) are reported in the following chapters to illustrate the power of this novel forensic tool as well as the need for careful and considered interpretation of the data.

III.5.2 The Skull from the Sea

How much or how little intelligence one can glean from isotope data very much depends on the hand one has been dealt in terms of type of material, amount and case circumstance/environmental conditions. The following is a point in case and as good a starting point as any to demonstrate how little conclusive information can be obtained from the forensic stable isotope analysis of human tissue in aid of human provenancing even though the isotopic profile seems to paint a compelling picture.

A partial human skull comprised of frontal, parietal and occipital bone was submitted to the Environmental Forensics and Human Health laboratory at Queen's University Belfast (United Kingdom) for stable isotope profiling by a senior officer with the Scottish police. The partial skull had been 'found' on the harbour wall next to the lifeboat station of a Scottish coastal town with the presumption being that it was put there deliberately by persons unknown. The skull was confirmed as very likely to be human by the casualty surgeon.

A further examination of the skull was undertaken by the forensic pathologist at Glasgow University (United Kingdom), who confirmed it to be a human skull and likely to have originated from a female of between 13 and 30 years of age. It was confirmed that no assaultive injury was evident and that silt was associated with the skull, suggesting that it had been in the sea. It was the opinion of the forensic pathologist that the skull had been in its present condition 'for a number of years'.

The skull was taken to the Scottish Universities Environmental Research Centre (SUERC)'s radio carbon laboratory (East Kilbride, Glasgow) for ^{14}C radioisotope analysis. Based on where the level of ^{14}C activity found in the skull coincided with the ^{14}C bomb-pulse curve (Shin et al., 2004; Wild et al., 2000), the skull could either have originated from someone who died in the late 1950s or, alternatively, from someone who died between 1990 and 2000, although the SUERC's scientist favoured the latter option.

The diagram shown in Figure III.23 represents the area of the skull that was present and available for isotope analysis. The upper and posterior parts of the cranial value were preserved, but all lateral elements and all facial and dental elements were absent. Bones present were frontal, right and left parietals, occipital, and the mastoid part of the left mastoid temporal. Recent cut marks were visible across both the left and right parietotemporal region, and we assumed these to have been made at SUERC to remove material for ^{14}C analysis.

The police force dealing with the case commissioned a report by Prof. Sue Black, a forensic anthropologist at the University of Dundee (UK) and what follows are two excerpts from her report.

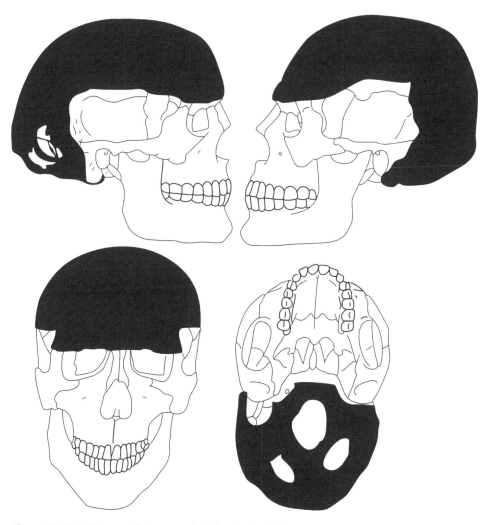

Figure III.23 Diagram of the area of skull submitted for isotope analysis (shaded in black). © 2005 S.M. Black, reproduced by kind permission.

Green algal deposits were evident on the lateral aspects of the skull. The bone surface was not consistent with a weathered appearance of having been exposed to the elements when for example having been washed up on a beach and exposed to alternations of wetting and drying through the action of sun, rain, and wind. There was minimal evidence of surface etchings which are again consistent with a sea bed borne environment.

This is a human skull. It originates from a young individual (most likely male) who is certainly younger than 18 years of age and may be older than 12 years. A likely range for inclusion would be 10–20 years and neither males nor females should be excluded as possible matches.

In the absence of any other source material and, hence, information that would have helped with identification by traditional means (teeth and/or DNA), bone samples were removed from the skull's temporal bone on either side with a diamond-wheel cutting tool. One set of samples was prepared for ^{18}O analysis by TC/EA-IRMS while the other set was used for strontium and lead isotope ratio analysis. Silver phosphate was precipitated from dissolved and decalcified bio-apatite, and six subsamples were analysed for ^{18}O isotopic composition. Concurrent with the samples, phosphate samples of known ^{18}O isotopic composition were analysed, and the raw data were corrected for any offset of measured values from accepted values and thus normalized to the international VSMOW scale.

Due to the various chemical processes involved during bone formation and bone mineralization, $\delta^{18}O$ values obtained from bone apatite (bone phosphate) have to be translated into corresponding $\delta^{18}O$ values for drinking water. Since this case work was carried out in 2005, the best two models available for determining ^{18}O isotope signatures of water taken up by a human body and used for bone mineralization came from studies in bio-archaeology and palaeoecology as proposed by Luz and Longinelli (D'Angela and Longinelli, 1990; Longinelli, 1984; Luz, Kolodny and Horowitz, 1984; Luz and Kolodny, 1985). We used both equations as proposed by Luz et al. and Longinelli to ensure we would not exclude potential geographic areas of origin (see Section I.5.2.1):

$$\delta^{18}O_{phosphate} = 0.64\, \delta^{18}O_{water} + 22.37 \text{ (Longinelli, 1984)} \tag{I.22}$$

$$\delta^{18}O_{phosphate} = 0.78\, \delta^{18}O_{water} + 22.7 \text{ (Luz and Kolodny, 1985)} \tag{III.1}$$

The averaged $\delta^{18}O$ value for bone phosphate prepared from the skull's temporal bone was 18.86‰ (1σ: 0.37‰), which after application of both correlation equations yielded a pooled mean of –6.4‰ for the $\delta^{18}O$ value of water with a 90th percentile range of –5.8 to –7.1‰. These $\delta^{18}O$ values are consistent with the western coastline of mainland United Kingdom, the Isle of Man and the east coast of Ireland (including Northern Ireland). It must be emphasized that in the absence of additional markers this does not exclude the possibility of the body originating from a different geographic location yet with similar isotope profile for water, such as parts of the west coast of France or the northwestern coast of Portugal, to name but two examples.

However, the story did not end there. Even though we did not have high expectations given the skull had been lying on the sea bed for more then 6 years, we submitted some temporal bone sample for strontium and lead isotope ratio analysis in the hope of hitting on an unusual or unexpected result. Similar to light element stable isotopes, trace element isotopes are introduced into the human body chiefly through dietary and water intake, but also through inhaled dust. Unlike light element isotopes, however, rare elements (or heavy metal) isotopes are not prone to isotopic fractionation since they are deposited in the metal ion form in which they have been ingested. We knew two compounding factors were conspiring against us (and the deceased):

(i) The skull had been in extensive contact with seawater for a considerable amount of time. Based on information given as to the result of ^{14}C dating, one has to assume this time frame to be of the order of greater than 6 years. During this time, leaching and exchange of strontium in the bone with strontium present in seawater could not be excluded.

(ii) Due to a combination of fertilizer use and globally sourced sea-food in the diet of modern people living in so-called First World countries, the strontium signature in tooth and bone mineral of modern people tends to become 'blurred', approaching the isotope ratio signature of near-surface seawater, and no longer being a direct reflection of soil and rock composition of a particular geographic location.

Lead isotope ratio analysis yielded a value of 1.164 ± 0.027 for the ^{206}Pb/^{207}Pb (corrected for NBS-982) – a value that was lower than that found in coal in the United Kingdom (1.18), but higher than that found in leaded fuel (1.07–1.08) used in the United Kingdom until 1987. In contrast, the corresponding isotope ratio value for leaded fuel used in the United States in the 1980s was 1.20. Since a considerable amount of households in the United Kingdom and Ireland still burn coal in domestic fireplaces, the finding of a ^{206}Pb/^{207}Pb isotope ratio of 1.164 would be consistent for a person who was living in the United Kingdom or Ireland for 12–18 years and who was initially exposed to lead from the combustion of two sources – leaded petrol and coal – with combustion of coal becoming the predominant source of lead nearer the time of death between 1990 and 2000.

The ^{87}Sr/^{86}Sr isotope ratio analysis yielded a value of 0.709480 ± 0.000023 (0.709437 corrected for NBS-987) – a value typical for hydrothermal dolomites and associated calcites, but also for present-day seawater due to the input of leachate from continental weathering. Thus, in this case the diagnostic value of the strontium isotopic signature was somewhat diminished because mineral exchange between bone and seawater could not be ruled out. Interestingly, this value did provide a good fit with some areas in the United Kingdom, with which the ^{18}O stable isotope data were also consistent such as the west coast of Scotland around Troon and Ayr and the Isle of Man, to name but two examples. However, in light of the compounding factors surrounding the evidentiary value of ^{87}Sr/^{86}Sr isotope ratios that were numerically similar to those found in seawater, we pointed out in our report to the Senior Investigating Officer that these findings should be treated with caution, and locations such as western coastal regions of France and Portugal should not be ruled out since δ^{18}O values obtained from bone were also consistent with these regions.

III.5.3 A Human Life Recorded in Hair

In this case a young man with Asian features was left in the Accident & Emergency department of a hospital in Wales (United Kingdom) by persons unknown and died shortly after of multiple stab wounds. A hit in the Interpol fingerprint database soon established the victim's identity, but there was no record of this individual ever having entered the United Kingdom – at least not officially. To help with their enquiries the

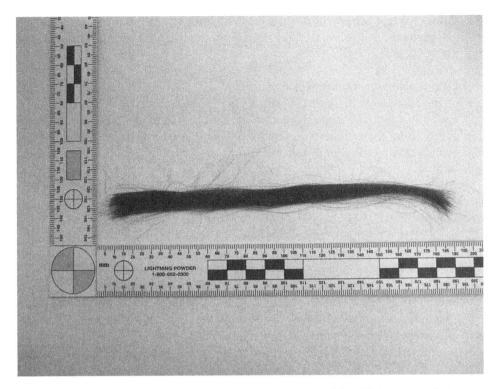

Figure III.24 Sample of scalp hair as submitted for sequential stable isotope analysis.

investigating team was interested to learn where the victim had lived prior to death and, if possible, for how long he had been in the United Kingdom and from where he had come before entering the country. A sample of scalp hair from the victim was delivered by a Scene of Crime Officer to the Environmental Forensics and Human Health Laboratory at Queen's University Belfast for stable isotope analysis (Figure III.24).

The lock of hair was cleaned and subsamples were analysed for isotopic composition of light elements (i.e. ^{2}H, ^{13}C, ^{15}N and ^{18}O). Isotope abundance of ^{2}H and ^{18}O reflect water intake (as discussed previously as liquid, but also as an integral part of food), which in turn is a reflection of geographic location. The isotope record of ^{13}C and ^{15}N is a reflection of diet and nutritional status and, hence 'lifestyle'. Given their association with geographic and ethnic differences in dietary intake, ^{13}C and ^{15}N data can also serve as qualifiers in cases where ^{2}H isotope data fit not just one particular region, but a cluster of disjointed geographic locations.

When looking at the ^{15}N data from the deceased's hair sample, two striking observation were made:

(i) The deceased's diet intake changed significantly in terms of protein consumption at three different points in time, with each change virtually coinciding with the three periods of residency, increasing from a low protein diet for period/location A to a protein-rich diet for period/location C.

(ii) There are also at least three periods where protein intake changed temporarily and dramatically either in terms of source or amount (or both), and these changes seemed to happen either immediately prior or concurrent with geographic movement from one location to the next.

Longitudinal studies carried out by the author's laboratory showed a $\delta^{15}N$ value of $9.09 \pm 0.31‰$ to be typical for an omnivore – a person whose protein intake is derived from both plant and animal sources (Fraser, Meier-Augenstein and Kalin, 2006; O'Connell and Hedges, 1999a; O'Connell *et al.*, 2001; Schoeninger and DeNiro, 1984; Schoeninger, DeNiro and Tauber, 1983). A $\delta^{15}N$ value of around 9–10‰ is usually indicative of an omnivore or an ovo-lacto vegetarian (who consumes eggs and/or dairy products), while a $\delta^{15}N$ value of around 7–8‰ is indicative of a vegan (who consumes no animal protein whatsoever). In contrast, $\delta^{15}N$ values in excess of around 10‰ are indicative of a diet relatively rich in meat (and fish). Examining the deceased's $\delta^{15}N$ profile over the 14.5 months prior to death showed three plateaus at 7.22, 8.23 and 9.95‰ that were more or less in line with the deceased's periods of residency at locations A, B and C, respectively (*cf*. Figure III.25).

What can complicate the otherwise straightforward interpretation of $\delta^{15}N$ values as indicators of diet in terms of protein source and intake is the fact that the human body can scramble the ^{15}N signal in times of physiological and metabolic stress. Conversely, sudden changes in ^{15}N isotope abundance of human tissue can provide valuable information on a person's physiological and metabolic status. Examples for such situations would be suffering from an infection or nutritional stress brought on by under-nourishment (including voluntary starvation as in a 'crash' diet). In the case of the latter, the body starts to satisfy its protein needs by breaking down muscle tissue, which leads to an increase of $\delta^{15}N$ values.

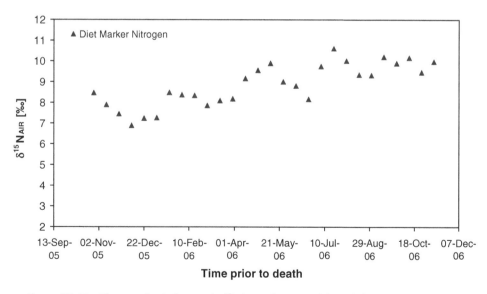

Figure III.25 Time-resolved changes in ^{15}N isotopic composition of the victim's scalp hair.

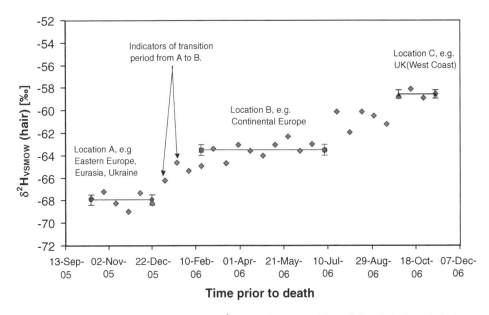

Figure III.26 Time-resolved changes in ^{2}H isotopic composition of the victim's scalp hair.

Such sharp increases or peaks in δ^{15}N values followed by a decay-like decline to a plateau were observed three times and they seemed to coincide or immediately precede periods of geographic movement. Interpreting these sharp rises in ^{15}N isotopic abundance as a consequence of a stark change in dietary habits (e.g. a crash diet) was supported by corresponding peaks and troughs in the ^{13}C isotope profile of the deceased's hair.

The results of the ^{2}H and ^{18}O isotope analysis of the victim's hair samples clearly indicated that the deceased was not resident in the United Kingdom for a prolonged period of time (i.e. more than 1 year), but had only lived in the United Kingdom for approximately 2–2.5 months prior to death. The least that could be said about the victim in terms of geographic life history was that he seemed to have lived in at least three distinctly different locations or regions during the 15 months prior to death in November 2006, with the last location being similar, if not identical, to the area where he died (Figure III.26).

A preliminary indication of geographic locations (using local water as a proxy) was derived from ^{2}H isotope analysis of hair cut into segments representing 2 week's worth of growth. Measured δ^2H values were broadly consistent and found to be typical for moderate, temperate climate zones as encountered between 40° and 55° Northern latitude in the United Kingdom, Central Europe and Eastern Europe. Measured δ^2H values from hair were converted into meteoric water δ^2H values using a correlation equation developed in and by our laboratory at Queen's University Belfast (Fraser and Meier-Augenstein, 2007). Calculated meteoric water δ^2H values showed exactly the same pattern as observed for measured δ^2H values from hair and confirmed the initial tentative interpretation of these data with regard to the geographic history of the deceased during the last 15 months of his life. The deceased had lived at three distinctly different geographic locations for defined periods of time over a period of 15 months prior to death. Isotope data also indicated two periods of movement or travel between the periods of temporary residency. Location A is characterized by a meteoric water δ^2H value of −66.8‰, which is indicative of an Eastern European location such as Poland or the Ukraine. Location B shows a significantly different meteoric water δ^2H value of −57.8‰, which is typical for the climate and latitude of Germany, the Czech Republic and the Alsace. Finally, location C is represented by a meteoric water δ^2H value of −47.6‰, which is typical for the Southwest and West of the United Kingdom, but can also point towards parts of Northern Ireland and the Republic of Ireland.

Moving from the oldest part of the hair to the most recently formed, the results of the isotope profiles were interpreted as follows. At 15 months prior to death, isotope records showed a geographic stationary period of 2.5–3 months spent in Eastern Europe (A, e.g. Poland or Ukraine) or parts of Eurasia (e.g. Eastern Turkey) followed by a direct move to Central Europe (B, e.g. Germany or the Czech Republic). After this almost instantaneous move from A to B, the victim lived in Central Europe for about 6–7 months from where he moved to the United Kingdom, eventually to arrive at his final location of residency on or near the west coast (C). In contrast to the move from Eastern Europe to Central Europe, this move was not a direct transition from B to C. This move either happened in stages over a period of 1.5–2 months interrupted by brief stays (less than 0.5 month) at different locations, or the period between leaving Central Europe and arriving at the final location was characterized by frequent travel and change of location not necessarily related to or influenced by the final area of residency.

Based on the match in the Interpol fingerprint database, investigations were carried out by the police in parallel to our analytical work and independently established that 1 year before his death the victim had entered Germany via the Ukraine. He had stayed in Germany for half a year (i.e. 6 months!) during which period he was arrested and fingerprinted in connection with a crime. Upon leaving Germany he had travelled to his final place of residence in the United Kingdom with stopovers in Dover, London and Birmingham.

III.5.4 Found in Newfoundland

This section deals with multi-isotope analysis undertaken on hair and dental enamel from the unidentified remains of an adult Caucasian male recovered at Minerals Road, Conception Bay South, Newfoundland (Canada) in 2001 by the Royal Newfoundland Constabulary (RNC). The investigating officer, Inspector John House (Officer in Charge, Crimes against Persons, RNC, Newfoundland, Canada) gave us the following background briefing.

> On May 17, 2001 a decomposed male human skull and a few associated vertebrae was discovered by hikers in a wooded area near Minerals Road, in the town of Conception Bay South, in the Canadian province of Newfoundland and Labrador. Conception Bay South is a bedroom community just outside the capital city of St. John's. A Forensic Identification Team, aided by the Chief Medical Examiner, carefully excavated the scene and collected evidence. Investigative procedures

employed in this case included (but were not limited to) the following: A Forensic anthropological examination was completed at the Canadian Museum of Civilization in Hull, Quebec. As a result of that examination, it was concluded that: (1) the skull and mandible were essentially intact and complete, and belonged to the same person; (2) the state of preservation suggested a possible time frame since death of between six months to a year, and five to ten years; (3) morphological features indicated a probable Caucasian male likely in his middle to late twenties at the time of death, but conceivably older. A craniofacial reconstruction has been completed by a Royal Canadian Mounted Police Forensic Artist. A DNA profile was generated through the assistance of the Bureau of Legal Dentistry in British Columbia. Extensive area canvassing (including specific targeting of local dentists), searches of missing person reports across Canada; and following up on various tips from the public have failed to assist in identifying the individual. The case quickly went cold.

In the autumn of 2007, Inspector John House began to explore other options that might help advance this investigation. This led him to explore possible contributions of stable isotope forensics and ^{14}C bomb-pulse analysis to the investigation (Hedges *et al.*, 2007; Shin *et al.*, 2004). For ^{14}C analysis, teeth were submitted to Dr Kirsty Spalding at the Karolinska Institute Department of Cell and Molecular Biology (Stockholm, Sweden). Strands of hair were also submitted to Dr Bruce Buchholz of the Lawrence Livermore National Laboratory Center for Accelerator Mass Spectrometry (Livermore, California, USA). Based on the ^{14}C bomb-pulse analysis of tooth enamel completed at the Karolinska Institute, it was estimated that the victim was born sometime around January 1958 ($\pm$2.3 years). Based on the corresponding analysis of hair shafts completed at Lawrence Livermore, it was estimated that the victim died around June 1995 ($\pm$1.7 years). In summary, it was believed that the victim in this case was a male who was born between late 1955 and early 1960.

After some initial discussions and consultation by e-mail and over the phone, Inspector John House decided to have a wide spectrum of isotope analyses done, and in the main these were carried out by Maria Hillier and Dr Vaughan Grimes at the Max Planck Institute for Evolutionary Anthropology (MPI EVA; Leipzig, Germany) on whose summary report this chapter draws a great deal. Stable isotope analysis of tooth enamel for ^{13}C of the carbonate fraction was carried out by the Stable Light Isotope Facility at the University of Bradford (United Kingdom), while ^{2}H isotope analysis of the victim's hair was carried out by our Stable Isotope Forensics Laboratory (Centre for Anatomy and Human Identification, University of Dundee, United Kingdom) located at the Scottish Crop Research Institute (SCRI; Invergowrie, Dundee, United Kingdom). Samples of the victim's remains comprising a third molar, an atlas and approximately 250 strands of scalp hair were originally submitted to Maria Hillier at the MPI EVA. Analysis of these samples occurred between January 2008 and February 2009.

Scalp hair was analysed for ^{2}H, ^{13}C, ^{15}N and ^{34}S to determine the victim's recent life history in terms of diet or geographic location, while tooth enamel was analysed for ^{13}C and ^{18}O to assess dietary influence and geographic marker during adolescence (8–15 years old) when mineralization of the third molar takes place (Moorrees, Fanning and

Hunt, 1963a, 1963b; Saunders *et al.*, 1993). In addition, strontium isotope ratios from tooth enamel were determined to serve as a second qualifier besides $\delta^{18}O$ values for geographic origin of the victim during the period of mineralization of the third molar. Once formed, dental enamel does not remodel and, as we have learned in Section I.5.2.3, is resistant to diagenetic changes such as the incorporation or exchange of oxygen or the uptake of exogenous strontium from the environment after death. This makes teeth an excellent proxy for the local geology during tooth mineralization, thus affording us a glimpse into a person's childhood (Koch, Tuross and Fogel, 1997; Lee-Thorp and Sponheimer, 2003a, 2003b).

The results from the different isotope analyses are summarized in Table III.5. Figure III.27 shows the changes in measured δ^2H values of the individual's scalp hair with tentative indications of corresponding geographic location consistent with all the other isotope data, but not necessarily the only possible, let alone true, interpretation of these results. Pooling the results of isotopic analysis from samples of scalp hair and dental enamel of a third molar to establish dietary and geographic life history of the deceased, a list of bullet points was drawn up to summarize findings, interpretations and conclusions:

- *Diet markers.* $\delta^{13}C$ and $\delta^{15}N$ values of the scalp hair indicate that, for approximately the last 19 months of his life, the individual consumed a diet consistent with most North American individuals subsisting on a diet with a slightly higher meat-derived protein intake compared to an occasional meat-eating omnivore or an ovo-lacto vegetarian. However, there was no indication in $\delta^{13}C$ and $\delta^{15}N$ values of the victim's hair, suggesting any change in dietary intake during the last 19 months prior to death indicative of a diet change imposed, for example, by a move to a country with a mainly C_3-plant-derived diet or by a change to a more vegetarian lifestyle or hinting at a change in nitrogen balance due to stark changes in metabolism (crash diet or infectious disease). The $\delta^{13}C$ values of the scalp hair were not consistent with $\delta^{13}C$ values expected for an individual residing in Newfoundland for the last 18 months of his life. This was a tentative conclusion, however, based solely on the latitude of Newfoundland and estimated C_4-plant abundance for the island since no reference $\delta^{13}C$ values of living humans, animals or plants from the island of Newfoundland were available for direct comparison.

- *Geo-markers.* Variations observed in δ^2H values of the individual's scalp hair were indicative of movement between and temporary residence in four distinctly different geographic locations. Changes in geographic location as suggested by δ^2H values from the individual's scalp hair were not reflected by similar changes in any of the dietary markers in his hair ($\delta^{13}C$ and $\delta^{15}N$ values), while $\delta^{13}C$ values of the individual's tooth enamel indicated that, from 8 to 15 years of age, the individual subsisted on a diet consistent with most North American individuals. The latter finding was corroborated by $\delta^{34}S$ values of the individual's scalp hair, which were consistent with those for individuals who reside in inland regions of North America. However, this was a tentative conclusion since no $\delta^{34}S$ values had been available for North Americans living in coastal regions. Strontium isotope ratio values of the individual's

Table III.5 Results of stable isotope analysis of the tissue samples studied in the case of the unidentified body found at Minerals Road, Conception Bay South, Newfoundland.

	$\delta^{13}C_{VPDB} \pm 1\sigma$	$\delta^{13}C_{VPDB}$ range	$\delta^{15}N_{AIR} \pm 1\sigma$	$\delta^{15}N_{AIR}$ range	$\delta^{18}O_{VSMOW} \pm 1\sigma$	$\delta^{18}O_{VSMOW}$ range	$\delta^{34}S_{VCDT} \pm 1\sigma$	$\delta^{34}S_{VCDT}$ range	$\delta^{2}H_{VSMOW}$ range ('true' $\delta^{2}H$)
Hair (measured)	−17.30 ± 0.17	−17.72 to −17.09	10.38 ± 0.05	10.31 to 10.51			2.09 ± 0.41	1.32 to 2.70	−70.8 to −65.1
Tooth (measured)	−9.36 ± 0.04				27.20 ± 0.12				
Tooth (adjusted)	−19.06[a]					−6.58 to −5.74[b]			

[a] Measured tooth carbonate $\delta^{13}C$ value has been adjusted for the equilibrium fraction between blood bicarbonate and bone carbonate to furnish a $\delta^{13}C$ value reflecting the ^{13}C isotopic composition of the victim's diet during adolescence.
[b] Measured tooth carbonate $\delta^{18}O$ value has been converted into a corresponding phosphate $\delta^{18}O$ value according to Iacumin et al. (1996), which was subsequently converted in $\delta^{18}O$ values for source water using the correlation equations published by Longinelli (1984) and Daux et al. (2008).
Based on data generated by Maria Hillier, Dr Vaughan Grimes and the author during this investigation.

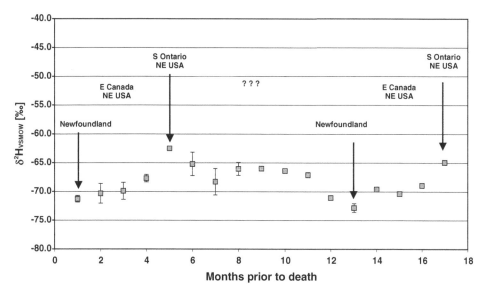

Figure III.27 Geographic life history for the last 17 months prior to death as gleaned from ^{2}H isotope analysis of scalp hair segments from the body found at Minerals Road, Conception Bay South, Newfoundland.

tooth enamel indicated that he may have resided during early childhood (8–15 years of age) in a region with a geological make-up similar to that found in the Minerals Road area. This inference was based solely on the consistency of the strontium isotope values from the third molar with geologically derived strontium values. Again, this was a tentative conclusion due to the absence of biologically available strontium data both in this area and the island of Newfoundland, in general. Oxygen isotope analysis from the tooth enamel carbonate yielded a range of δ^{18}O values (–5.82 to –6.58‰), suggesting that during tooth mineralization the individual resided in a region where drinking water containing a higher level of ^{18}O than would be expected for the island of Newfoundland (range for δ^{18}O$_{expected}$: –10.5 to –12.9‰). The range of δ^{18}O values for source water derived from tooth enamel did not match any of the geographic markers derived from ^{2}H analysis of the individual's scalp hair, either suggesting that neither of the geographic locations he might have been during the last 19 months of his life were consistent with the place where he lived between age 8 and 15. Given the contrasting results of ^{18}O isotope and strontium isotope ratio analysis for the individual's tooth enamel it would seem unlikely for the individual to have lived in Newfoundland between the age of 8 and 15 years.

Based on comparison of the isotope profile generated from the individual's scalp hair and tooth enamel with published data and individually held data it was concluded that the following regions could most likely be excluded as being potential candidates for the individual's recent provenance: Australia, Central Europe including the United Kingdom, Russia and Scandinavia. While consistent with movement between parts of

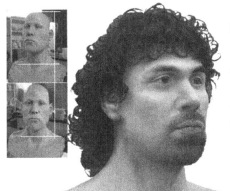

Figure III.28 Poster based on information derived from, amongst other sources, stable isotope analysis for a public appeal for information regarding the murder victim found at Minerals Road, Conception Bay South, Newfoundland. Reproduced with permission of Inspector John House (Royal Newfoundland Constabulary).

East Canada, South Ontario and the Northeast United States, the profile of geographic movement and temporary residence during the last 19 months of his life based on $\delta^2 H$ values from the individual's scalp hair could also be consistent with a recent geographic life history played out in other parts of the United States and/or Canada.

At the time of writing this book no further information was available as to whether the isotope profile of the deceased individual found in Newfoundland had helped to advance the investigation or if a public appeal for information (*cf*. Figure III.28) based on information obtained from stable isotope analysis and other techniques had brought Inspector House any closer to identifying the body.

III.5.5 The Case of 'The Scissor Sisters'

This chapter gives an account of a case that has already been published by us as case report in the forensic journal *Science & Justice* (Meier-Augenstein and Fraser, 2008). In this case forensic intelligence provided from a stable isotope profile helped to identify a murder victim whose body was severely mutilated to conceal both the victim's and his killers' identity. Owing to the particular circumstances of this case the forensic intelligence gained from stable isotope analysis concerning the victim's provenance and life history also proved instrumental in the apprehension of his killers. Due to the extraordinary circumstances of this crime, this case soon gained a certain notoriety in the country where the crime had been committed and, once the perpetrators' identities were known, the media soon dubbed them first 'The Butcher Sisters' and later on 'The Scissor Sisters'. The case background as provided to us by one of the investigating officers, Detective Garda Mike Smyth, was as follows.

> The body of a male was discovered on a March evening in 2005 by a group of schoolboys playing along the Royal Canal at Ballybough, in the North Inner City of Dublin. The body had been dismembered with the torso being cut in two, whilst the arms had been separated from the trunk at the shoulders and the legs had been removed at the groin and dismembered further at the knees. The head was missing from the lower neck up and was never recovered. Interestingly enough the penis had been removed in a very particular manner. The penis and the scrotum skin had been removed but leaving the gonads *in situ*. The torso bore twenty-one stab wounds, of which five or six were capable of being fatal. The torso also bore angular slice marks on the upper back, which appeared to be somewhat symbolic and because of this seemingly ritualistic context police even travelled to West Africa in pursuance of their original enquiries. The body was that of a middle-aged man of African, Afro-Caribbean or Afro-American descent although this was not apparent at first when the body was found. Pigment disintegration had occurred on a massive scale leading to exposed skin appearing as almost completely white. Only during the body's examination by the forensic pathologist it became apparent that the un-exposed skin was of a dark complexion and that the pubic hair was consistent with Africoid characteristics.

Apart from the aforementioned details, the investigating team of the Garda (An Garda Síochána, Dublin, Republic of Ireland) had no other information about the body, its point of origin or its true ethnicity, let alone a name since no missing person report had been filed fitting the victim's description. Fingerprints were recovered from the dismembered arms, but they did not match anybody on either the national or Interpol's fingerprint database. The investigators were in the process of manually checking the fingerprints of every immigrant into Ireland, a dataset which in 2005 already ran into the hundreds of thousands. Since this work would have meant spending thousands of man-hours, the investigating officer decided to contact our laboratory to see if isotope analysis of various body tissues might yield information to help him bring focus to the investigation if nothing else but by a process of elimination. As it turned out later, checking all the immigrants' fingerprints on record would have meant pursuing a fruitless avenue because the deceased had entered the Republic of Ireland prior to compulsory fingerprinting of immigrants being introduced in this country.

As a result of our discussions with Detective Garda Mike Smyth, a complete set of fingernails from one hand, several strands of pubic hair and a ring-shaped slice cut from the femur of the deceased were submitted to us for stable isotope analysis together with volunteered fingernail clippings and hair samples from a male control person whose recent geographic life history as a resident of Dublin was known.

Sample preparation and comparative stable isotope analysis of hair and fingernail samples from both the victim and the control person were carried according to published methods and principles (Bowen *et al.*, 2005; Farmer, Meier-Augenstein and Kalin, 2005, 2006; Fraser and Meier-Augenstein, 2007; Sponheimer *et al.*, 2003). The isotope signatures for ^{2}H, ^{13}C, ^{15}N and ^{18}O obtained from the victim's hair and nail showed no significant differences to those obtained from the control person or our local group of volunteers for a longitudinal study into isotopic signatures of modern human hair and nail (Fraser, Meier-Augenstein and Kalin, 2006). Based on the assumption that both victim and control person were in a comparable good state of health, a δ^{15}N value of $10.9 \pm 0.4‰$ on average suggested the victim's diet was slightly more meat protein-rich, whereas a δ^{15}N value of $9.26 \pm 0.2‰$ suggested the control person to be an omnivore eating a balanced diet with a moderate intake of meat protein. The δ^{13}C values for the victim's nails and hair of -22.34 ± 0.25 and $-22.16 \pm 0.1‰$, respectively, showed no sign of C_4-plant-derived carbon sources in his diet, thus excluding a recent North American or African point of origin (Meier-Augenstein and Liu, 2004). Both δ^2H and δ^{18}O values of the victim's hair and nail where in line with those obtained from the control person. The δ^2H value of $-66.24 \pm 2.66‰$ from the victim's hair was all but identical to that of $-66.2 \pm 2.08‰$ obtained from the control person's hair, and the former value was consistent with known δ^2H values of $-57.4‰$ for tap water in Dublin and surrounding areas including the east coast of Northern Ireland as established by systematic longitudinal studies carried by my then PhD student Dr Isla Fraser (Fraser, Meier-Augenstein and Kalin, 2006, 2008; Fraser and Meier-Augenstein, 2007). We therefore concluded that in all probability the victim lived in this part of Ireland for at least the last 7 months prior to death. This time line was based on the known growth rate for finger nails of approximately 3 mm/month (Wilson and Gilbert, 2007) and the length of the nails submitted for analysis (on average 19 mm).

To see if ^{18}O isotope data from the phosphate fraction of bone bio-apatite would provide more information about the victim's past life history, particularly about his point of origin, a segment was cut from the slice of femur, and subsampled further from its innermost and outermost parts. Bio-apatite from powdered bone samples was extracted and its phosphate fraction precipitated as silver phosphate for ^{18}O isotope analysis using a modified procedure of that originally described by Stephan (2000). The complete protocol for the preparation of silver phosphate from bio-apatite is given in Appendix III.B.

Due to the various chemical processes involved during bone formation and bone mineralization, δ^{18}O values obtained from bone apatite have to be translated into corresponding δ^{18}O values for drinking water. At the time of working this case, the best two models we had for determining ^{18}O isotope signatures of water taken up by a human body and used for bone mineralization came from studies in bio-archaeology and palaeoecology as proposed by Luz and Longinelli (see Section I.5.2.1) (D'Angela and Longinelli, 1990; Longinelli, 1984; Luz, Kolodny and Horowitz, 1984; Luz and Kolodny, 1985). We used both equations to ensure we would not exclude potential geographic areas of origin:

$$\delta^{18}O_{phosphate} = 0.64\ \delta^{18}O_{water} + 22.37 \text{(Longinelli, 1984)} \tag{I.22}$$

$$\delta^{18}O_{phosphate} = 0.78\ \delta^{18}O_{water} + 22.7 \text{(Luz and Kolodny, 1985)} \tag{III.1}$$

The inner and, hence, more recently formed part yielded a δ^{18}O value of $21.21 \pm 0.46\text{‰}$. Given the results of the Luz and the Longinelli correlations where all but identical within the error of measurement, we pooled the correlation data resulting in a mean oxygen signature of -3.04‰ (95% confidence interval range: -2.36 to -3.72‰). Since the femur remodels approximately every 25 years, this signature has to be a composite of the ^{18}O signature of the drinking water consumed by the victim in recent times and of the water consumed prior to his arrival in the Republic of Ireland. If the victim would have lived in Dublin for longer than 25 years, one would have expected to find a δ^{18}O value in the region of -7.3 to -8.3‰. In contrast, the outer and, hence, 'older' part of the femur yielded an ^{18}O signature of $22.34 \pm 0.47\text{‰}$. Following the same approach as above, the corresponding source water ^{18}O signature was calculated to yield a δ^{18}O value of -1.43‰, with a 95% confidence interval range of -2.11 to -0.75‰ (Meier-Augenstein and Fraser, 2008).

An ^{18}O isotopic signature of source water as close to the δ^{18}O value of the international standard VSMOW ($\delta^{18}O = 0\text{‰}$) is very unusual. Broadly speaking, it is indicative of a hot, low-altitude coastal region near the equator. Based on the latest data available at the time as released by the IAEA from its GNIP for observations up to December 2001, we could only find five regions worldwide with consistent ^{18}O signatures. These regions included part of the east coast of Brazil (Salvador (Bahia) to Recife (Pernambuco)), the Windward Islands (Lesser Antilles), the Horn of Africa, the United Arab Emirates and part of Oman, and the area between the Gulf of Kachchh and the Gulf of Khambat on the west coast of India.

Thus far our findings suggested that the victim lived in an area/location consistent with Dublin or the east coast of Ireland in general for a prolonged period of time prior to death, but ultimately originated from a different part of the world and quite likely from one of the aforementioned five regions. Based on an estimated remodelling time for femoral bone of 25 years (Hedges *et al.*, 2007), we tried to determine for how long the victim had lived in Ireland for the inner section of the femur to reflect a $\delta^{18}O$ value of −3.04‰. Based on a simple linear two-pool mixing model, a linear regression line was determined connecting the $\delta^{18}O$ values for the victim's geographic point of origin at time $t_O = 0$ years and Dublin at time $t_D = 25$ years. Using the solution of the linear regression analysis for that line, the point in time was calculated for which the $\delta^{18}O$ value from the inner part of the femur would fall on that line (*cf*. Figure III.29). The 95% confidence limit determined for the $\delta^{18}O$ value was translated into a corresponding uncertainty interval around the time point by calculating times for $\delta^{18}O$ values at either end of the 95% confidence range. As a result we concluded the victim had come to Ireland approximately 6.3 ± 2.9 years prior to death (Meier-Augenstein and Fraser, 2008).

Taking this time frame into consideration, we looked to interpret our findings in the context of GNIP ^{18}O data in meteoric precipitation prior to 1999, but also in the context of the victim's complexion and the nature or characteristics of his pubic hair. On the basis of this contextual information we decided to rank the Horn of Africa (i.e. countries such as Ethiopia, Somalia, Kenya, Eritrea, Uganda and the eastern part of Sudan) as the most likely point of origin out of the aforementioned five possible areas.

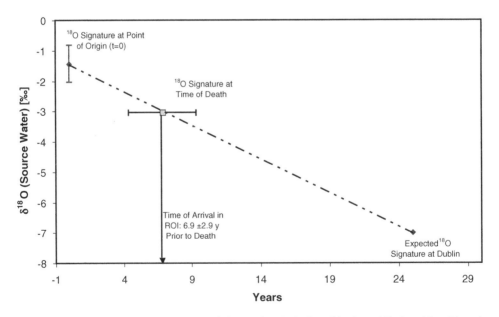

Figure III.29 Geographic life trajectory of the murder victim found in the Dublin Royal Canal based on ^{18}O isotope analysis of bone phosphate extracted from his femur. ROI, Republic of Ireland. © 2008 Elsevier, reproduced with permission from Meier-Augenstein and Fraser (2008).

The above findings and conclusions were reported to the investigating officer, who at the time of receiving the report had developed several potential leads, with one particular line of investigation that required obtaining legal and financial authorization for requesting DNA samples from a woman and her child presumed to have been fathered by the deceased and to have a DNA cross-matching procedure carried out. According to the investigating officer, the results of the forensic stable isotope analysis provided him with the justification needed to pursue the DNA testing of both the child believed to have been fathered by the victim and the mother of this child. The DNA cross-match showed the victim to indeed be the father of this child, which in turn helped the police to establish the victim's true identity. The deceased, the live-in boyfriend of the mother of his eventual killers was a 39-year-old man of African descent, originally from Kenya who immigrated to Ireland in 1998 (i.e. 7 years prior to his death!). Once the victim's identity was confirmed the two sisters were quickly identified as most likely suspects. Both women eventually confessed to having committed this crime, leading to their convictions for murder and manslaughter on 28 October 2006, and eventual sentencing to life and 15 years' imprisonment, respectively, on 4 December 2006. At the Central Criminal Court, Justice Paul Carney said the details of the victim's dismemberment were the most grotesque of any case that he had dealt with in recent times.

III.5.6 Conclusions

Even though equally applicable to all areas of forensic work, mentioning the following imperatives seems to me particularly appropriate in the context of working with human remains. All case work must be carried out in a fume hood within a dedicated laboratory. Fresh disposable gloves, hats and laboratory coats should be worn at all times, and sterile or new equipment (such as scalpel blades and bench cover) must be used to prevent contamination. Evidence received from the police must be receipted and logged into the laboratory sample register using an established chain of custody procedure. Signatures of the police or scene of crime officer delivering the samples and the recipient of the samples must be collected. All descriptions and procedures must be documented contemporaneously using laboratory journals, documentation log sheets and examination sheets.

One has to appreciate that extracting information from stable isotope signatures of human tissue samples of unknown history with the view of forming hypotheses about potential provenance and life circumstance of a hitherto unidentified person is naturally a combination of generating good quality data and contextual as well as intuitive interpretation. Certain assumptions have to be made based on what is known or presumed to be known at the time about the science behind stable isotope records in different human tissue and global or regional differences in stable isotope distribution. Attention is therefore drawn to the following. Data on ^{13}C and ^{15}N isotopic composition of human tissue provide information on a person's diet and, hence, lifestyle or life circumstance and changes therein. These data can also be used as qualifiers when interpreting ^{2}H isotope data with regard to geographic point of origin or change in geographic location in instances when observed δ^2H values are not unique to one particular region, but are consistent with several regions in the world. However, even in conjunction with information on dietary intake, it may not always be possible to narrow down provenance of a person to a handful of regions in the world with isotopic profiles that are consistent with the profile obtained from a victim's tissue samples.

The four cases described in this chapter on human provenancing demonstrate that stable isotope profiles or signatures of recently deceased people can unlock valuable information recorded in human tissue to recreate a person's life history or geographic life trajectory if interpreted in context and with cognisance of potentially compounding factors that may skew results. These cases also illustrate how life history trajectories based on stable isotope signatures can provide investigative focus to scenarios where identification of a body using traditional methods is hampered by circumstances such as mutilation to prevent identification, severe deterioration of the body or mass disasters, be they due to natural causes or the result of a terrorist attack, for example.

There are several scenarios and situations that will benefit greatly from any complementary or corroborating information on a person's identity that can be gleaned from multivariate analysis of data derived from human remains. These scenarios include post-bomb scene management, fire and explosion investigation, murder enquiries, missing person enquiries (including abandoned babies), and disaster victim identification management.

If you, dear reader, should find yourself involved in a case of human provenancing based on isotopic signatures, do not set your sights on 'identifying' this person or this person's point of origin. From an investigative point of view, reducing the size of the haystack is already a significant step forward and, hence, a success – finding the needle is a bonus! To what extent stable isotope profiling can help in narrowing down where an unidentified person had come from is aptly illustrated by the two maps shown in Figure III.30. Inferences made about a person's life history and geographic life trajectory can be further constrained by analysis of, usually, tooth and bone samples for trace element abundance and composition as well as trace element isotope ratios. For example, while two disparate regions may share the same δ^{18}O isobar, they may differ in their geology and the more likely candidate of the two regions may emerge on the basis of the resultant ^{87}Sr/^{86}Sr isotope ratio record in bone bio-apatite. Based on this information, traditional methods of identification such as a search through dental records or the search for ante-mortem DNA samples or next of kin can be carried out in a much more focused way, thus increasing the chances of a successful identification.

Let me close this chapter with a personal remark. Being part of a team or process aiming to assert a person's basic human right to their own identity has always and will always be a very gratifying but equally challenging aspect of my life. I have found working with or on human remains to be a delicate balancing act. In order to be objective one has to be detached, but not too detached lest one forgets to deal with the remains in a respectful manner; yet there is also an element of passion involved fuelled by the desire to preserve the dignity of the deceased by giving them back their identity. However, understandable as it is, one also has to curb this passion so as not to loose objectivity.

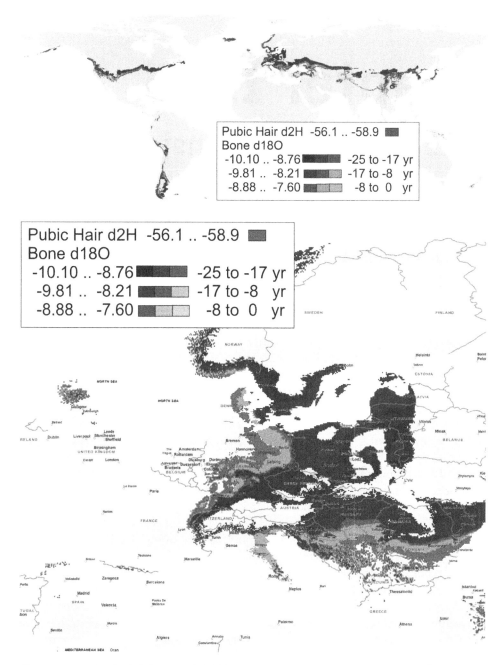

Figure III.30 (Top) Global map with highlighted areas with model predictions for $\delta^{18}O$ values in precipitation ranging from −10.1 to −7.6‰ illustrating the constraining power of stable isotope profiling in aid of human provenancing. (Bottom) Zoomed-in version of the global map focusing on Central Europe and the United Kingdom. © 2009 J. Hoogewerf, maps provided courtesy of and reproduced with permission from J. Hoogewerff, University of East Anglia, United Kingdom. Maps are based on the $\delta^{2}H$ and $\delta^{18}O$ model reported in van der Veer et al. (2009).

Chapter III.6
Stable Isotope Forensics of other Physical Evidence

III.6.1 Microbial Isotope Forensics

The anthrax attacks of 2001 in the United States, also known by their FBI case name 'Amerithrax', highlighted the urgent need for methods to trace the origins of microbial agents. With bioterrorism having thus become a real threat (as opposed to a hypothetical threat scenario), scientists at the University of Utah investigated if stable isotope signatures might help constrain the range of possible regions where microbiological agents used in a bioterrorism attack might have been cultivated. While DNA analysis can reveal information specific to a particular bacterial strain, it does not provide direct intelligence about the environment in which a particular batch of micro-organisms was grown. Stable isotope signatures were seen as a potential forensic tool to complement genetics-based investigations. The research team at the University of Utah hypothesized that stable isotope composition of cellular components of bacteria might provide a record of both nutrients and water used in preparing their culture media, and could thus be used to provide information about their growth environment. If relationships between growth environment and cellular isotope composition could be established one could draw conclusions about the waters in which the microbes had been grown and, hence, make inferences about geographic location of the clandestine laboratory where the bacteria had been cultivated.

In a series of studies, Helen Kreuzer-Martin and colleagues looked systematically at the various sources of forensic intelligence, namely the bacterium itself, its product the spores and the media in which it was cultured. Initially, they investigated the correlation of ^{2}H and ^{18}O isotopic composition of cells and spores from *Bacillus subtilis* – a Gram-positive, endospore-forming soil bacterium with the isotopic composition of culture media water (Kreuzer-Martin *et al.*, 2003). Even though their studies showed that contrary to the common assumption for intracellular water to be isotopically identical to extracellular water these two pools of water could be quite distinctly different, Kreuzer-Martin and colleagues could demonstrate that the ^{2}H and ^{18}O isotopic composition of microbes was linearly related to the corresponding isotopic composition of the water with which their culture mediums had been prepared, thus providing intelligence on potential points of geographic origin. Based on the calculated linear relationship they predicted δ^2H and δ^{18}O values of spores produced in nutritionally identical media but prepared with local water sourced from five different locations around the United States. Each of the spore δ^2H and δ^{18}O values matched predicted values within a 95% confidence interval, suggesting that stable isotope signatures could be useful for tracing the geographic point of origin of bacterial spores.

Seized culture media used to produce microbiological pathogens could be a potentially useful source of information revealing links between a bacterium or its spores and the clandestine laboratory in which they were grown. Thus, in a further study, Helen Kreuzer-Martin and colleagues set out to determine general relationships between ^{2}H, ^{13}C and ^{15}N isotopic composition of bacteriological culture media and spores of *B. subtilis*. In numerous media varying both in nutrient composition and water isotopic make-up, medium to spore enrichment in ^{13}C and ^{15}N was found to be 0.3 ± 2.0 and $4.5 \pm 0.7‰$, respectively. Mass balance for the contribution of hydrogen isotopes from nutrients (70%) and water (30%) to spores was achieved in a series of independent experiments by varying the isotopic composition of either nutrients or water (Kreuzer-Martin et al., 2004). Ultimately, this research group surveyed 516 samples of bacteriological culture media and recorded their corresponding δ^2H, δ^{13}C and δ^{15}N values. Observed variations in δ^2H, δ^{13}C and δ^{15}N values were consistent with expected isotopic variation in the plant and animal products upon which the media were based. Differences in observed stable isotope signatures between culture media were reflected in stable isotope signatures of bacterial cultures grown on these media, and were of sufficient magnitude to suggest that investigations of clandestine bio-weapon manufacture would be possible if links existed between seized culture media and produced bacteria or spores.

III.6.2 Paper, Plastic (Bags) and Parcel Tape

III.6.2.1 Paper

Forensic examination of mass-produced articles such as printer or photocopier paper, plastic materials, or adhesive tape is a challenging task since analytical techniques such as Raman or infrared spectroscopy are not able to distinguish between materials of the same chemical make-up. Similarly, by comparing such materials on the basis of physical characteristics such as refractive index, tensile strength, thickness or weight, one is not able to match a recovered to a control material with a high enough degree of certainty. Establishing a link by visual comparison between, for example, a piece of adhesive tape and a roll of tape is usually only really successful when the tear pattern on either end can be shown to match up. Clearly in such instances analytical techniques are required that can pick up on subtle differences in material make-up that are not detectable by traditional spectroscopic techniques such as infrared, Raman or ultraviolet/visible spectroscopy or even organic MS. Analytical techniques able to detect subtle differences in trace elemental composition, isotopic composition of trace element and isotopic composition of light elements are X-ray fluorescence spectrometry (XRF), ICP-MS or multicollector (MC)-ICP-MS, and, last but not least, CF-IRMS.

Forensic scientists at the Netherlands Forensic Institute (NFI) came to the same conclusion, and decided to investigate if it was possible to discriminate mass-produced document paper by XRF, laser ablation (LA)-ICP-MS and CF-IRMS using multivariate statistical techniques (see Chapter II.4), such as PCA, cluster analysis, and discriminant analysis for statistical data evaluation (van Es *et al.*, 2006; van Es, de Koeijer and van der Peijl, 2007a, 2007b). In total, 25 different types of standard white 80 g/m^2 document paper from 16 different European manufacturers or wholesalers were analysed by XRF, LA-ICP-MS and CF-IRMS with the following results (van Es, de Koeijer and van der Peijl, 2009). Evaluating data from XRF spectra of 13 chemical elements using cluster analysis (single linkage; Euclidean distance) separated 22 out of the 25 paper types. Evaluating measured data of 51 elemental isotopes by LA-ICP-MS in the same way yielded separation of 23 out of 25 paper types. When combining δ^2H, δ^{13}C and δ^{18}O values, 21 out of 25 paper types could be fully discriminated by CF-IRMS alone. In other words, three-element isotopic signatures of document paper obtained by CF-IRMS exhibited a power of discrimination (84%) almost as good as the discriminatory power of XRF based on 13 elements (88%). Measured δ^2H, δ^{13}C and δ^{18}O values for all 25 papers ranged from –110 to –42, –26.5 to –21.9 and 20.5 to 30.5‰, respectively.

Given the already very good results obtained by either technique in isolation it should not come as a great surprise that any combination of two of the three techniques resulted in full discrimination of all 25 paper types, whereas the combination of all three techniques was the most discriminating, adding further discriminatory power over and above all other combinations of these techniques. Noteworthy from a practical point of view (i.e. instrument purchase cost) is the fact that a combination of XRF intensities for sulfur (S_{XRF}) with δ^{13}C values together with δ^2H/δ^{13}C and δ^{18}O/δ^{13}C data pairs obtained by CF-IRMS discriminated all 25 paper types.

III.6.2.2 Plastic and Plastic Bags

Plastic materials such as cling film or plastic bags frequently form part of seized physical evidence since they are a convenient means to package drugs, explosives or money. As a result, forensic casework quite often requires comparison of plastic bags found at different locations. This is currently achieved almost exclusively through exploiting physical properties such as double refraction indices. Double refraction or birefringence is observed in transparent materials that possess different indices of refraction for light travelling in different directions through this material. However, small fragments are not amenable to physical comparisons and unrelated plastic bags or pieces thereof may exist that may exhibit all but identical birefringence patterns. One could argue the situation encountered here is not too dissimilar to that when forensically comparing wooden matchsticks (see Chapter III.4) where unrelated matchsticks might be indistinguishable by thin-section microscopy, yet could be distinguished on the basis of their different stable isotope signatures. By the same token, case circumstances may be such that plastic bags with similar birefringence patterns exhibit different isotopic signatures. In other words, we are looking at yet another example of forensic stable isotope analysis being able to provide information complementary to an existing, well-established forensic technique.

The first report of a comparative examination of pieces of cling film by stable isotope analysis formed part of a criminal investigation concerning illegal drugs where police (Avon and Somerset Constabulary, United Kingdom) had seized five 'wraps' of heroin (a 'wrap' is a small quantity of heroin, typically less than 100 mg, bound in cling film) within a short time frame in a small area (Carter *et al.*, 2005). One of these five samples was found to comprise a very small quantity of heroin (approximately 2 mg) bound with a large quantity of cling film making it appear larger. While trivariate isotope profiles based on δ^2H, $\delta^{13}C$ and $\delta^{15}N$ values suggested that the 2-mg wrap was related to one of the other four wraps (#4) but not any of the other three seizures (#1, #2 and #3), trivariate $^2H/^{13}C/^{18}O$ isotope profiles of the cling film samples revealed that the three seizures #1, #2 and #3 formed distinct individual groups, while the cling film from the 2-mg wrap was isotopically indistinguishable from seizure #4.

In a study published in 2008, 16 grip-seal plastic bags from a wide range of sources were analysed for their ^2H and ^{13}C isotopic composition, although two out of this group of 16 were believed to have come from the same source (Taylor *et al.*, 2008). Average δ^2H and $\delta^{13}C$ values obtained from analysis of subsamples taken from the top, middle, seam and grip-seal of each bag ranged from −105 to −48 and −32.1 to −27.2‰, respectively. Classical discriminant analysis using the Mahalanobis distance applied to the δ^2H and $\delta^{13}C$ data alone was able to correctly identify 13 of the 16 samples as originating from a specific bag. This compares well with a comparison based solely on physical data, size and weight, which was able to distinguish 14 of the 16 bags. Thirteen correct classifications were the same for either comparison method. Classical discriminant analysis based on stable isotope data alone only misclassified one pair of bags that had no physical characteristics in common.

Also included in this study were six boxes containing 24 resealable food storage bags each, all of the same brand, but from four different batches. In all cases, $\delta^{13}C$ values of the coloured grip-seals were significantly different from subsamples taken from top, middle, white panel and seam of the bags, which yielded consistent $\delta^{13}C$ and δ^2H values. A simple bivariate scatter plot of δ^2H versus $\delta^{13}C$ values showed grouping of isotope data according to batches and, hence, correct batch discrimination although in this plot values for batches 2 and 4 were in close vicinity, suggesting a possible similarity. However, a scatter plot of $\delta^{13}C$ values from samples of the red part of the grip-seal versus $\delta^{13}C$ values from samples of the blue part clearly distinguished between all four batches. This finding was corroborated by the different birefringence patterns exhibited by bags from batch 2 and batch 4.

An exciting extension of the forensic examination of plastic materials by stable isotope analytical techniques has been reported with the characterization of post-blast plastic debris from IEDs (Quirk *et al.*, 2009). Studying δ^2H and $\delta^{13}C$ values of pre- and post-blast plastic fragments from four matching pairs of commercially available two-way, hand-held radios of the type considered to be representative of the kind of two-way radio known to have been used to initiate detonation of an IED, it was found that the rate of correct associations by classical discriminant analysis of both δ^2H and δ^{13}C data of 'receiver radio' fragments following detonation with plastic explosive with the still intact 'sender radio' increased with number of samples taken from different sampling points on the radios such as antenna, main casing and screen display cover. For example, for all pairs of radio sets δ^{13}C values of 'mode' and 'send/receive' switch material (sampling points 2 and 5) were all but identical, yet significantly different from the δ^{13}C values of the antenna or display screen materials (sampling points 1 and 3). The striking similarities observed for δ^{13}C values of 'mode' and 'send/receive' switch material from all pairs of radios was reflected by the close similarity of the Raman spectra of these samples, suggesting the switch material was of the same chemical composition for either switch. However, for three of the four pairs of radios δ^2H values for the 'mode' switch material were different from those of the 'send/receive' switch material, highlighting again the importance of multivariate isotope analysis and its added value as analytical technique complementary to traditional forensic analytical techniques.

III.6.2.3 Parcel Tape

Pressure-sensitive adhesive (PSA) tapes, better known, depending on application, under their more common names sticky tape, duct tape, Scotch® tape or brown parcel tape, to name but a few, are encountered in many criminal investigations. PSA tapes are often used to restrain individuals during, for example, a robbery and other offences against the person, or to package drugs or indeed drug money (Carter *et al.*, 2005). Like paper and plastic materials, PSA tapes are mass-produced, ubiquitous items of modern everyday life, thus posing serious challenges to any forensic examination aiming to distinguish between products from different manufacturers or to link material recovered during a criminal investigation to a potential source. To determine if stable isotope signatures would be of benefit to the forensic examination, scientists at the NFI (Dobney *et al.*, 2003a; van der Peijl, Dobney and Wiarda, 2004) and at MSA have engaged in systematic studies of PSA tapes and their stable isotopic composition. (Incidentally, like our stable isotope forensics laboratory, both NFI and MSA are members of the FIRMS Network; see Section II.5.1.) Virtually running parallel, however, these two studies had slightly different aims. While work at MSA focussed exclusively on how useful stable isotope profiling could be used as a forensic tool (Carter *et al.*, 2004), work at NFI was building on previous studies involving LA-ICP-MS (Dobney *et al.*, 2002, 2003a) and was also concerned with investigating what advantages, if any, CF-IRMS analysis would have over traditional techniques for PSA tape examination such as visible/FTIR and XRF (Montero *et al.*, 2005).

For the MSA study, five rolls of commonly available brands of brown parcel tape designated P1–P5 were analysed for their ^{2}H, ^{13}C and ^{18}O isotopic composition in conjunction with a further five samples of brown parcel tape designated P6–P10, which had been removed from parcels delivered to the laboratory at various points in time prior to the study experimental work. Samples P6–P10 were carefully removed from the parcels to ensure that no paper or other contaminants adhered to the adhesive or the backing material. In addition to analysing samples of whole or intact tape (designated 'untreated'), subsamples were treated with a solvent wash to remove adhesive material for subsequent stable isotope analysis of the backing material only (designated 'treated'). Looking at $\delta^{13}C$ values of untreated samples in isolation, all parcel tapes but P5 and P10 could be distinguished on the basis of their $\delta^{13}C$ values alone. When combining $\delta^{13}C$ data of untreated samples with $\delta^{13}C$ data of treated samples (i.e. adhesive removed), samples P5 and P10 could also be distinguished. The same was true for bivariate plots of δ^2H versus $\delta^{13}C$ values from all parcel tapes. Analysis of untreated samples for their ^{18}O isotopic composition revealed samples P3, P6 and P7 contained no measureable amount of oxygen. Since none of the treated samples contained any measurable amount of oxygen, the presence or absence of oxygen in untreated PSA tape samples could be regarded as an additional discriminator since it is indicative of the type of adhesive (acrylic versus natural rubber). A triangular or ternary plot of δ^2H versus $\delta^{13}C$ versus $\delta^{18}O$ values for untreated samples of the seven tapes containing oxygen showed that all seven data clusters could be readily distinguished. Probably the most surprising finding was that short- to medium-term storage, even including storage in a water-filled cistern, had no significant influence on isotopic composition of PSA tape and it was concluded that stable isotope profiling might have the ability to link control samples of PSA tape to recovered samples originating from the same batch following a period of use (Carter *et al.*, 2004).

In the study carried out by the NFI, in total eight samples of three different brands of brown parcel tape (Ruban Embal, TESA and V & D) were purchased from different outlets at different times. Parcel tapes from these three brands have in common the use of polypropylene as backing material. Examining these samples by traditional techniques such as 'visual/microscopic inspection', FTIR and XRF established that acrylic and natural rubber adhesives were used by TESA/V & D and Ruban Embal, respectively. The traditional techniques were able to distinguish between the three brands but save for a partial discrimination within one brand, were not able to distinguish between different samples from the same brand. In contrast, the new techniques of ICP-MS and CF-IRMS were both able to distinguish between all eight tape samples based on either trace element ratios such as Mn/Co or V/^{53}Cr or measured δ^2H and $\delta^{13}C$ values, respectively (Dobney *et al.*, 2002, 2003a, 2003b). Intact tape samples from the brands using acrylic adhesive (TESA and V & D) could also all be distinguished using $\delta^{13}C$ and $\delta^{18}O$ values (Montero *et al.*, 2005). Some of the stable isotope data generated by the NFI study alongside XRF and ICP-MS data are presented in Figure III.31 to illustrate what level of discrimination that could be achieved on the basis of δ^2H and $\delta^{13}C$ values from CF-IRMS analysis of brown parcel tape. However, this is merely an illustration and should not be interpreted as a recommendation or endorsement to rely exclusively on stable isotope data for the forensic examination of PSA tapes.

Even though the brands of parcel tapes used in the MSA study were different from those examined by the NFI study, I think it is fair to conclude that both studies established independently of each other that stable isotope profiling of PSA tapes can provide very powerful forensic information and may potentially even be able to link a recovered to a control tape sample even when they cannot be matched up by their tear patterns.

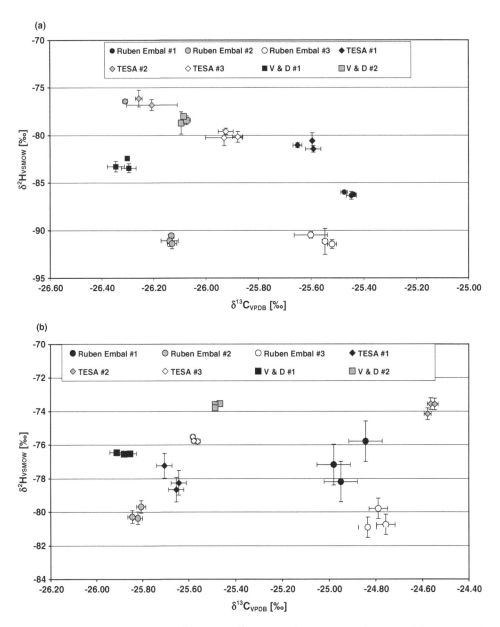

Figure III.31 Bivariate plots of δ^2H versus δ^{13}C values of (top) intact (untreated) brown parcel tape samples and (bottom) treated brown parcel tape samples (i.e. backing material only). Graphs are based on data provided courtesy of Dr Gerard van der Peijl (NFI).

Finally, it should be noted that XRF and ICP-MS are inherently more sensitive to contamination, and due to the low abundance of trace metals within PSA tape these techniques might require relatively large samples of up to 10 cm² in area in conjunction with special handling conditions to avoid contamination. With sample requirements of approximately 1 mm² in area for a single analysis, CF-IRMS compared favourably in this instance (Carter *et al.*, 2004).

III.6.3 Conclusions

A brief summing up of what has been presented in this chapter could and should be never to listen to people whose entire expertise seems to consist of phrases such as 'it will never work' or 'I don't believe this has any chances of success' or 'I can't really see how X can be measured let alone successfully applied to Y'. The sad truth of course is that this kind of 'expert' is listened to when decisions on funding and support of novel and visionary research are being made, which goes some way to explain why it was not possible to report more progress in this chapter on the use of stable isotope techniques as analytical tools for mass-produced everyday items such as paper, plastic, paint, adhesive tape or synthetic and natural fibres.

Clearly, the data and resulting information presented in this chapter and others (*cf.* my then PhD student Dr Nicola Farmer's work on white paint and wooden matchsticks described in Sections II.4.2 and III.3.4) demonstrate the great potential of stable isotope analytical techniques to discriminate between different brands, manufacturers, batches and provenance of consumer products, mass produced under well-controlled conditions. In addition, we learned that information from stable isotope profiles was not necessarily better than that from other advanced analytical techniques, although it could be obtained more readily or required smaller amounts of sample than other techniques. Similar to XRD spectra of strike-heads of matchsticks providing independent corroboration of information obtained from stable isotope profiles of matchstick wood (Farmer *et al.*, 2007), in the NFI study of PSA tapes the FTIR, XRF and ICP-MS data were used to gauge and corroborate the quality of findings gained by CF-IRMS. Thus, another take-home message from this chapter on stable isotope analysis of other physical evidence is a repetition of the take-home message from the chapter on drugs, namely to combine stable isotope profiles whenever possible or feasible with data from independent analytical techniques such as XRF or GC-MS.

Finally, I would hypothesise that one key element of the remarkable findings or insights gained thus far from stable isotope profiles of mass produced items is the very fact that these materials are manufactured by well-controlled processes on an industrial scale where batch-to-batch variations in isotopic composition are most likely a reflection of changes in isotopic composition of precursor and raw materials, and less a reflection of changes in manufacturing conditions, unlike the case of clandestine production of illicit drugs.

Chapter III.7
Summary

Having read this book I hope you, dear reader, will have come to appreciate the role stable isotope signatures or profiles already play in everyday life, in areas such as food standards or environmental protection, and the powerful contribution they can make towards forensic examination of physical evidence. Scrutinizing the examples of actual case work and scientific research presented in this part of the book, I think it is fair to say that stable isotope signatures or profiles in conjunction with data from other analytical techniques such as FTIR, ICP-MS, XRD, XRF and GC-MS will significantly increase the probative power of established analytical techniques for the forensic examination of drugs, flammable liquids, explosives, fibres, textiles, paints, paper, plastics, adhesive tapes and organic materials, in general.

That being said, from what has been discussed both here and Chapter III.2 on drugs it would appear that bivariate, but in particular trivariate, stable isotope profiles or signatures can yield powerful forensic information of up to 84% specificity, thus providing invaluable support to intelligence-led policing and focused deployment of police resources. Stable isotope signatures already show great potential when used on their own; however, from an evidence point of view they are of even greater benefit when used in conjunction with complementary and corroborating data from other forensic analytical techniques. For example, multidimensional isotope analysis (or multivariate isotope analysis) can provide information on whether drug samples from two different seizures that are chemically indistinguishable are likely to share sample provenance or sample history (e.g. the same source and/or geographical origin), thus revealing the 'recreational' user to be a dealer or confirm links in a suspected drug trafficking chain. This information becomes even more powerful or compelling when combined with data from independent analytical techniques probing different compound aspects or characteristics. The combination of GC-MS data with IRMS data and their analysis by multivariate statistical techniques has been shown to provide answers to forensic questions that otherwise would have been elusive. Being able to come to a conclusive assessment about sample provenance will be a significant advance in forensic science, intelligence gathering, and crime detection and reduction. To deliver on its potential for 'real-time forensics', intelligence-led policing and more secure convictions, three major conditions must be met:

(i) All the methods employed in the pursuance of forensic stable isotope analysis must be proven to be fit-for-purpose and all potential pitfalls must be known where reasonably practical. Analytical methods from seemingly similar techniques such as GC-MS adopted or adapted for use in stable isotope analysis must, however, be properly validated in their entirety to demonstrate fitness-for-purpose.

(ii) Databases of multivariate isotope signatures for each type of physical evidence need to be established to determine the probative value or discriminatory power of such isotope signatures when coming to a decision of how likely it is for two samples A and B to share a common provenance or sample history, should their isotopic signatures be indistinguishable.

(iii) Stable isotope datasets as mentioned in (ii) should be linked to physical and spectroscopic datasets of the same samples recorded contemporaneously whenever and wherever it is possible and feasible to do so, with the aim to compile a multilayered, multivariate database of independent complementary profiles so that the 'uniqueness' of such a profile can be properly assessed by forensic statistical methods.

More and better financial and logistical support by governments, government agencies and funding bodies is still required to address the challenges that continue to exist along the road to realizing this exciting technique's full forensic potential to combat terrorism and serious and organized crime, and to safeguard a basic human right – a person's right to their identity.

Appendix III.A
'Play True?': Stable Isotopes in Anti-doping Control or *Quis custodiet ipsos custodes*?[1]

The inclusion of anti-doping control in this book on the basis of the application of ^{13}C-CSIA of testosterone metabolites could be regarded as a misplaced example since anti-doping control, although often billed as one, is not a true forensic application of analytical science since in most countries it is not used for the pursuance of criminal law in a court of justice. However, it is a good example nonetheless, namely of how easy it is to discredit a good analytical technique, and with it the reputation of potentially innocent people, when the methods applied are not fit-for-purpose. Unfortunately, this can and does happen when methods are devised without an independent review process testing not only administrative procedures but also proficiency, and if such methods are applied uncritically, inappropriately and without any deeper understanding of the underlying science.

So, let us use this allegedly 'forensic' application of stable isotope analysis to test what we have learned in the course of reading this book and find out if ^{13}C-CSIA of testosterone metabolites as certified by the World Anti-Doping Agency (WADA) is indeed a robust forensic analytical method that is fit-for-purpose.

Before going into the details of ^{13}C-CSIA of testosterone metabolites in anti-doping control it must be noted that there are virtually as many different analytical protocols and result-reporting procedures as there are anti-doping laboratories. This results in the rather unique situation that an athlete's urine sample would be declared positive for testosterone abuse by laboratory A, while the same sample if analysed elsewhere

---

[1] Quote from the Roman poet/satirist Juvenal (*Satire* 6.346–348), usually translated as 'Who watches the watchers?'.

may be declared negative or at least inconclusive. According to protocols issued by WADA, $\delta^{13}C$ values are determined for four testosterone breakdown products as well as two endogenous reference compounds (that are not part of the biochemical pathway of testosterone), from which subsequently four net $\delta^{13}C$ values or $\Delta\delta^{13}C$ values are determined by calculating the difference between $\delta^{13}C$ values of key testosterone metabolites and $\delta^{13}C$ values of their chosen endogenous reference compounds. However, what constitutes a positive test (proof of exogenous testosterone doping)? The anti-doping laboratory dealing with the case described in this chapter considers the test to be positive if any *one* of the $\Delta\delta^{13}C$ values for the four metabolites is abnormal. However other WADA accredited laboratories such as the anti-doping laboratories at the University of California – Los Angeles (UCLA) and at the National Measurement Institute, Australia require that the $\Delta\delta^{13}C$ values for at least *two* metabolites be abnormal. How can these laboratories all be accredited by WADA for the same analysis with the same purpose and yet apply different criteria for what constitutes a failed test?

As we have seen in all the examples of forensic or quasi-forensic applications of stable isotope techniques, the chemical and thermodynamical processes leading to isotopic fractionation and, hence, the subtle differences in isotopic composition we can exploit *do not occur at random*. The fractionation factor α for a given process or reaction is a reaction-specific constant that in certain circumstances even enables us to predict the isotopic composition of a product of a certain process or of the precursor used to manufacture the product. In other words, a sound working knowledge of the underlying (bio)chemical principles and pathways pertaining to a compound under investigation can, and almost invariably will, provide useful diagnostic information to check if results are qualitatively meaningful. Simply put, if we have information, such as melting point, density, behaviour against acid and so on, on a yellowish metal that all agree for this material to be gold then we have a reasonable expectation for the outcome of yet another more modern analytical method to declare this metal to be gold too. Let us say the modern method is ICP-MS and this sophisticated MS system proclaims the metal to be platinum. Either the traditional chemical information is wrong or the ICP-MS analysis has delivered an incorrect result. Given that platinum is not yellow, but the atomic masses of platinum and gold only differ by 1.88 amu, one possible conclusion could be the ICP-MS was not properly mass calibrated. The alternative conclusion would be that one trusts the result from the modern instrument (merely because it is modern), and overrules the outcome from the tried and tested methods. Strictly speaking, both approaches are incorrect. If the nature of a material has to be confirmed by n methods then all n methods must yield consistent results. The moment contradictory results are obtained, all n methods have to be scrutinized, checked, validated and recalibrated, and a new sequence of analysis has to be initiated and recorded by an unbroken audit trail.

How does this help us in stable isotope forensics? Let us have a look at the case of an athlete who stands accused of testosterone abuse on grounds of a failed ^{13}C isotope drug test based on only one abnormal $\Delta\delta^{13}C$ value. It should be noted that in this particular case the failings of the analytical laboratory involved started with the cardinal sin a forensic laboratory should never commit, namely to have no proper chain-of-custody record and, hence, no traceability of sample movement or transfer because samples had been transferred from sample collection to the laboratory without a delivery document. To make matters worse the official transfer record document did not list the sample ID number pertaining to the athlete, although it did list an ID number numerically close (i.e. similar to the sample ID number issued at collection). Thus, in all probability any evidence produced by this laboratory would have already been declared inadmissible in a court of law on these grounds.

III.A.1 Testosterone Metabolism and ^{13}C Isotopic Composition

Let us now have a look at the results reported by this laboratory and the analytical methods employed for analysis. The laboratory dealing with the analysis of the athlete's urine samples reported the following δ^{13}C values for the four key testosterone metabolites in sample A:

5β-Androstanediol	−23.76‰
Etiocholanolone	−23.83‰
5α-Androstanediol	−27.72‰
Androsterone	−25.05‰

In addition, differences in δ^{13}C values between testosterone metabolites and endogenous reference compounds were reported as $\Delta\delta^{13}$C values. In the case of the two androstanediols, the endogenous reference compound is 5β-pregnanediol. The $\Delta\delta^{13}$C values for [5β-pregnanediol – 5β-androstanediol] and [5β-pregnanediol – 5α-androstanediol] were 2.15 and 6.14‰, respectively. It was chiefly on the basis of the $\Delta\delta^{13}$C value for [5β-pregnanediol – 5α-androstanediol] of 6.14‰ that the laboratory reported an adverse finding and accused the athlete of testosterone abuse.

When looking at the metabolic pathway of testosterone (Figure III.A.1) one can appreciate the wisdom of those laboratories that decided to declare a test positive only if δ^{13}C values or $\Delta\delta^{13}$C values for at least two testosterone metabolites were abnormal. Considering that during the conversion of testosterone to androsterone (or etiocholanolone) via 5α-androstanediol (or 5β-androstanediol) the number of carbons remains unchanged and no carbon–carbon bond is involved during the conversion, it is reasonable to assume that the difference in δ^{13}C values between two paired metabolites such as 5α-androstanediol/5β-androstanediol will be small, and only reflect minute differences in the kinetics of saturating the double bond between carbons 4 and 5. Indeed, published data show this to be the case. Mean difference in δ^{13}C values between 5β-androstanediol and 5α-androstanediol reported from longitudinal volunteer studies ($n = 73$) was 0.66‰, with 5α-androstanediol being more depleted in ^{13}C than 5β-androstanediol under normal circumstances (i.e. in the absence of exogenous testosterone administration) (Aguilera et al., 2001). At the higher end of the spectrum the reported differences in δ^{13}C values between 5β-androstanediol

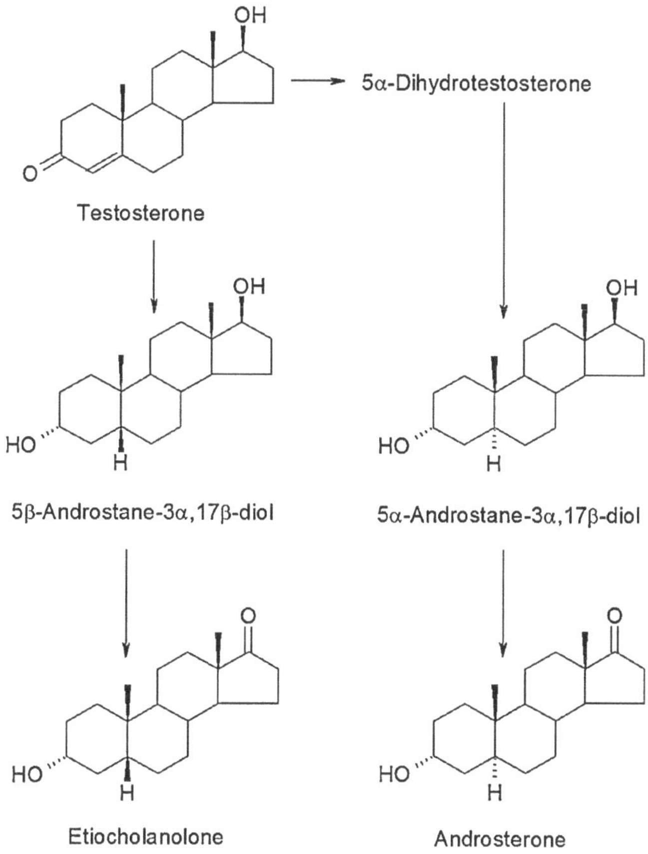

Figure III.A.1 Schematic of the metabolic pathway of testosterone.

and 5α-androstanediol ranged from 1.39 to 1.74‰ for individual controls, while for athletes to whom testosterone had been administered the differences in $\delta^{13}C$ values between 5β-androstanediol and 5α-androstanediol ranged from 0.66 to 2.76‰ with a mean of 2.13 ± 0.78‰ (Aguilera et al., 2001; Shackleton et al., 1997). Using the results reported by the anti-doping laboratory at the centre of this case, the difference in $\delta^{13}C$ values between 5β-androstanediol and 5α-androstanediol from the athlete's sample as reported by the laboratory was 3.96‰, which is highly unusual, even when considering samples were taken approximately half way through during an endurance competition. Such a stark difference in $\delta^{13}C$ values between 5β-androstanediol and 5α-androstanediol has not been reported even for administration of testosterone by means of a testosterone-containing gel (Testogel®) (Piper et al., 2008). In the Testogel study the maximum differences in $\delta^{13}C$ values between 5β-androstanediol and 5α-androstanediol were 3.3 and 2.4‰ for intermittent and continued Testogel application, respectively, although the difference value of 3.3‰ was only observed once after the initial administration of Testogel with all subsequent analyses yielding differences of below 3‰. In addition, the results reported for the Testogel study were data obtained from individual subjects ($n=1$) and not from a trial with a sufficiently large control group, thus no conclusions could be drawn regarding how typical or normal these individual data were or what their population standard deviation might have been.

Since the endogenous reference compound 5β-pregnanediol is not a part of the testosterone pathway and could almost be considered to be an individual-specific constant it is not surprising for the $\Delta\delta^{13}C$ values for 5β-pregnanediol and the two 5-androstanediols to track each other perfectly, too, irrespective of study design or method of testosterone application. This means the theoretical difference between $\Delta\delta^{13}C$ [5β-pregnanediol−5β-androstanediol] and $\Delta\delta^{13}C$ [5β-pregnanediol−5α-androstanediol] can be as little as zero and there are published data to support this (Shackleton *et al.*, 1997). However, even though a difference of approximately 2.5‰ has been observed on one occasion, differences between $\Delta\delta^{13}C$ [5β-pregnanediol−5β-androstanediol] and $\Delta\delta^{13}C$ [5β-pregnanediol−5α-androstanediol] in athletes after testosterone administration were typically of the order 2.1‰ (Aguilera *et al.*, 2001). Table III.A.1 provides a summary of the above and also includes the corresponding athlete's data together with data from the study into the administration of Testogel (Piper *et al.*, 2008).

Even on the basis of a difference in $\Delta\delta^{13}C$ values of approximately 2.5‰ observed once in one volunteer, the differences in $\Delta\delta^{13}C$ values of 3.99 and 3.74‰ for the athlete's A and B sample, respectively, are uncommonly large. For what should be highly correlated data, such differences are inconsistent with the results of all the published studies involving statistically meaningful volunteer populations and representative population mean $\delta^{13}C$ values of testosterone metabolites.

It is also interesting to note that the anti-doping laboratory involved in this case has chosen to report *any* abnormal $\Delta\delta^{13}C$ value as *proof positive* of testosterone abuse. One has to ask if only one $\Delta\delta^{13}C$ value is regarded as sufficient proof of testosterone abuse why the $\Delta\delta^{13}C$ value of [5β-pregnanediol−5α-androstanediol] was chosen and not the $\Delta\delta^{13}$ value of [5β-pregnanediol−5β-androstanediol], considering one of the most respected and cited scientific papers (Aguilera *et al.*, 2001) recommends just that?

> The data on mean z-scores ... reveal that both the differences and ratios are better indices of testosterone use than the absolute $\delta^{13}C$ values for 5βA and 5αA. This is not unexpected given that 5βP correlates with both 5βA and 5αA. Furthermore, the difference 5βP−5βA and the ratio 5βA/5βP are more robust indicators of testosterone use than the corresponding difference 5βP−5αA and the ratio 5αA/5βP.

Question III.A.1: What would you consider is the appropriate course of action to take when analytical results are seemingly at odds with what is scientifically known about biochemical pathways and isotopic fractionation during chemical reactions concerning the particular analyte/s concerned?

Table III.A.1 Athlete's versus reported $\Delta\delta^{13}C$ values for pathway-linked testosterone metabolites.

	$\Delta\delta^{13}C$ [5β-pregnanediol – 5β-androstanediol]	$\Delta\delta^{13}C$ [5β-pregnanediol – 5α-androstanediol]	Difference between $\Delta\delta^{13}C$	Ratio of $\delta^{13}C$ (5β-androstanediol)/$\delta^{13}C$ (5β-pregnanediol)	Ratio of $\delta^{13}C$ (5α-androstanediol)/$\delta^{13}C$ (5β-pregnanediol)	Ratio-of-ratios
Athlete's A sample	2.15	6.14	3.99	1.08	1.21	1.12
Athlete's B sample	2.65	6.39	3.74	1.10	1.22	1.11
Aguilera et al.[a] (controls)	1.40	2.10	0.70	1.06	1.09	1.03
Aguilera et al.[b] (testosterone administration)	5.90	7.90	2.00	1.25	1.33	1.06
Shackleton et al.[c] (testosterone administration)	3.92	3.90	0.02	1.145	1.149	1.003
Testosterone gel study – week 6 of administration[d]	3.85	6.27	2.42	1.17	1.27	1.08

[a] Pooled mean values from three control athletes (Aguilera et al., 2001). Within their control group of 73 volunteers (non-athletes) the maximum values for $\Delta\delta^{13}C$ [5β-pregnanediol – 5β-androstanediol] and $\Delta\delta^{13}C$ [5β-pregnanediol – 5α-androstanediol] were 3.17 and 3.72, respectively.
[b] Pooled mean values from three athletes after testosterone administration (Aguilera et al., 2001).
[c] Nine day mean values for one volunteer after testosterone administration; day-to-day absolute difference ranged from 0 to 1.37‰, although for another volunteer an absolute difference of around 2.5‰ was observed on day 10 after testosterone administration (Shackleton et al., 1997).
[d] Values are maximum values observed for one volunteer during week 6 of continuous application of Testogel; $\Delta\delta^{13}C$ values for [5β-pregnanediol – 5β-androstanediol] and [5β-pregnanediol – 5α-androstanediol] during the 6 weeks increased from 2.9 to 3.1 and from 4.29 to 5.29‰, respectively, yet the absolute difference between the two $\Delta\delta^{13}C$ values did not exceed 2.19‰ apart for that one sampling point shown in the table (Piper et al., 2008).

III.A.2 Analytical Methodology: Gas Chromatography and Peak Identification

GC/C-IRMS measurement of urinary steroids has become a potent analytical method in doping control, although scrupulous attention to detail is necessary. Successful implementation of the method requires strict adherence to quality control and quality assurance protocols. When it is coupled with strict system-suitability and batch-acceptance criteria, it can be used in a routine fashion to obtain valid $\delta^{13}C$ data.

The aforementioned is the conclusion the UCLA Olympic Analytical Laboratory has come to and this need for scrupulous attention to detail has been echoed at the Second Annual United States Anti-Doping Agency (USADA) Symposium on Anti-Doping Science, who in their published symposium proceedings stated that 'characterization of peak envelope – width, height, resolution, shoulders, asymmetry – and development of criteria for acceptable peak characteristics are important' (Bowers, Hilderbrand and Symanski, 2003).

Let us now consider how the anti-doping laboratory in this case measures up in meeting the requirements stated above.

Since it is impossible to establish compound identity of a peak on a GC/C-IRMS instrument, as we have learned compound peaks eluting from the gas chromatograph's column are on-line combusted to yield CO_2 (plus N_2 and H_2O), one is stuck with the problem of how to associate the CO_2 peak detected by the IRMS instrument with its parent organic compound? In addition to the dilemma of how to prove peak identity one is also faced with the problem how to prove the peak was caused by a single compound and not an overlap of two or more? The obvious answer is of course given in Section II.2.4.1, namely the use of a GC(-MS)/C-IRMS hybrid system that permits unambiguous compound identification and peak interrogation by means of organic MS. This technology has been available since the mid-1990s and the question has to be asked why most of the laboratories engaged in anti-doping control have not embraced this technology even though they have adopted CSIA as the method of choice?

Thus, instead of choosing to use available instrumentation that would provide unambiguous compound identification of peaks analysed for their ^{13}C isotopic composition, the laboratory involved in this case elected to employ peak matching by peak retention characteristics – a method that is quite often and mistakenly billed as peak identification by peak retention characteristics. As any analytical scientist involved in GC will know, matching retention characteristics are by no means a cast-iron guarantee that two peaks were caused by the same chemical compound or indeed one compound. That being said, many a routine gas chromatographic analysis run on a simple GC-FID system, that like an IRMS system only detects peaks, is based on a GC-MS method that established retention characteristics for all peaks and linked this information to compound identity and peak integrity per peak determined by their mass spectra.

Question III.A.2: If technology and methods exist that would enable you to measure compound- or material-characterizing properties by a destructive technique and to determine compound or material identity simultaneously under the same experimental conditions, would you as the analyst or as the laboratory director consider it acceptable to risk a case or a person's life and reputation on the mere comparison of retention times from two different chromatographic systems?

If we have a closer look at the matter of GC method development for routine analysis by peak detection based on an existing GC-MS method we find that this is a well-researched and well-understood area of GC, with research into retention parameters of organic compounds in relation to polarity of the GC column (or, to be more accurate, the polarity of the stationary phase with which the GC column is packed or coated), film thickness of the stationary phase, column dimensions, choice of carrier gas (H_2, He or N_2), carrier gas flow rate and column temperature going back to the 1950s. For the successful set-up of a GC method for peak detection without compound identification that permits peak matching to peaks identified by an established GC-MS method, several retention parameters must be known or be determined and several prerequisites must be met (http://www.shimadzu.com/products/lab/ms/tutorial/oh80jt0000007e8m.html[2]; last accessed 22 July 2009):

(i) We need to know the 'gas hold-up time' t_0 ('dead time' of the system): the time required by a compound not retained by the stationary phase of the GC column from injection to reach the detector.

(ii) We need to know the 'adjusted retention time' t'_{IS} of an internal standard: the time difference ($t_{IS} - t_0$) between the absolute retention time of the peak of an unretained compound and the retention time of the peak of the internal standard.

(iii) From condition (ii) follows the need for an internal standard and this internal standard compound needs to satisfy the condition that its peak must elute in a part of the chromatogram completely free of any other compound peaks.

(iv) Since absolute retention times are affected by many operational parameters, retention parameters less dependent on column length and carrier gas flow have to be calculated. They are expressed by the relative relation of adjusted retention time between the standard sample internal standard and the unknown sample: relative retention and retention index.

The advantage of relative retention is that it depends only on the ratio of distribution coefficients and *the effects from some parameters, such as column length and carrier gas flow, are basically cancelled out*. However, relative retention only makes allowances for different column length and different carrier gas flow in *otherwise identical set-ups* (i.e. identical stationary phase and identical temperature programme). Only if one analyses the same sample on the same GC columns using the same analytical method can one expect to see the same chromatogram. The following two prerequisites must be met in

---

[2] Click on 'Click here for details' beneath the last bullet point '• Identification Using LRI'; in the new window scroll through the presentation using '≫'. Do be sure to open the tabs on the right hand side of each screen.

order to satisfy the 'principle of identical treatment':

(i) The polarity and film thickness of the GC columns used in either GC system (i.e. the GC detection system and the GC-MS system) must match.

(ii) The temperature programmes used in either GC system must be identical.

Strictly speaking GC methods based on temperature programming must be set up using $n+1$ internal standards for n target compound peaks so linear retention indices (LRIs), also known as Kovats indices, can be employed.

The International Union of Pure and Applied Chemistry (IUPAC) defines relative retention (r) as follows (http://old.iupac.org/publications/analytical_compendium/ TOC_cha9.html and in particular document http://old.iupac.org/publications/ analytical_compendium/Cha09sec237.pdf; last accessed 23 July 2009):

$$r = t'_{rs}/t'_{IS} = (t_{rs} - t_0)/(t_{IS} - t_0)$$

For comparison, here is the procedure to determine retention parameters employed by the laboratory in this case:

> Next an internal standard (5-α-androstanol acetate) is added for a purpose that will be explained below. The first element of compound identification is the GC 'retention time (RT)' and the second one is the molecular fingerprint recorded by the MS, which fragments the molecule into ions. A parameter that is even better than the retention time is the relative retention time (RRT). It relies on the internal standard that was added to each tube during sample preparation. The internal standard has its own characteristic retention time. The relative retention time of any other compound is simply (RT of other compound)/(RT of internal standard). This makes comparisons of retention times easier because it normalizes them. Peaks and the sequence of peaks from the GC-MS and the GC/C-IRMS are compared to identify metabolites and the endogenous reference compounds.

The above procedure clearly relies on retention parameters of GC compound peaks for compound identification, and thus aims to establish a link between retention parameters of a GC compound peak and its mass spectrometric fingerprint obtained by GC-MS. However, while an internal standard has been used, there is no mention of the system's gas hold-up time being determined, which means the relative retention r could not have been calculated. Consequently, a parameter dubbed relative retention time is, one is tempted to say, made up since it is not to be found in IUPAC's compendium of terms related to the chromatographic process or any textbook on theory and practice of GC. In addition, the term relative retention time for a ratio of retention times is a misnomer since such a ratio would result in a number without a unit.

Furthermore, in the athlete's samples anything from six to up to 11 (!) unidentified peaks were present in close proximity to the internal standard with at least two of these peaks being inside a ± 0.2 min time window of the internal standard (i.e. within the time window specified by WADA's own Technical Document *TD2003IDCR* for identification of a given single peak by retention time or relative retention).

Question III.A.3: Since it is impossible to verify peak identity in GC/C-IRMS other than by matching its relative retention against that established by GC-MS, how did the laboratory determine which of the three peaks sharing the 'same' retention time window of ± 0.2 min in the GC/C-IRMS chromatogram was that of their internal standard?

Setting aside the matter of correctly applied chromatographic theory and terminology, the issue of chromatographic conditions is as crucial, if not more so. Recapitulating what has been said above, only if one analyses a subsample of the same sample on the same GC column using the same analytical method can one expect to see the same chromatogram because retention times depend on many factors: analytical conditions such as temperature programme, type of column (stationary phases and film thickness) and age of column (column degradation and presence of active spots). Agilent, a manufacturer of GC-MS systems and GC columns, advises users in their *GC Column Selection Guide* (http://www.chem.agilent.com/Library/selectionguide/Public/5989-6159.pdf; p. 146):

> ... when changing a temperature program, confirmation of peak identities in the new chromatogram is essential. *Peak retention orders can shift upon a change in the temperature program (called peak inversions)*. Peak misidentifications or an apparent loss of a peak (actually co-eluting with another peak) are common results of undetected peak inversions. This is especially true for the most polar stationary phases.

The laboratory in this case used the following conditions for their GC-MS method with which they identified compound peaks by molecular MS fingerprints and determined the corresponding retention parameters of these peaks.

GC column: *Agilent 19091A-433*; 30 m × 0.25 mm; film thickness 0.25 μm. According to Agilent's part number 19091A-433 this column is called Ultra-1, a *non-polar* column coated with a stationary phase comprised of *100% methyl-polysiloxane*. Agilent lists as equivalent columns BP-1, HP-1, Rtx-1 and *DB1* (*GC Column Selection Guide*, p. 83)

Method MAN_52	Temperature (°C)	Hold time (min)	Temperature ramp (°C/min)
Step 1	70	1	30
Step 2	270	12	10
Step 3	300	3	—

For comparison here are the conditions used by the laboratory to analyse subsamples of the same samples anchored by their GC-MS for ^{13}C-CSIA isotope analysis on their GC/C-IRMS instrument.

GC column: *DB-17MS*; 30 m × 0.25 mm; film thickness 0.25 μm. Agilent classifies this column as a *mid-polar* column coated with a stationary phase comprised of *50% phenyl, 50% methyl-polysiloxane*. Agilent lists as equivalent columns BPX-50, HP-50+ and Rtx-50. The Agilent part number of the DB-17MS is *122-4732* (*GC Column Selection Guide*, p. 83)

Method MAN_41	Temperature (°C)	Hold time (min)	Temperature ramp (°C/min)
Step 1	70	1	30
Step 2	*271*	*0*	*0.6*
Step 3	*281*	*3*	*5*
Step 4	300	5	—

Question III.A.4: In how many ways does the GC/C-IRMS method differ from the GC-MS method that is supposed to provide information on compound peak identity by matching peaks through peak retention parameters?

Not surprisingly when comparing GC-MS chromatograms with GC/C-IRMS chromatograms generated by the laboratory tasked with the analysis of urine samples for testosterone abuse in this case, minor peaks present and eluting very close to peaks of target key metabolites during GC-MS analysis were no longer visible in the corresponding GC/C-IRMS chromatograms. For both the athlete's A and B sample, the GC-MS chromatograms showed a minor peak within –0.2 min of the 5α-androstanediol peak, yet this peak is absent from either GC/C-IRMS chromatogram. Similarly, while the peaks for etiocholanolone and androsterone showed a minor peak within +0.2 min of androsterone in the GC-MS chromatograms, in the GC/C-IRMS chromatograms this peak is no longer visible. Since compound peaks do not vanish into thin air one plausible conclusion might be that they were now co-eluting (i.e. were overlapping to 100%) with peaks of any one of the target key metabolites due to the changes in polarity and, hence, selectivity of the stationary phase and/or the changes in the temperature programme. The effect of peak overlap on measured $\delta^{13}C$ values of compound peaks has been documented by one the groups pioneering and developing GC/C-IRMS. Overlapping peaks in GC/C-IRMS result in a loss of accuracy, which depending on the isotopic composition of either parent peak and the degree of overlaps can yield measured $\delta^{13}C$ values that are 0.5–2‰ too negative or too positive than the true $\delta^{13}C$ value of the clean compound peak (Ricci *et al.*, 1994).

Question III.A.6: Taking all of the above into account, is it possible to come up with a definitive answer as to whether this athlete was guilty of testosterone abuse?

Question III.A.7: Taking all of the above into account, is it possible to come up with a definitive answer as to whether the data generated by the laboratory involved in this case would have been deemed admissible in a court of law?

Question III.A.8: What lessons should be learned from this case for the application of CSIA in a forensic context?

For the avoidance of doubt, the criticism implied in this chapter is not indiscriminately levelled at all analytical laboratories engaged in anti-doping testing. Anti-doping laboratories do exist whose published methods clearly demonstrate basic tenets of analytical and chromatographic science are being met, such as the use of identical GC columns in both the GC/C-IRMS system and the GC-MS system used for compound peak identification (Aguilera, Chapman and Catlin, 2000; Piper *et al.*, 2008). Unfortunately, this does not help the athletes whose samples are not being sent to one of these laboratories but to the one at the centre of this chapter.

Question III.A.9: Is a laboratory lottery really an acceptable way to combat doping where athletes' lives and reputations depend on which laboratory will deal with their samples?

I would sum up my private conclusion from this case as follows – what must be avoided at all costs and at all times when applying stable isotope analytical techniques in a forensic context is what we in our stable isotope forensics laboratory have coined a 'lackadaisical non-descript data determination' approach.

Appendix III.B
Sample Preparation Procedures

III.B.1 Preparing Silver Phosphate from Bio-apatite for ^{18}O Isotope Analysis

The procedure outlined in the following is essentially based on a published procedure (Stephan, 2000), but with some modifications of our own.

Day 1

- A sample of 40 mg of powdered bone is placed in a test tube.

- An aliquot of 2 ml of 2.5% sodium hypochlorite solution is added and left at room temperature with gentle agitation for 24 h. This is based on Fluka product #71696; a 10% solution. Dilute 2.5 ml of Fluka #71696 with 7.5 ml of water.

Day 2

- The solution is divided between two Eppendorf centrifuge caps and centrifuged for 5 min at 10 000 rpm.

- The supernatant (including the dissolved organic substances) is removed.

- The pellet is washed four or five times to neutrality with doubly distilled water. Each step requires centrifugation prior to pH check (Merck indicator strips) of the supernatant. The final supernatant should also be checked for the presence of chloride ions by adding a few drops of the supernatant to 0.5 ml of a 1 M $AgNO_3$ solution (1.7 g in 10 ml of water; keep away from light and store in fridge; not to be confused with the buffered silver nitrate solution required on day 5 to precipitate silver phosphate); the presence of chloride is indicated by the formation of a milky precipitation. The final supernatant must be free of chloride ions!

- An aliquot of 2 ml of 0.125 M sodium hydroxide solution is added to dissolve humic substances for 48 h at room temperature with gentle agitation; to prepare a 0.125 M NaOH solution, dissolve 0.5 g NaOH in 100 ml of water.

Day 4

- The solution is centrifuged for 5 min at 10 000 rpm.

- The supernatant is removed.

- The pellet is washed five times to neutrality with doubly distilled water. Each step requires centrifugation prior to pH check (Merck indicator strips) of supernatant.

- An aliquot of 2 ml of 2 M hydrofluoric acid is added and left for 24 h at room temperature. The 2 M hydrofluoric acid would be equivalent to 2 g of 100% hydrofluoric acid in 50 ml of water. However, the typical concentration of commercial products is 49–51%, therefore use 4 g (= 3.5 ml @ 1.15 g/ml) of the same added to 46.5 ml of water.

Day 5

- A 200-ml glass beaker has to be washed with approximately 30% nitric acid. Mix 30 g of concentrated HNO_3 into 70 ml of water. This is based on Fluka products #78005 or #84392; both are greater than 99.5% in HNO_3.

- The sample treated with 2 M hydrofluoric acid is centrifuged for 5 min at 10 000 rpm.

- The supernatant (containing the dissolved apatite) is removed, pipetted into the 200-ml beaker and 3 ml of 2 M potassium hydroxide is added. To prepare 2 M KOH dissolve 11.22 g of KOH in 100 ml of water.

- The pellet is washed with 2 ml doubly distilled water, centrifuged and the supernatant also pipetted into the beaker containing the first aliquot of apatite solution.

- Doubly distilled water is added to make up the apatite solution to a total volume of 200 ml.

- An aliquot of 30 ml of buffered silver nitrate solution is added. *Note*: The published procedure has been modified due to a discrepancy in the units used in the original reference (Stephan, 2000). The buffered silver nitrate solution used here is an aqueous solution, being 0.2 M $AgNO_3$, 0.35 M NH_4NO_3 and 0.75 M concentrated NH_4OH (around 25% NH_3 in water; 0.9 g/ml).

- The solution is gradually warmed up to 70 °C and the temperature held for 3 h without adding water to compensate for loss due to evaporation.

- The solution is allowed to cool down slowly to room temperature.

- At the start of this process the solution pH should be 10; at the end of this process the pH should be 7.

- The crystals formed are filtered on a pre-weighed 0.2-μm filter, washed several times with distilled water and dried overnight at 50 °C.

Day 6

- The apatite precipitate (Ag_3PO_4) is weighed into silver capsules (approximately 0.2 mg per capsule). Crimped silver capsules are stored in a 96-well plate and kept in a sealed and evacuated desiccator over P_4O_{10} for 7 days prior to analysis by TC/EA-IRMS.

To make 0.75 M NH_4OH one adds 26.29 ml of 25% NH_4OH to 223.71 ml of water to yield 250 ml of 0.75 N NH_4OH (= 0.75 M). Add to this 8.5 g of $AgNO_3$ and 7 g of NH_4NO_3 to make the buffered silver nitrate solution. An alternative recipe is based on commercially available 5 M NH_4OH: 26.29 ml of 5 M NH_4OH (Aldrich #318612) is added to 148.96 ml of water to yield 175.25 ml of 0.75 M NH_4OH.

The formula weights for silver nitrate, ammonium nitrate and ammonia required when preparing the buffered silver nitrate solution are:

1 mol $AgNO_3$ = 170 g; 0.2 mol = 34 g
1 mol NH_4NO_3 = 80 g; 0.35 mol = 28 g
1 mol concentrated NH_4OH (around 25%) = around 7.13 M

III.B.2 Acid Digest of Carbonate from Bio-apatite for ^{13}C and ^{18}O Isotope Analysis

For a triplicate analysis, at least 12 mg of tooth enamel or bone bio-apatite has to be drilled from the tooth or the bone using a dental drill whilst wearing nitrile latex-free gloves, cleaning the drill thoroughly with deionized water between samples. Powdered samples are first immersed in 0.1 M NaOCl for 2 h to remove organic material and then rinsed three times in deionized water before being washed with 5% acetic acid for 4 h to remove diagenetic carbonates. After further rinsing with deionized water the samples are dried for 48 h at below 45 °C in a drying oven. Prior to analysis samples are stored for 7 days in an evacuated desiccator using P_2O_5 as desiccant.

Approximately 4 mg of sample is weighed into an Exetainer® (Labco, High Wycombe, United Kingdom) and subjected to the procedure summarized below. At least two sets of three Exetainers filled with 0.5 mg each either of the international reference material NBS-19 ($\delta^{13}C_{VPDB} = +1.95‰$; $\delta^{18}O_{VPDB} = -2.20‰$) or the international reference material LSVEC ($\delta^{13}C_{VPDB} = -46.6‰$; $\delta^{18}O_{VPDB} = -26.7‰$) are put through the same sample preparation procedure contemporaneously to scale-anchor and quality-control sample results. Exetainers are then flushed with N5.7 grade nitrogen (N_2-BIP; Air Products, Crewe, United Kingdom) for 10 min to remove all traces of atmospheric CO_2. Once flushed, 0.5 ml of absolute (water-free) sulfuric acid (99.999%) or freshly distilled, water-free *ortho*-phosphoric acid is added by injection through the Exetainer's septum. Acid digest is carried out by placing the prepared Exetainers into a thermostatically controlled heater block at 50 °C for 6 h. Samples are allowed to cool at room temperature for at least 6 h (typically for 12 h overnight) prior to stable isotope analysis of the evolved CO_2 for $\delta^{13}C$ and $\delta^{18}O$ isotopic composition.

In human tooth enamel, as in human bone, oxygen is present in both the phosphate and carbonate fractions of bio-apatite, and, in the absence of diagenetic changes, $\delta^{18}O$ values from either fraction may be used to calculate corresponding $\delta^{18}O$ values for source water using appropriate correlation equations. If carbonate is to be used for this purpose one needs to consider that $\delta^{18}O$ values for ^{18}O in carbonate are anchored on the VPDB scale. Prior to conversion of $\delta^{18}O_{carbonate}$ values into $\delta^{18}O_{phosphate}$ values, $\delta^{18}O_{carbonate}$ values versus VPDB must be adjusted to the VSMOW scale on which $\delta^{18}O_{phosphate}$ values are traditionally reported.

The conversion of a $\delta^{18}O_{VPDB}$ value of a given carbonate sample X into a $\delta^{18}O_{VSMOW}$ value is given by (Friedman and O'Neil, 1977):

$$\delta^{18}O_{VSMOW}(X) = 1.03086 \, \delta^{18}O_{VPDB}(X) + 30.86 \quad (I.21)$$

Once $\delta^{18}O_{carbonate}$ values have been adjusted to the VSMOW scale they can be converted into $\delta^{18}O_{phosphate}$ values by virtue of:

$$\delta^{18}O_{phosphate} = 0.98 \, \delta^{18}O_{carbonate} - 8.5 \; (\text{Iacumin } et \, al., 1996) \quad (I.25)$$

The original correlation between $\delta^{18}O$ values of human bone phosphate and $\delta^{18}O$ values of source water (Equation I.22) was based on the least-squares best fit of measured $\delta^{18}O$ values and was obtained using bone belonging to people who died between the end of the nineteenth century and 1950 (Longinelli, 1984). In my opinion this equation has still some relevance today, especially if one considers that a unified correlation equation (Equation I.23) proposed in 2008 (Daux *et al.*, 2008) is not too dissimilar to the equation proposed by Longinelli a quarter of a century ago:

$$\delta^{18}O_{phosphate} = 0.64 \, \delta^{18}O_{water} + 22.37 \quad (I.22)$$
$$\delta^{18}O_{phosphate} = 0.65 \, \delta^{18}O_{water} + 21.89 \quad (I.23)$$

When applying this technique to determining the potential geographic life history or trajectory of a people I would recommend making use of both equations and to compare results.

III.B.3 Standard Protocol for Preparing Hair Samples for ^{2}H Isotope Analysis

The following protocol will of course have to be adapted for use in your own laboratory, chiefly as far as the two different waters used for equilibration are concerned. Absolute δ^2H values for the two waters are not important, but the difference between their δ^2H values should be approximately 100‰ or more. Equally important is how to store water samples, be it for the purpose of equilibrating samples or be it to serve as reference/standard materials. A stock solution of 'standard' water should be sterile filtrated using 0.2-μm but at least 0.25 μm filters and dispensed into 2-ml, 11-mm amber autosampler vials. Vials should be filled to overflow to leave no head-space volume, their rims wiped dry and subsequently crimp-sealed. This way samples can be stored in a normal fridge almost indefinitely. Water for equilibration purposes should be treated similarly, with the only difference being the size of the crimp-seal vial. Here I would suggest using 22 ml crimp-seal vials.

Case hair sample(s)

(i) Once cleaned, oven-dried hair samples are finely cut (chopped) using a stainless steel scalpel blade on a polished granite slab.

(ii) The cut hair sample is placed in glass vial, sealed with Parafilm into which a few holes have been punched and stored in an evacuated desiccator containing self-indicating phosphorous pentoxide (Sicapent®) for 7 days. This drying step removes residual moisture traces from the sample.

(iii) To monitor and later correct for hydrogen exchange, multiple subsamples of the sample (at least $n=3$) are weighed out into silver capsules, which are slightly crimped and subsequently stored in a sealed desiccator designated for laboratory tap water (LTW) use only together with a Petri dish containing SCRI LTW (δ^2H$_{measured}$: −54.1‰) for 4 days so labile hydrogen atoms prone to exchange will all reflect the same ^{2}H abundance level.

(iv) Should sufficient case material be available, multiple subsamples of the sample (at least $n=3$) are weighed out into silver capsules, which are slightly crimped and subsequently stored in a sealed desiccator designed for Canadian Calgary water (CCW) use only together with a Petri dish containing CCW (δ^2H$_{measured}$: −141.9‰) for 4 days so labile hydrogen atoms prone to exchange will all reflect the ^{2}H abundance of Calgary water.

(v) The hydrogen-equilibrated hair samples are removed from the equilibration desiccator(s) and the silver capsules are crimped shut.

(vi) Crimped silver capsules containing the hair sample are placed in a 96-well plate and sample positions are concurrently entered into a well plate template sample sheet.

(vii) The well plate is covered with its lid. Together with the folded well plate template sample sheet, the lid is secured to the well plate by a rubber band.

(viii) The sample well plate is placed in a desiccator over Sicapent and kept there under vacuum until the samples are analysed (but at least for 7 days). Samples equilibrated with LTW and CCW are kept isolated from each other, and are placed in separate desiccators to avoid cross-exchange during drying down.

Control hair sample for contemporaneous determination of hydrogen exchange

- An aliquot of 'standard' hair (e.g. ID F07BT235NE) is prepared contemporaneously with the case sample(s) to permit calculation of hydrogen exchange rate and, if required, subsequent conversion of measured $\delta^2 H_{total}$ values into $\delta^2 H_{true}$ values. This control procedure is carried out even if sufficient case sample material is available.

- Following points (i) and (ii) above, four replicates of a subsample of standard hair are treated as described in point (iii), while four replicates of a second subsample are treated as described in point (iii) but placed in a separate CCW desiccator together with a Petri dish containing CCW ($\delta^2 H_{measured}$: −141.9‰) for 4 days.

- The standard hair sample equilibrated with CCW is subsequently prepared as described in points (v)–(viii) above but for the fact that for storage and further drying a CCW desiccator is used to avoid any cross-contamination between CCW standard hair samples and case samples as well as LTW standard hair samples.

- Case hair samples and standard hair samples are therefore analysed on a 'like-for-like' basis, in accordance with the 'principle of identical treatment'.

The exchange mole fraction f_{Hxch} of exchangeable hydrogen is calculated from the ratio of the difference in observed $\delta^2 H$ values for standard hair to the difference in known $\delta^2 H$ values of waters LTW and CCW used for equilibration:

$$f_{Hxch} = \frac{\delta^2 H_{std-hair,\ CCW} - \delta^2 H_{std-hair,\ LTW}}{\delta^2 H_{water,\ CCW} - \delta^2 H_{water,\ LTW}}$$

'True' $\delta^2 H$ values for hair samples can then be calculated by inserting numerical values obtained for f_{Hxch}, $\delta^2 H_{hair,\ total}$ and $\delta^2 H_{water,\ LTW}$ into Equation II.12 (Section II.6.3.2.1).

The above protocol is based on published methods and is subject to continuous refinement and revision (Bowen *et al.*, 2005; Chesson *et al.*, 2009; Fraser and Meier-Augenstein, 2007; Sauer *et al.*, 2009; Schimmelmann, 1991; Sharp *et al.*, 2003).

References Part III

Ader, M., Coleman, M.L., Doyle, S.P., Stroud, M. and Wakelin, D. (2001) Methods for the stable isotopic analysis of chlorine in chlorate and perchlorate compounds. *Analytical Chemistry*, **73**, 4946–4950.

Aguilera, R., Chapman, T.E. and Catlin, D.H. (2000) A rapid screening assay for measuring urinary androsterone and etiocholanolone delta C-13 (parts per thousand) values by gas chromatography/combustion/isotope ratio mass spectrometry. *Rapid Communications in Mass Spectrometry*, **14**, 2294–2299.

Aguilera, R., Chapman, T.E., Starcevic, B., Hatton, C.K. and Catlin, D.H. (2001) Performance characteristics of a carbon isotope ratio method for detecting doping with testosterone based on urine diols: controls and athletes with elevated testosterone/epitestosterone ratios. *Clinical Chemistry*, **47**, 292–300.

Barbour, M.M., Andrews, T.J. and Farquhar, G.D. (2001) Correlations between oxygen isotope ratios of wood constituents of *Quercus* and *Pinus* samples from around the world. *Australian Journal of Plant Physiology*, **28**, 335–348.

Begley, I.S., Coleman, M., Beadah, A., Doyle, S., Wakelin, D., Stroud, M., Ader, M., Isaacs, M., Poswa, A., Brookes, S. and Belanger, C. (2003) Stable isotope ratio analysis of explosives – a new type of evidence. *Forensic Science International*, **136**, 140.

Benson, S.J. (2009) Introduction of isotope ratio mass spectrometry (IRMS) for the forensic analysis of explosives. PhD Thesis. University of Technology, Sydney.

Benson, S.J., Lennard, C.J., Maynard, P., Hill, D.M., Andrew, A.S. and Roux, C. (2009a) Forensic analysis of explosives using isotope ratio mass spectrometry (IRMS) discrimination of ammonium nitrate sources. *Science & Justice*, **49**, 73–80.

Benson, S.J., Lennard, C.J., Maynard, P., Hill, D.M., Andrew, A.S. and Roux, C. (2009b) Forensic analysis of explosives using isotope ratio mass spectrometry (IRMS) – preliminary study on TATP and PETN. *Science & Justice*, **49**, 81–86.

Besacier, F., Guilluy, R., Brazier, J.L., ChaudronThozet, H., Girard, J. and Lamotte, A. (1997) Isotopic analysis of C-13 as a tool for comparison and origin assignment of seized heroin samples. *Journal of Forensic Sciences*, **42**, 429–433.

Billault, I., Courant, F., Pasquereau, L., Derrien, S., Robins, R.J. and Naulet, N. (2007) Correlation between the synthetic origin of methamphetamine samples and their N-15 and C-13 stable isotope ratios. *Analytica Chimica Acta*, **593**, 20–29.

Black, S. and Thompson, T. (2007) Body modification, in *Forensic Human Identification – An Introduction* (eds T.J.T. Thomson and S.M. Black), CRC Press, Boca Raton, FL, pp. 379–399.

Boettger, T., Haupt, M., Knoller, K., Weise, S.M., Waterhouse, J.S., Rinne, K.T., Loader, N.J., Sonninen, E., Jungner, H., Masson-Delmotte, V., Stievenard, M., Guillemin, M.T., Pierre, M., Pazdur, A., Leuenberger, M., Filot, M., Saurer, M., Reynolds, C.E., Helle, G. and Schleser, G.H. (2007) Wood cellulose preparation methods and mass spectrometric analyses of delta C-13, delta O-18, and nonexchangeable delta H-2 values in cellulose, sugar, and starch: an interlaboratory comparison. *Analytical Chemistry*, **79**, 4603–4612.

Bohlke, J.K., Sturchio, N.C., Gu, B.H., Horita, J., Brown, G.M., Jackson, W.A., Batista, J. and Hatzinger, P.B. (2005) Perchlorate isotope forensics. *Analytical Chemistry*, **77**, 7838–7842.

Bol, R. and Pflieger, C. (2002) Stable isotope (C-13, N-15 and S-34) analysis of the hair of modern humans and their domestic animals. *Rapid Communications in Mass Spectrometry*, **16**, 2195–2200.

Bol, R., Marsh, J. and Heaton, T.H.E. (2007) Multiple stable isotope (O-18, C-13, N-15 and S-34) analysis of human hair to identify the recent migrants in a rural community in SW England. *Rapid Communications in Mass Spectrometry*, **21**, 2951–2954.

Bolck, A., Weyermann, C., Dujourdy, L., Esseiva, P. and van den Berg, J. (2009) Different likelihood ratio approaches to evaluate the strength of evidence of MDMA tablet comparisons. *Forensic Science International*, **191**, 42–51.

Bowen, G.J. and Wilkinson, B. (2002) Spatial distribution of delta O-18 in meteoric precipitation. *Geology*, **30**, 315–318.

Bowen, G.J., Chesson, L., Nielson, K., Cerling, T.E. and Ehleringer, J.R. (2005) Treatment methods for the determination of delta H-2 and delta O-18 of hair keratin by continuous-flow isotope-ratio mass spectrometry. *Rapid Communications in Mass Spectrometry*, **19**, 2371–2378.

Bowers, L.D., Hilderbrand, R.L. and Symanski, E.J. (eds) (2003) *Application of Gas Chromatography-Combustion-Isotope Ratio Mass Spectrometry to Doping Control (Second Annual USADA Symposium on Anti-Doping Science)*, USADA, Colorado Springs, CO.

Brand, W.A., Coplen, T.B., Aerts-Bijma, A.T., Boehlke, J.K., Gehre, M., Geilmann, H., Groning, M., Jansen, H.G., Meijer, H.A.J., Mroczkowski, S.J., Qi, H.P., Soergel, K., Stuart-Williams, H., Weise, S.M. and Werner, R.A. (2009a) Comprehensive inter-laboratory calibration of reference materials for $d^{18}O$ versus VSMOW using various on-line high-temperature conversion techniques. *Rapid Communications in Mass Spectrometry*, **23**, 999–1019.

Brand, W.A., Huang, L., Mukai, H., Chivulescu, A., Richter, J. and Rothe, M. (2009b) How well do we know VPDB? Variability of $d^{13}C$ and $d^{18}O$ in CO_2 generated from NBS19-calcite. *Rapid Communications in Mass Spectrometry*, **23**, 915–926.

Brooks, J.R., Buchmann, N., Phillips, S., Ehleringer, B., Evans, R.D., Lott, M., Martinelli, L.A., Pockman, W.T., Sandquist, D., Sparks, J.P., Sperry, L., Williams, D. and Ehleringer, J.R. (2002) Heavy and light beer: a carbon isotope approach to detect C-4 carbon in beers of different origins, styles, and prices. *Journal of Agricultural and Food Chemistry*, **50**, 6413–6418.

Buchanan, H.A.S. (2009) An evaluation of isotope ratio mass spectrometry for the profiling of 3,4-methylenedioxy-methamphetamine. PhD Thesis. University of Strathclyde.

Buchanan, H.A.S., Nic Daéid, N., Meier-Augenstein, W., Kemp, H.F., Kerr, W.J. and Middleditch, M. (2008) Emerging use of isotope ratio mass spectrometry as a tool for discrimination of 3,4-methylenedioxymethamphetamine by synthetic route. *Analytical Chemistry*, **80**, 3350–3356.

Carter, J.F., Titterton, E.L., Grant, H. and Sleeman, R. (2002a) Isotopic changes during the synthesis of amphetamines. *Chemical Communications*, **21**, 2590–2591.

Carter, J.F., Titterton, E.L., Murray, M. and Sleeman, R. (2002b) Isotopic characterisation of 3,4-methylenedioxyamphetamine and 3,4-methylenedioxymethylamphetamine (ecstasy). *Analyst*, **127**, 830–833.

Carter, J.F., Grundy, P.L., Hill, J.C., Ronan, N.C., Titterton, E.L. and Sleeman, R. (2004) Forensic isotope ratio mass spectrometry of packaging tapes. *Analyst*, **129**, 1206–1210.

Carter, J.F., Sleeman, R., Hill, J.C., Idoine, F. and Titterton, E.L. (2005) Isotope ratio mass spectrometry as a tool for forensic investigation (examples from recent studies). *Science & Justice*, **45**, 141–149.

Casale, J., Casale, E., Collins, M., Morello, D., Cathapermal, S. and Panicker, S. (2006) Stable isotope analyses of heroin seized from the merchant vessel *Pong Su*. *Journal of Forensic Sciences*, **51**, 603–606.

Casale, J.F., Ehleringer, J.R., Morello, D.R. and Lott, M.J. (2005) Isotopic fractionation of carbon and nitrogen during the illicit processing of cocaine and heroin in South America. *Journal of Forensic Sciences*, **50**, 1315–1321.

Cerling, T.E., Ehleringer, J.R., West, A., Stange, E. and Dorigan, J. (2003) Forensic applications of stable isotopes in hair. *Forensic Science International*, **136**, 172.

Chesson, L.A., Podlesak, D.W., Cerling, T.E. and Ehleringer, J.R. (2009) Evaluating uncertainty in the calculation of non-exchangeable hydrogen fractions within organic materials. *Rapid Communications in Mass Spectrometry*, **23**, 1275–1280.

Collins, M., Cawley, A.T., Heagney, A.C., Kissane, L., Robertson, J. and Salouros, H. (2009) delta C-13, delta N-15 and delta H-2 isotope ratio mass spectrometry of ephedrine and pseudoephedrine: application to methylamphetamine profiling. *Rapid Communications in Mass Spectrometry*, **23**, 2003–2010.

D'Angela, D. and Longinelli, A. (1990) Oxygen isotopes in living mammals bone phosphate – further results. *Chemical Geology: Isotope Geoscience Section*, **86**, 75–82.

Daux, V., Lecuyer, C., Heran, M.A., Amiot, R., Simon, L., Fourel, F., Martineau, F., Lynnerup, N., Reychler, H. and Escarguel, G. (2008) Oxygen isotope fractionation between human phosphate and water revisited. *Journal of Human Evolution*, **55**, 1138–1147.

DeNiro, M.J. and Epstein, S. (1979) Relationship between the oxygen isotope ratios of terrestrial plant cellulose, carbon dioxide, and water. *Science*, **204**, 51–53.

Desage, M., Guilluy, R., Brazier, J.L., Chaudron, H., Girard, J., Cherpin, H. and Jumeau, J. (1991) Gas chromatography with mass spectrometry or isotope-ratio mass spectrometry in studying the geographical origin of heroin. *Analytica Chimica Acta*, **247**, 249–254.

Dickson, J.H., Oeggl, K., Holden, T.G., Handley, L.L., O'Connell, T.C. and Preston, T. (2000) The omnivorous Tyrolean Iceman: colon contents (meat, cereals, pollen, moss and whipworm) and stable isotope analyses. *Philosophical Transactions of the Royal Society of London Series B Biological Sciences*, **355**, 1843–1849.

Dobney, A.M., Wiarda, W., de Joode, P. and van der Peijl, G.J.Q. (2002) Sector field ICPMS applied to the forensic analysis of commercially available adhesive packaging tapes. *Journal of Analytical Atomic Spectrometry*, **17**, 478–484.

Dobney, A.M., Wiarda, W., de Joode, P. and van der Peijl, G.J.Q. (2003a) Forensic comparison of packaging tapes – an elemental and isotopic perspective. *Forensic Science International*, **136**, 352–353.

Dobney, A.M., Wiarda, W., de Joode, P. and van der Peijl, G.J.Q. (2003b) ICPMS for forensic investigations: the NFI experience. *Forensic Science International*, **136**, 360–361.

Dupras, T.L. and Schwarcz, H.P. (2001) Strangers in a strange land: stable isotope evidence for human migration in the Dakhleh Oasis, Egypt. *Journal of Archaeological Science*, **28**, 1199–1208.

Ehleringer, J.R., Casale, J.F., Lott, M.J. and Ford, V.L. (2000) Tracing the geographical origin of cocaine. *Nature*, **408**, 311–312.

Ehleringer, J.R., Bowen, G.J., Chesson, L.A., West, A.G., Podlesak, D.W. and Cerling, T.E. (2008) Hydrogen and oxygen isotope ratios in human hair are related to geography. *Proceedings of the National Academy of Sciences of the United States of America*, **105**, 2788–2793.

Farmer, N., Curran, J., Lucy, D., Nic Daéid, N. and Meier-Augenstein, W. (2009) Stable isotope profiling of burnt wooden safety matches. *Science & Justice*, **49**, 107–113.

Farmer, N., Meier-Augenstein, W. and Lucy, D. (2009) Stable isotope analysis of white paints and likelihood ratios. *Science & Justice*, **49**, 114–119.

Farmer, N.L., Meier-Augenstein, W. and Kalin, R.M. (2005) Stable isotope analysis of safety matches using isotope ratio mass spectrometry – a forensic case study. *Rapid Communications in Mass Spectrometry*, **19**, 3182–3186.

Farmer, N.L., Ruffell, A., Meier-Augenstein, W., Meneely, J. and Kalin, R.M. (2007) Forensic analysis of wooden safety matches – a case study. *Science & Justice*, **47**, 88–98.

Fogel, M.L. and Tuross, N. (2003) Extending the limits of paleodietary studies of humans with compound specific carbon isotope analysis of amino acids. *Journal of Archaeological Science*, **30**, 535–545.

Fraser, I. and Meier-Augenstein, W. (2007) Stable ^{2}H isotope analysis of human hair and nails can aid forensic human identification. *Rapid Communications in Mass Spectrometry*, **21**, 3279–3285.

Fraser, I., Meier-Augenstein, W. and Kalin, R.M. (2006) The role of stable isotopes in human identification: a longitudinal study into the variability of isotopic signals in human hair and nails. *Rapid Communications in Mass Spectrometry*, **20**, 1109–1116.

Fraser, I., Meier-Augenstein, W. and Kalin, R.M. (2008) Stable isotope analysis of human hair and nail samples: the effects of storage on samples. *Journal of Forensic Sciences*, **53**, 95–99.

Friedman, I. and O'Neil, J.R. (1977) *Compilation of Stable Isotope Fractionation Factors of Geochemical Interest*, Vol. 440-KK, US Geological Survey, Reston, VA.

Fuller, B.T., Richards, M.P. and Mays, S.A. (2003) Stable carbon and nitrogen isotope variations in tooth dentine serial sections from Wharram Percy. *Journal of Archaeological Science*, **30**, 1673–1684.

Fuller, B.T., Fuller, J.L., Sage, N.E., Harris, D.A., O'Connell, T.C. and Hedges, R.E.M. (2004) Nitrogen balance and delta N-15: why you're not what you eat during pregnancy. *Rapid Communications in Mass Spectrometry*, **18**, 2889–2896.

Fuller, B.T., Fuller, J.L., Sage, N.E., Harris, D.A., O'Connell, T.C. and Hedges, R.E.M. (2005) Nitrogen balance and delta N-15: why you're not what you eat during nutritional stress. *Rapid Communications in Mass Spectrometry*, **19**, 2497–2506.

Galimov, E.M., Sevastyanov, V.S., Kulbachevskaya, E.V. and Golyavin, A.A. (2005) Isotope ratio mass spectrometry: $d^{13}C$ and $d^{15}N$ analysis for tracing the origin of illicit drugs. *Rapid Communications in Mass Spectrometry*, **19**, 1213–1216.

Gentile, N., Siegwolf, R.T.W. and Delemont, O. (2009) Study of isotopic variations in black powder: reflections on the use of stable isotopes in forensic science for source inference. *Rapid Communications in Mass Spectrometry*, **23**, 2559–2567.

Grynpas, M. (1993) Age and disease-related changes in the mineral of bone. *Calcified Tissue International*, **53**, S57–S64.

Hedges, R.E.M., Clement, J.G., Thomas, C.D.L. and O'Connell, T.C. (2007) Collagen turnover in the adult femoral mid-shaft: modeled from anthropogenic radiocarbon tracer measurements. *American Journal of Physical Anthropology*, **133**, 808–816.

Hill, P.A. (1998) Bone remodelling. *British Journal of Orthopaedics*, **25**, 101–107.

Holden, J.L., Clement, J.G. and Phakey, P.P. (1995) Age and temperature-related changes to the ultrastructure and composition of human bone-mineral. *Journal of Bone and Mineral Research*, **10**, 1400–1409.

Hoogewerff, J., Papesch, W., Kralik, M., Berner, M., Vroon, P., Miesbauer, H., Gaber, O., Kunzel, K.H. and Kleinjans, J. (2001) The last domicile of the Iceman from Hauslabjoch: a geochemical approach using Sr, C and O isotopes and trace element signatures. *Journal of Archaeological Science*, **28**, 983–989.

Iacumin, P., Bocherens, H., Mariotti, A. and Longinelli, A. (1996) An isotopic palaeoenvironmental study of human skeletal remains from the Nile Valley. *Palaeogeography Palaeoclimatology Palaeoecology*, **126**, 15–30.

Keppler, F., Harper, D.B., Kalin, R.M., Meier-Augenstein, W., Farmer, N., Davis, S., Schmidt, H.L., Brown, D.M. and Hamilton, J.T.G. (2007) Stable hydrogen isotope ratios of lignin methoxyl groups as a paleoclimate proxy and constraint of the geographical origin of wood. *New Phytologist*, **176**, 600–609.

Koch, P.L., Tuross, N. and Fogel, M.L. (1997) The effects of sample treatment and diagenesis on the isotopic integrity of carbonate in biogenic hydroxylapatite. *Journal of Archaeological Science*, **24**, 417–429.

Kreuzer-Martin, H.W., Lott, M.J., Dorigan, J. and Ehleringer, J.R. (2003) Microbe forensics: oxygen and hydrogen stable isotope ratios in *Bacillus subtilis* cells and spores. *Proceedings of the National Academy of Sciences of the United States of America*, **100**, 815–819.

Kreuzer-Martin, H.W., Chesson, L.A., Lott, M.J., Dorigan, J.V. and Ehleringer, J.R. (2004) Stable isotope ratios as a tool in microbial forensics – part 1. Microbial isotopic composition as a function of growth medium. *Journal of Forensic Sciences*, **49**, 954–960.

Kurashima, N., Makino, Y., Sekita, S., Urano, Y. and Nagano, T. (2004) Determination of origin of ephedrine used as precursor for illicit methamphetamine by carbon and nitrogen stable isotope ratio analysis. *Analytical Chemistry*, **76**, 4233–4236.

Kurashima, N., Makino, Y., Urano, Y., Sanuki, K., Ikehara, Y. and Nagano, T. (2009) Use of stable isotope ratios for profiling of industrial ephedrine samples: application of hydrogen isotope ratios in combination with carbon and nitrogen. *Forensic Science International*, **189**, 14–18.

Lee-Thorp, J. and Sponheimer, M. (2003a) Three case studies used to reassess the reliability of fossil bone and enamel isotope signals for paleodietary studies. *Journal of Anthropological Archaeology*, **22**, 208–216.

Lee-Thorp, J. and Sponheimer, M. (2003b) Tooth enamel remains a virtually closed system for stable light isotope and trace element archives in fossils [Abstract]. *American Journal of Physical Anthropology*, **120** (Suppl. 36), 138.

Loader, N.J., Robertson, I. and McCarroll, D. (2003) Comparison of stable carbon isotope ratios in the whole wood, cellulose and lignin of oak tree-rings. *Palaeogeography Palaeoclimatology Palaeoecology*, **196**, 395–407.

Lock, C.M. (2009) [Title withheld]. PhD Thesis. Queen's University Belfast.

Lock, C.M. and Meier-Augenstein, W. (2008) Investigation of isotopic linkage between precursor and product in the synthesis of a high explosive. *Forensic Science International*, **179**, 157–162.

Longinelli, A. (1984) Oxygen isotopes in mammal bone phosphate – a new tool for paleohydrological and paleoclimatological research. *Geochimica et Cosmochimica Acta*, **48**, 385–390.

Luz, B. and Kolodny, Y. (1985) Oxygen isotope variations in phosphate of biogenic apatites. 4. Mammal teeth and bones. *Earth and Planetary Science Letters*, **75**, 29–36.

Luz, B., Kolodny, Y. and Horowitz, M. (1984) Fractionation of oxygen isotopes between mammalian bone-phosphate and environmental drinking water. *Geochimica et Cosmochimica Acta*, **48**, 1689–1693.

Macko, S.A., Engel, M.H., Andrusevich, V., Lubec, G., O'Connell, T.C. and Hedges, R.E.M. (1999) Documenting the diet in ancient human populations through stable isotope analysis of hair. *Philosophical Transactions of the Royal Society of London Series B-Biological Sciences*, **354**, 65–75.

Mas, F., Beemsterboer, B., Veltkamp, A.C. and Verweij, A.M.A. (1995) Determination of common-batch members in a set of confiscated 3,4-(methylendioxy) methylamphetamine samples by measuring the natural isotope abundances: a preliminary study. *Forensic Science International*, **71**, 225–231.

Mays, S.A., Richards, M.P. and Fuller, B.T. (2002) Bone stable isotope evidence for infant feeding in mediaeval England. *Antiquity*, **76**, 654–656.

McCarroll, D. and Loader, N.J. (2006) Isotopes in tree rings, in *Isotopes in Palaeoenvironmental Reseach* (ed. M.J. Leng), Springer, Berlin, pp. 67–115.

Meier-Augenstein, W. (1999) Use of gas chromatography-combustion-isotope ratio mass spectrometry in nutrition and metabolic research. *Current Opinion in Clinical Nutrition and Metabolic Care*, **2**, 465–470.

Meier-Augenstein, W. (2007) Stable isotope fingerprinting – chemical element 'DNA'?, in *Forensic Human Identification – An Introduction* (eds T.J.T. Thomson and S.M. Black), CRC Press, Boca Raton, FL, pp. 29–53.

Meier-Augenstein, W. and Fraser, I. (2008) Forensic isotope analysis leads to identification of a mutilated murder victim. *Science & Justice*, **48**, 153–159.

Meier-Augenstein, W. and Liu, R.H. (2004) Forensic applications of isotope ratio mass spectrometry, in *Advances in Forensic Applications of Mass Spectrometry* (ed. J. Yinon), CRC Press, Boca Raton, FL, pp. 149–180.

Meier-Augenstein, W., Kemp, H.F. and Lock, C.M. (2009) N_2: a potential pitfall for bulk ^2H isotope analysis of explosives and other nitrogen-rich compounds by continuous-flow isotope-ratio mass spectrometry. *Rapid Communications in Mass Spectrometry*, **23**, 2011–2016.

Montero, S., Wiarda, W., de Joode, P. and van der Peijl, G. (2005) LA-ICP-MS and IRMS investigations on packaging and duct tapes, presented at the Forensic Isotope Ratio Mass Spectrometry (FIRMS) Network Conference, Brands Hatch, http://www.forensic-isotopes.org/conference/conf2005/Van Der Peijl Presentation.pdf.

Moorrees, C.F., Fanning, E.A. and Hunt, E.E. (1963a) Formation and resorption of 3 deciduous teeth in children. *American Journal of Physical Anthropology*, **21**, 205–213.

Moorrees, C.F., Fanning, E.A. and Hunt, E.E. (1963b) Age variation of formation stages for ten permanent teeth. *Journal of Dental Research*, **42**, 1490–1502.

Nic Daéid, N. and Meier-Augenstein, W. (2008) Feasibility of source identification of seized street drug samples by exploiting differences in isotopic composition at natural abundance level by GC/MS as compared to isotope ratio mass spectrometry (IRMS). *Forensic Science International*, **174**, 259–261.

O'Connell, T.C. and Hedges, R.E.M. (1999a) Investigations into the effect of diet on modern human hair isotopic values. *American Journal of Physical Anthropology*, **108**, 409–425.

O'Connell, T.C. and Hedges, R.E.M. (1999b) Isotopic comparison of hair and bone: archaeological analyses. *Journal of Archaeological Science*, **26**, 661–665.

O'Connell, T.C., Hedges, R.E.M., Healey, M.A. and Simpson, A.H.R. (2001) Isotopic comparison of hair, nail and bone: modern analyses. *Journal of Archaeological Science*, **28**, 1247–1255.

O'Reilly, W. (2007) The 'Adam' case, London, in *Forensic Human Identification – An Introduction* (eds T.J.T. Thomson and S.M. Black), CRC Press, Boca Raton, FL, pp 473–484.

Palhol, F., Lamoureux, C. and Naulet, N. (2003) N-15 isotopic analyses: a powerful tool to establish links between seized 3,4-methylenedioxymethamphetamine (MDMA) tablets. *Analytical and Bioanalytical Chemistry*, **376**, 486–490.

Palhol, F., Lamoureux, C., Chabrillat, M. and Naulet, N. (2004) N-15/N-14 isotopic ratio and statistical analysis: an efficient way of linking seized ecstasy. *Analytica Chimica Acta*, **510**, 1–8.

Petzke, K.J. and Lemke, S. (2009) Hair protein and amino acid ^{13}C and ^{15}N abundances take more than 4 weeks to clearly prove influences of animal protein intake in young women with a habitual daily protein consumption of more than 1 g per kg body weight. *Rapid Communications in Mass Spectrometry*, **23**, 2411–2420.

Phillips, S.A., Doyle, S., Philp, L. and Coleman, M. (2003) Network developing forensic applications of stable isotope ratio mass spectrometry conference 2002 – abstracts. *Science & Justice*, **43**, 153–160.

Pierrini, G., Doyle, S., Champod, C., Taroni, F., Wakelin, D. and Lock, C. (2007) Evaluation of preliminary isotopic analysis (C-13 and N-15) of explosives: a likelihood ratio approach to assess the links between Semtex samples. *Forensic Science International*, **167**, 43–48.

Piper, T., Mareck, U., Geyer, H., Flenker, U., Thevis, M., Platen, P. and Schanzer, W. (2008) Determination of C-13/C-12 ratios of endogenous urinary steroids: method validation, reference population and application to doping control purposes. *Rapid Communications in Mass Spectrometry*, **22**, 2161–2175.

Quirk, A.T., Bellerby, J.M., Carter, J.F., Thomas, F.A. and Hill, J.C. (2009) An initial evaluation of stable isotopic characterisation of post-blast plastic debris from improvised explosive devices. *Science & Justice*, **49**, 87–93.

Rauch, E., Rummel, S., Lehn, C. and Buettner, A. (2007) Origin assignment of unidentified corpses by use of stable isotope ratios of light (bio-) and heavy (geo-) elements – a case report. *Journal of Forensic Sciences*, **168**, 215–218.

Ricci, M.P., Merritt, D.A., Freeman, K.H. and Hayes, J.M. (1994) Acquisition and processing of data for isotope-ratio-monitoring mass spectrometry. *Organic Geochemistry*, **21**, 561–571.

Richards, M.P., Fuller, B.T. and Hedges, R.E.M. (2001) Sulphur isotopic variation in ancient bone collagen from Europe: implications for human palaeodiet, residence mobility, and modern pollutant studies. *Earth and Planetary Science Letters*, **191**, 185–190.

Richards, M.P., Mays, S. and Fuller, B.T. (2002) Stable carbon and nitrogen isotope values of bone and teeth reflect weaning age at the Medieval Wharram Percy site, Yorkshire, UK. *American Journal of Physical Anthropology*, **119**, 205–210.

Richards, M.P., Fuller, B.T., Sponheimer, M., Robinson, T. and Ayliffe, L. (2003) Sulphur isotopes in palaeodietary studies: a review and results from a controlled feeding experiment. *International Journal of Osteoarchaeology*, **13**, 37–45.

Rieley, G. (1994) Derivatization of organic compounds prior to gas chromatographic-combustion-isotope ratio mass spectrometric analysis – identification of isotope fractionation processes. *Analyst*, **119**, 915–919.

Saudan, C., Kamber, M., Barbati, G., Robinson, N., Desmarchelier, A., Mangin, P. and Saugy, M. (2006) Longitudinal profiling of urinary steroids by gas chromatography/combustion/isotope ratio mass spectrometry: diet change may result in carbon isotopic variations. *Journal of Chromatography B*, **831**, 324–327.

Sauer, P.E., Schimmelmann, A., Sessions, A.L. and Topalov, K. (2009) Simplified batch equilibration for D/H determination of non-exchangeable hydrogen in solid organic material. *Rapid Communications in Mass Spectrometry*, **23**, 949–956.

Saunders, S., Devito, C., Herring, A., Southern, R. and Hoppa, R. (1993) Accuracy tests of tooth formation age estimations for human skeletal remains. *American Journal of Physical Anthropology*, **92**, 173–188.

Saurer, M. (2003) The influence of climate on the oxygen isotopes in tree rings. *Isotopes in Environmental and Health Studies*, **39**, 105–112.

Saurer, M., Cherubini, P., Bonani, G. and Siegwolf, R. (2003) Tracing carbon uptake from a natural CO_2 spring into tree rings: an isotope approach. *Tree Physiology*, **23**, 997–1004.

Schimmelmann, A. (1991) Determination of the concentration and stable isotopic composition of nonexchangeable hydrogen in organic matter. *Analytical Chemistry*, **63**, 2456–2459.

Schmidt, H.L., Werner, R.A. and Rossmann, A. (2001) O-18 pattern and biosynthesis of natural plant products. *Phytochemistry*, **58**, 9–32.

Schneiders, S., Holdermann, T. and Dahlenburg, R. (2009) Comparative analysis of 1-phenyl-2-propanone (P2P), an amphetamine-type stimulant precursor, using stable isotope ratio mass spectrometry. *Science & Justice*, **49**, 94–101.

Schoeninger, M.J. and DeNiro, M.J. (1984) Nitrogen and carbon isotopic composition of bone-collagen from marine and terrestrial animals. *Geochimica et Cosmochimica Acta*, **48**, 625–639.

Schoeninger, M.J., DeNiro, M.J. and Tauber, H. (1983) Stable nitrogen isotope ratios of bone-collagen reflect marine and terrestrial components of prehistoric human diet. *Science*, **220**, 1381–1383.

Schwarcz, H.P. and Walker, P.L. (2006) Characterization of a murder victim using stable isotopic analyses [Abstract]. *American Journal of Physical Anthropology*, **129** (Suppl. 42), 160.

Sewenig, S., Fichtner, S., Holdermann, T., Fritschi, G. and Neumann, H. (2007) Determination of delta C-13(V-PDB) and delta N-15(AIR) values of cocaine from a big seizure in Germany by stable isotope ratio mass spectrometry. *Isotopes in Environmental and Health Studies*, **43**, 275–280.

Shackleton, C.H.L., Phillips, A., Chang, T. and Li, Y. (1997) Confirming testosterone administration by isotope ratio mass spectrometric analysis of urinary androstanediols. *Steroids*, **62**, 379–387.

Sharma, S.P. and Lahiri, S.C. (2005) Characterization and identification of explosives and explosive residues using GC-MS, an FTIR microscope, and HPTLC. *Journal of Energetic Materials*, **23**, 239–264.

Sharma, S.P., Purkait, B.C. and Lahiri, S.C. (2005) Qualitative and quantitative analysis of seized street drug samples and identification of source. *Forensic Science International*, **152**, 235–240.

Sharp, Z.D., Atudorei, V., Panarello, H.O., Fernandez, J. and Douthitt, C. (2003) Hydrogen isotope systematics of hair: archeological and forensic applications. *Journal of Archaeological Science*, **30**, 1709–1716.

Shibuya, E.K., Sarkis, J.E.S., Negrini-Neto, O., Moreira, M.Z. and Victoria, R.L. (2006) Sourcing Brazilian marijuana by applying IRMS analysis to seized samples. *Forensic Science International*, **160**, 35–43.

Shibuya, E.K., Sarkis, J.E.S., Negrini-Neto, O. and Ometto, J.P.H.B. (2007) Multivariate classification based on chemical and stable isotopic profiles in sourcing the origin of marijuana samples seized in Brazil. *Journal of the Brazilian Chemical Society*, **18**, 205–214.

Shin, J.Y., O'Connell, T., Black, S. and Hedges, R. (2004) Differentiating bone osteonal turnover rates by density fractionation; validation using the bomb C-14 atmospheric pulse. *Radiocarbon*, **46**, 853–861.

Simmons, E.D., Pritzker, K.P.H. and Grynpas, M.D. (1991) Age-related changes in the human femoral cortex. *Journal of Orthopaedic Research*, **9**, 155–167.

Sponheimer, M., Robinson, T., Ayliffe, L., Roeder, B., Hammer, J., Passey, B., West, A., Cerling, T., Dearing, D. and Ehleringer, J. (2003) Nitrogen isotopes in mammalian herbivores: hair delta N-15 values from a controlled feeding study. *International Journal of Osteoarchaeology*, **13**, 80–87.

Stephan, E. (2000) Oxygen isotope analysis of animal bone phosphate: method refinement, influence of consolidants, and reconstruction of palaeotemperatures for Holocene sites. *Journal of Archaeological Science*, **27**, 523–535.

Sturchio, N.C., Bohlke, J.K., Beloso, A.D., Streger, S.H., Heraty, L.J. and Hatzinger, P.B. (2007) Oxygen and chlorine isotopic fractionation during perchlorate biodegradation: laboratory results and implications for forensics and natural attenuation studies. *Environmental Science and Technology*, **41**, 2796–2802.

Taylor, E., Carter, J.F., Hill, J.C., Morton, C., Nic Daéid, N. and Sleeman, R. (2008) Stable isotope ratio mass spectrometry and physical comparison for the forensic examination of grip-seal plastic bags. *Forensic Science International*, **177**, 214–220.

Teitelbaum, S.L. (2000) Bone resorption by osteoclasts. *Science*, **289**, 1504–1508.

van der Peijl, G.J.Q., Dobney, A.M. and Wiarda, W. (2004) IRMS and ICPMS studies on packaging tapes, in Proceedings of the 56th Annual Meeting of the American Academy of Forensic Sciences, Dallas, TX, pp. 106–107.

van der Veer, G., Voerkelius, S., Lorentz, G., Heiss, G. and Hoogewerff, J.A. (2009) Spatial interpolation of the deuterium and oxygen-18 composition of global precipitation using temperature as ancillary variable. *Journal of Geochemical Exploration*, **101**, 175–184.

van Deursen, M.M., Lock, E. and Poortman-van der Meer, A.J. (2006) Organic impurity profiling of 3,4-methylenedioxymethamphetamine (MDMA) tablets seized in the Netherlands. *Science & Justice*, **46**, 135–152.

van Es, A., de Koeijer, J. and van der Peijl, G. (2009) Discrimination of document paper by XRF, LA-ICP-MS and IRMS using multivariate statistical techniques. *Science & Justice*, **49**, 120–126.

van Es, A.J.J., Montero, S., Wiarda, W., de Joode, P. and van der Peijl, G.J.Q. (2006) Casework investigations for tapes, polymers, ink and paper using IRMS and (LA-)ICPMS studies, in Proceedings of the 58th Annual Meeting of the American Academy of Forensic Sciences, Seattle, pp. 60–61.

van Es, A.J.J., de Koeijer, J.A. and van der Peijl, G.J.Q. (2007a) Discrimination of document paper by laser ablation inductively coupled mass spectrometry (LA-ICP-MS), in Proceedings of the 59th Annual Meeting of the American Academy of Forensic Sciences, San Antonio, TX, p. 406.

van Es, A.J.J., de Koeijer, J.A. and van der Peijl, G.J.Q. (2007b) Discrimination of document paper by inorganic analysis and multivariate statistical techniques, in Proceedings of the 59th Annual Meeting of the American Academy of Forensic Sciences, San Antonio, TX, pp. 406–407.

Wada, E. and Hattori, A. (1990) *Nitrogen in the Sea: Forms, Abundance and Rate Processes*; CRC Press, Boca Raton, FL.

Werner, R.A. and Brand, W.A. (2001) Referencing strategies and techniques in stable isotope ratio analysis. *Rapid Communications in Mass Spectrometry*, **15**, 501–519.

West, J.B., Hurley, J.M. and Ehleringer, J.R. (2009) Stable isotope ratios of marijuana. I. Carbon and nitrogen stable isotopes describe growth conditions. *Journal of Forensic Sciences*, **54**, 84–89.

Weyermann, C., Marquis, R., Delaporte, C., Esseiva, P., Lock, E., Aalberg, L., Bozenko, J.S., Dieckmann, S., Dujourdy, L. and Zrcek, F. (2008) Drug intelligence based on MDMA tablets data – I. Organic impurities profiling. *Forensic Science International*, **177**, 11–16.

Widory, D. (2006) Combustibles, fuels and their combustion products: a view through carbon isotopes. *Combustion Theory and Modelling*, **10**, 831–841.

Widory, D. (2009) Sourcing explosives: a multi-isotope approach. *Science & Justice*, **49**, 62–72.

Wild, E.M., Arlamovsky, K.A., Golser, R., Kutschera, W., Priller, A., Puchegger, S., Rom, W., Steier, P. and Vycudilik, W. (2000) ^{14}C dating with the bomb peak: an application to forensic medicine. *Nuclear Instruments and Methods in Physics Research B: Beam Interactions with Materials and Atoms*, **172**, 944–950.

Wilkinson, C. (2007) Facial anthropology and reconstruction, in *Forensic Human Identification – An Introduction* (eds T.J.T. Thomson and S.M. Black), CRC Press, Boca Raton, FL, pp. 231–270.

Wilson, F.G., Forster, A. and Roberts, E. (1950) US Patent 2 525 252.

Wilson, A.S. and Gilbert, M.T.P. (2007) Hair and nail, in *Forensic Human Identification – An Introduction* (eds T.J.T. Thomson and S.M. Black), CRC Press, Boca Raton, FL, pp. 147–174.

Wopenka, B. and Pasteris, J.D. (2005) A mineralogical perspective on the apatite in bone. *Materials Science and Engineering C*, **25**, 131–143.

Government Agencies and Institutes with Dedicated Stable Isotope Laboratories

- Alcohol/Tobacco Tax & Trade Bureau; Beltsville, USA.

- Australian Federal Police Forensic Service; Canberra, Australia.

- Bundeskriminalamt (BKA); Wiesbaden, Germany.

- Bureau of Alcohol, Tobacco, Firearms and Explosives (BATF); Ammendale, USA.

- Canada Border Services Agency; Ottawa, Canada.

- Counter Terrorism Forensic Science Research Unit (CTFSRU) of the Federal Bureau of Investigation (FBI); Quantico, USA.

- Customs and Border Protection (CBP); Los Angeles, USA.

- Customs and Border Protection (CBP); San Francisco, USA.

- Customs and Border Protection (CBP); Savannah, USA.

- Customs and Border Protection (CBP); Springfield, USA.

- Department of Homeland Security, Oak Ridge National Laboratories; Oak Ridge, USA.

- Drug Enforcement Agency (DEA); Dulles, Washington, USA.

- Environmental Protection Agency (EPA); Athens, USA.

- Forensic Explosives Laboratory (FEL); Fort Halstead, UK.

- Forensic Institute; Lausanne, Switzerland.

- Netherlands Forensic Institute (NFI); The Hague, Netherlands.

- Ustredni Celne-Technika Customs Laboratory; Prague, Czech Republic.

- VPOP Customs Laboratory; Budapest, Hungary.

Acknowledgements

Behind every book stands an author, but behind every author stands a multitude of people without whom the book would have never seen the light of day. Since it is hard to single out any one person without being unjust to the others, I have decided to adopt a chronological and alphabetical approach for giving thanks. I am indebted to my parents who encouraged and supported my going to university every step of the way; to Nico Giulini, best friend and best man; to my teacher and subsequent PhD supervisor, the late Hermann Schildknecht, whose pioneering work in semiochemistry inspired me to pursue an academic scientific career; to Ben V. Burger who taught me everything there is to know about good analytical chemistry, in general, and gas chromatography, in particular, and whose infectious enthusiasm for ecological chemistry proved irresistible; to Dietz Rating whose pioneering work in non-invasive breath tests introduced me to the world of stable isotopes; to Willi Brand, Steven Brookes, Stewart Craig, Yvo Ghoos, Karleugen Habfast, David Halliday, Andreas Hilkert, Tom Preston, Mike J. Rennie and Charlie Scrimgeour who all played their part in my getting involved with instrument and method development for stable isotope analytical techniques, thus helping me to understand stable isotopes and stable isotope instrumentation from the bottom up; to Sue Black, Sean Doyle and Robert Kalin for the parts they played in my ever-increasing engagement with developing stable isotope techniques into a new forensic investigative tool; to Dave Barclay, Mark Harrison and Roy McComb who were the first law enforcement officers in the United Kingdom to put their trust in stable isotope forensics as well as me; to Sarah Benson, Jim Carter, Jim Ehleringer, Martin Grimes, Gerard Hamill, Jurian Hoogewerff, Jennifer McKinley, Niamh Nic Daéid, Alastair Ruffell, Gerard van der Peijl, Andrew Wilson and Patricia Wiltshire for their collegiate friendship, generosity and support; to my PhD students Nicola Farmer, Isla Fraser, Claire Lock and Lisa Reidy; to Nicola McGirr and Fiona Woods at John Wiley & Sons for all their support and help, especially when a downturn in my health and the passing away of my mother threatened the entire book project, and of course to all the people at John Wiley & Sons who worked tirelessly behind the scenes; and last, but not least, to my wife Helen who was and still is always there to comfort, support and help, and who, as an academic professional in her own right, is also an integral and essential part of the UK stable isotope forensics team.

Stable Isotope Forensics: An Introduction to the Forensic Application of Stable Isotope Analysis Wolfram Meier-Augenstein
© 2010 John Wiley & Sons, Ltd

Recommended Reading

When you start reading this book it is not unlikely that stable isotopes and associated terms will be all Greek to you, and what you read or learn here will still be but a small facet of a larger spectrum. Fortunately, there are a number of excellent books and reviews available to which I would like to draw your attention.

Books

Blackledge, R.D. (2007) *Forensic Analysis on the Cutting Edge*, John Wiley & Sons, Inc., New York.
Brereton, R.G. (2006) *Chemometrics – Data Analysis for the Laboratory and Chemical Plant*, John Wiley & Sons, Ltd, Chichester, UK.
Criss, R.E. (1999) *Principles of Stable Isotope Distribution*, Oxford University Press, New York.
Dawson, T.E. and Siegwolf, R.T.W. (2007) *Stable Isotopes as Indicators of Ecological Change*, Terrestrial Ecology Series 1, Elsevier, London.
de Groot, P.A. (2004) *Handbook of Stable Isotope Analytical Techniques*, vol. 1, Elsevier, Amsterdam.
Ehleringer, J.R., Hall, A.J. and Farquhar, G.D. (1993) *Stable Isotopes and Plant Carbon–Water Relations*, Physiological Ecology Series, Elsevier, London.
Faure, G. and Mensing, T.M. (2005) *Isotopes: Principles and Applications*, John Wiley & Sons, Inc., New York.
Fry, B. (2006) *Stable Isotope Ecology*, Springer, New York.
Griffiths, H. (1997) *Stable Isotopes: The Integration of Biological, Ecological and Geological Processes*, Garland Science, New York.
Hobson, K.A. and Wassenaar, L.I. (2008) *Tracking Animal Migration with Stable Isotopes*, Terrestrial Ecology Series 2, Elsevier, London.
Lajtha, K. and Michener, R. (2007) *Stable Isotopes in Ecology and Environmental Science*, 2nd edn, Methods in Ecology, Wiley-Blackwell, Oxford.
Lucy, D. (2005) *Introduction to Statistics for Forensic Scientists*, John Wiley & Sons, Ltd, Chichester.
Mook, W.G. (2006) *Introduction to Isotope Hydrology*, Taylor & Francis/Balkema, London.
Nic Daeid, N. (2009) *Fifty Years of Forensic Science*, John Wiley & Sons, Ltd, Chichester.
Platzner, I.T. (1997) *Modern Isotope Ratio Mass Spectrometry*, John Wiley & Sons, Ltd, Chichester.
Pye, K. (2007) *Geological and Soil Evidence*, CRC Press, Boca Raton, FL.

Stable Isotope Forensics: An Introduction to the Forensic Application of Stable Isotope Analysis Wolfram Meier-Augenstein
© 2010 John Wiley & Sons, Ltd

Pye, K. and Croft, D.J. (2004) *Forensic Geoscience: Principles, Techniques and Applications*, Special Publications 232, Geological Society, London.

Ritz, K., Dawson, L. and Miller, D. (2009) *Criminal and Environmental Soil Forensics*, Springer Science + Business Media, Dordrecht.

Ruffell, A. and McKinley, J. (2008) *Geoforensics*, John Wiley & Sons, Ltd, Chichester.

Sharp, Z.D. (2007) *Principles of Stable Isotope Geochemistry*, Pearson Prentice Hall, Upper Saddle River, NJ.

Taroni, F., Aitken, C., Garbolino, P. and Biedermann, A. (2006) *Bayesian Networks and Probabilistic Inference in Forensic Science*, John Wiley & Sons, Ltd, Chichester.

Thomson, T.J.T. and Black, S.M. (2007) *Forensic Human Identification*, CRC Press, Boca Raton, FL.

West, J.B., Bowen, G.J., Dawson, T.E. and Tu, K.P. (2009) *Isoscapes: Understanding Movement, Pattern, and Process on Earth through Isotope Mapping*, Springer Science + Business Media, Dordrecht.

Yinon, J. (2004) *Advances in Forensic Applications of Mass Spectrometry*, CRC Press, Boca Raton, FL.

Reviews

Adams, M.A. and Grierson, P.F. (2001) Stable isotopes at natural abundance in terrestrial plant ecology and ecophysiology: an update. *Plant Biology*, **3**, 299–310.

Asche, S., Michaud, A.L. and Brenna, J.T. (2003) Sourcing organic compounds based on natural isotopic variations measured by high precision isotope ratio mass spectrometry. *Current Organic Chemistry*, **7**, 1527–1543.

Barbour, M.M. (2007) Stable oxygen isotope composition of plant tissue: a review. *Functional Plant Biology*, **34**, 83–94.

Benson, S., Lennard, C., Maynard, P. and Roux, C. (2006) Forensic applications of isotope ratio mass spectrometry – a review. *Forensic Science International*, **157**, 1–22.

Blessing, M., Jochmann, M.A. and Schmidt, T.C. (2008) Pitfalls in compound-specific isotope analysis of environmental samples. *Analytical and Bioanalytical Chemistry*, **390**, 591–603.

Boschker, H.T.S. and Middelburg, J.J. (2002) Stable isotopes and biomarkers in microbial ecology. *FEMS Microbiology Ecology*, **40**, 85–95.

Brand, W.A. (1996) High precision isotope ratio monitoring techniques in mass spectrometry. *Journal of Mass Spectrometry*, **31**, 225–235.

Brenna, J.T. (1994) High-precision gas isotope ratio mass-spectrometry – recent advances in instrumentation and biomedical applications. *Accounts of Chemical Research*, **27**, 340–346.

Brenna, J.T., Corso, T.N., Tobias, H.J. and Caimi, R.J. (1997) High-precision continuous-flow isotope ratio mass spectrometry. *Mass Spectrometry Reviews*, **16**, 227–258.

Brettell, T.A., Butler, J.M. and Almirall, J.R. (2007) Forensic science. *Analytical Chemistry*, **79**, 4365–4384.

Calderone, G., Guillou, C. and Naulet, N. (2003) [Official methods based on stable isotope techniques for analysis of food. Ten years of European experience]. *L'Actualite Chimique*, (8/9), 22–24.

Collins, M.J., Nielsen-Marsh, C.M., Hiller, J., Smith, C.I., Roberts, J.P., Prigodich, R.V., Weiss, T.J., Csapo, J., Millard, A.R. and Turner-Walker, G. (2002) The survival of organic matter in bone: a review. *Archaeometry*, **44**, 383–394.

Dawson, T.E., Mambelli, S., Plamboeck, A.H., Templer, P.H. and Tu, K.P. (2002) Stable isotopes in plant ecology. *Annual Review of Ecology and Systematics*, **33**, 507–559.

Ehleringer, J.R., Rundel, P.W. and Nagy, K.A. (1986) Stable isotopes in physiological ecology and food web research. *Trends in Ecology and Evolution*, **1**, 42–45.

Evershed, R.P., Dudd, S.N., Charters, S., Mottram, H., Stott, A.W., Raven, A., van Bergen, P.F. and Bland, H.A. (1999) Lipids as carriers of anthropogenic signals from prehistory. *Philosophical Transactions of the Royal Society of London Series B Biological Sciences*, **354**, 19–31.

Evershed, R.P., Dudd, S.N., Copley, M.S., Berstan, R., Stott, A.W., Mottram, H., Buckley, S.A. and Crossman, Z. (2002) Chemistry of archaeological animal fats. *Accounts of Chemical Research*, **35**, 660–668.

Farquhar, G.D., Ehleringer, J.R. and Hubick, K.T. (1989) Carbon isotope discrimination and photosynthesis. *Annual Review of Plant Physiology and Plant Molecular Biology*, **40**, 503–537.

Fox, T. and Bearhop, S. (2008) The use of stable-isotope ratios in ornithology. *British Birds*, **101**, 112–130.

Glaser, B. (2005) Compound-specific stable-isotope (delta C-13) analysis in soil science, *Journal of Plant Nutrition and Soil Science/Zeitschrift fur Pflanzenernahrung und Bodenkunde*, **168**, 633–648.

Handley, L.L. and Raven, J.A. (1992) The use of natural abundance of nitrogen isotopes in plant physiology and ecology. *Plant Cell And Environment*, **15**, 965–985.

Handley, L.L. and Scrimgeour, C. (1997) Terrestrial plant ecology and ^{15}N natural abundance: the present limits to interpretation for uncultivated systems with original data from a Scottish old field. *Advances in Ecological Research*, **27**, 134–213.

Hobbie, E.A. and Werner, R.A. (2004) Intramolecular, compound-specific and bulk carbon isotope patterns in C-3 and C-4 plants: a review and synthesis. *New Phytologist*, **161**, 371–385.

Hobson, K.A. and Wassenaar, L.I. (1999) Stable isotope ecology: an introduction. *Oecologia*, **120**, 312–313.

Kelly, S., Heaton, K. and Hoogewerff, J. (2005) Tracing the geographical origin of food: the application of multi-element and multi-isotope analysis. *Trends in Food Science and Technology*, **16**, 555–567.

Kennedy, B.V. and Krouse, H.R. (1990) Isotope fractionation by plants and animals – implications for nutrition research. *Canadian Journal of Physiology and Pharmacology*, **68**, 960–972.

Koch, P.L. (1998) Isotopic reconstruction of past continental environments. *Annual Review of Earth and Planetary Sciences*, **26**, 573–613.

Lee-Thorp, J.A. (2008) On isotopes and old bones. *Archaeometry*, **50**, 925–950.

Lichtfouse, E. (2000) Compound-specific isotope analysis. Application to archaeology, biomedical sciences, biosynthesis, environment, extraterrestrial chemistry, food science, forensic science, humic substances, microbiology, organic geochemistry, soil science and sport. *Rapid Communications in Mass Spectrometry*, **14**, 1337–1344.

Macko, S.A., Engel, M.H., Andrusevich, V., Lubec, G., O'Connell, T.C. and Hedges, R.E. M. (1999) Documenting the diet in ancient human populations through stable isotope analysis of hair. *Philosophical Transactions of the Royal Society of London Series B Biological Sciences*, **354**, 65–75.

Macko, S.A., Engel, M.H. and Qian, Y.R. (1994) Early diagenesis and organic matter preservation – a molecular stable carbon-isotope perspective. *Chemical Geology*, **114**, 365–379.

Meier-Augenstein, W. (1999a) Applied gas chromatography coupled to isotope ratio mass spectrometry. *Journal of Chromatography A*, **842**, 351–371.

Meier-Augenstein, W. (1999b) Use of gas chromatography-combustion-isotope ratio mass spectrometry in nutrition and metabolic research. *Current Opinion in Clinical Nutrition and Metabolic Care*, **2**, 465–470.

Meier-Augenstein, W. (2002) Stable isotope analysis of fatty acids by gas chromatography-isotope ratio mass spectrometry. *Analytica Chimica Acta*, **465**, 63–79.

Meier-Augenstein, W. (2007) Stable isotope fingerprinting – chemical element 'DNA'?, in *Forensic Human Identification – An Introduction* (eds T.J.T. Thomson and S.M. Black), CRC Press, Boca Raton, FL, pp. 29–53.

Meier-Augenstein, W. and Liu, R.H. (2004) Forensic applications of isotope ratio mass spectrometry, in *Advances in Forensic Applications of Mass Spectrometry* (ed. J. Yinon), CRC Press, Boca Raton, FL, pp. 149–180.

Pearson, O.M. and Lieberman, D.E. (2004) The aging of Wolff's law: ontogeny and responses to mechanical loading cortical bone. *American Journal of Physical Anthropology*, **47**, 63–99.

Peterson, B.J. and Fry, B. (1987) Stable isotopes in ecosystem studies. *Annual Review of Ecology and Systematics*, **18**, 293–320.

Philp, R.P. (2007) The emergence of stable isotopes in environmental and forensic geochemistry studies: a review. *Environmental Chemistry Letters*, **5**, 57–66.

Preston, T. and Barrie, A. (1991) Recent progress in continuous-flow isotope ratio mass-spectrometry. *American Laboratory*, **23**, H32.

Raven, J.A. and Handley, L.L. (1987) Transport processes and water relations. *New Phytologist*, **106**, 217–233.

Richards, M.P., Fuller, B.T., Sponheimer, M., Robinson, T. and Ayliffe, L. (2003) Sulphur isotopes in palaeodietary studies: a review and results from a controlled feeding experiment. *International Journal of Osteoarchaeology*, **13**, 37–45.

Ruffell, A. and McKinley, J. (2005) Forensic geoscience: applications of geology, geomorphology and geophysics to criminal investigations. *Earth-Science Reviews*, **69**, 235–247.

Schmidt, H.L., Werner, R.A. and Rossmann, A. (2001) O-18 pattern and biosynthesis of natural plant products. *Phytochemistry*, **58**, 9–32.

Schmidt, H.L., Rossmann, A., Stockigt, D. and Christoph, N. (2005) [Stable isotope analysis. Origin and authenticity of foodstuffs]. *Chemie in Unserer Zeit*, **39**, 90–99.

Schmidt, T.C., Zwank, L., Elsner, M., Berg, M., Meckenstock, R.U. and Haderlein, S.B. (2004) Compound-specific stable isotope analysis of organic contaminants in natural environments: a critical review of the state of the art, prospects, and future challenges. *Analytical and Bioanalytical Chemistry*, **378**, 283–300.

Slater, C., Preston, T. and Weaver, L.T. (2001) Stable isotopes and the international system of units. *Rapid Communications in Mass Spectrometry*, **15**, 1270–1273.

Slater, G.F. (2003) Stable isotope forensics – when isotopes work. *Environmental Forensics*, **4**, 13–23.

Taylor, A., Branch, S., Day, M.P., Patriarca, M. and White, M. (2007) Clinical and biological materials, foods and beverages. *Journal of Analytical Atomic Spectrometry*, **22**, 415–456.

Wada, E., Mizutani, H. and Minagawa, M. (1991) The use of stable isotopes for food web analysis. *Critical Reviews In Food Science and Nutrition*, **30**, 361–371.

Werner, R.A. and Brand, W.A. (2001) Referencing strategies and techniques in stable isotope ratio analysis. *Rapid Communications in Mass Spectrometry*, **15**, 501–519.

Woodhead, J.D. (2006) Isotope ratio determination in the earth and environmental sciences: developments and applications in 2004/2005. *Geostandards and Geoanalytical Research*, **30**, 187–196.

Author's Biography

Dr Meier-Augenstein is a Senior Lecturer (Associate Professor) in Stable Isotope Forensics at the Centre for Anatomy & Human Identification at the University of Dundee. He is also the Principal Scientist for Stable Isotopes at the Scottish Crop Research Institute (Invergowrie, Scotland, UK). Dr Meier-Augenstein is a registered expert advisor with the National Policing Improvement Agency (NPIA, UK) and holds a Diplom-Chemiker degree (MChem) and a Doctorate in Bio-organic Chemistry (PhD) both awarded by the University of Heidelberg (Germany). After time spent in 1990/91 as Feodor-von-Lynen Fellow at the University of Stellenbosch (R.S.A.) he initially worked at the University Children's Hospital in Heidelberg before moving to Dundee in 1994. In 2003 he joined The Queen's University Belfast but returned to Dundee in 2008. He is a Fellow of the Royal Society of Chemistry (FRSC) and has over 18 years of experience in the field of stable isotope technology having made contributions to instrument design and method development as well as to several textbooks on the subject of stable isotope analytical techniques including their use in forensic science. Dr Meier-Augenstein has assisted police forces and coroners' offices in the UK and abroad in murder enquiries involving unidentified and sometimes mutilated victims of serious crime with provision of Stable Isotope Profiles or 'Signatures' to aid victim identification. He is an executive director of the Forensic Isotope Ratio Mass Spectrometry (FIRMS) Network, an organisation with international membership involved in quality control and quality assurance of stable isotope data generated for and used in a forensic context. He also serves as a council member of the British Association for Human Identification (BAHID) as well as member of the editorial boards for the scientific journals *Science & Justice* and *Bioanalysis*.

Dr Wolfram Meier-Augenstein, CChem, FRSC
Centre for Anatomy & Human Identification, University of Dundee, UK
Scottish Crop Research Institute, Invergowrie, Dundee, UK

Stable Isotope Forensics: An Introduction to the Forensic Application of Stable Isotope Analysis   Wolfram Meier-Augenstein
© 2010 John Wiley & Sons, Ltd

Index

Note: Figures and Table are indicated by *italic page numbers*, Boxes by **emboldened numbers**; IRMS = isotope ratio mass spectrometry.

accreditation 101
'Adam' case 190
AIR (atmospheric air) standard 6
   $^{15}N/^{14}N$ isotope ratio 6
n-alkanes, IRMS secondary standards *132*
amino acids, racemization of 33
ammonium nitrate
   bulk isotope analysis 114, 116, 171–2, 182–3
   hygroscopicity 108, *109*, 183
   total δ^2H versus true δ^2H values 106, 108
amphetamines 154–7
   $\delta^{15}N/\delta^{13}C$ bivariate plots *157*
ANFO (ammonium nitrate + fuel oil) 171
animals, hydrogen sources 19
anthrax attacks 214
anthrogenic activity, $\delta^{15}N$ in soil affected by 28
anti-doping control 47–8, 224–35
arson attacks
   data on 188
   matchstick evidence 184–9
atmospheric gases, monitoring of 45
atomic number 3
atomic percent 5
   approximation to δ notation 6
atomic structure 3–4
Australian Federal Police, Forensic Operations Laboratory, explosives studies *109*, 169, 170, 172, 182–3

Barclay, Dave 147
Bayesian analysis 94–9

BEVABS (European Office for Wine, Alcohol and Spirit Drinks) 38, 49
beverages, authenticity and provenance 39–41
bio-apatite 193
   acid digest of carbonate from 238–9
   diagenetic changes 24–5
   silver phosphate prepared from 196, 209, 236–8
   and source water 20–2, 192
biochemical processes, isotopic fractionation in 11–12, **15**, 20, 186
bioterrorism 214
birefringence pattern analysis, plastic bags 216, *217*
bivariate plots
   advantages 222
   drugs *44*, *157*
   explosives *171*, 172–3, *174*
   matchsticks *188*
   paints *97*
   parcel tapes 219, *220*
   plastic bags 217
   wines *41*
BKA (Bundeskriminalamt, Weisbaden, Germany) 154, 155
bone bio-apatite, acid digest of carbonate from 238–9
bone carbonate, diagenetic changes 24
bone collagen, diagenetic changes 32
bone diagenesis 24–5

bone phosphate (bio-apatite)
    and ^{18}O in source water 19–23, 196
    sample preparation for ^{18}O isotope analysis 196, 209, 236–9
bone remodelling 23, 193, 210
Brand, Willi 81, 103
Brazil, cannabis/marijuana 149–50
Brenna, Tom 103
BSIA (bulk stable isotope analysis) 77
    compared with CSIA 81
    drugs 161, *162*
    explosives 114, 116, 170–8
    foodstuffs 42
    generic considerations 104–5
    isobaric interference 104–5
    matchsticks 185–8
    MDMA synthesis products 161, *162*
    particular considerations for ^{2}H-BSIA 105–16
    see also ^{13}C-BSIA; EA-IRMS; ^{2}H-BSIA; TC/EA-IRMS
bulk isotope analysis *see* BSIA

C$_3$-plants 25
    and δ^{13}C values 40, 203
    see also grapes; sugar beet
C$_4$-plants 25
    and δ^{13}C values 40, 203, 208
    see also corn (maize); sugar cane
^{13}C-BSIA 77
    matchsticks 185–6
^{13}C-CSIA 79
    air pollutants 45
    anti-doping control 47–8, 224
    drugs 152, 154
^{14}C bomb-pulse analysis 194, 202
^{14}C radioisotope analysis 194
calibration, in CF-IRMS 85
calibration materials 5–6, 74, 85
Calvin–Benson cycle 25
cannabis 149–50, *153*
carbon
    fixation processes 25, 26
    isotopes 4, 5, *6*
    sources 25, 26, 192–3
    stable isotopic distribution 25–7
carbon-12 (^{12}C) 4
    natural abundance *6*
carbon-13 (^{13}C) 4
    abundance analysis 69, 71
    calibration and reference material(s) 5, *7*, 85, *89*

isotope analysis 46, 71, 77, 88, 104, 149–50, 185–6, 238–9
    natural abundance *6*
    organic reference materials *7*, *89*
    in single-seed vegetable oils *39*
    see also δ^{13}C
carbonated beverages 39–40
cellulose nitrate, ^{2}H-BSIA 116
CF-IRMS (continuous flow isotope ratio mass spectrometry) 33–4, 74–6
    applications 45, 216, 218, 219
    basis 69
    calibration in 85
    contrasted with dual-inlet IRMS 118–19
    key dates in development *76*
    laboratory set-up 123–35
    quality control in 85–6
CF-IRMS laboratory
    equipment and tools *130*, *131*
    forensic laboratory considerations 129–30
    gas supply 126–9
    infrastructure and services 130–1
    initial considerations 123
    location 124
    power supply 125–6
    pre-installation requirements 124
    setting up 123–35
    space allocation 125
    temperature control 124, 125
chain-of-custody documentation 158, 211
    lacking in anti-doping laboratory 226
CHAMP profiling method 163–4
chemical processes, isotopic fractionation in 11–12, **15**
chemically identical compounds, isotopic differences 8, *9*
chemometrics 91–4
chlorates 170
chromatographic isotope effect, in GC/C-IRMS 118–20
chromatography *see* gas chromatography; GC/C-IRMS; liquid chromatography
cling film, forensic examination of 217
cluster analysis 91, 162
    see also hierarchical cluster analysis
cocaine 152–4
collagen, stable isotopic make-up 32
compound-specific isotope analysis *see* CSIA
continuous flow isotope ratio mass spectrometry *see* CF-IRMS
Coplen, Tyler 103

copper sulfate 105–6
  anhydrous 106
corn syrup 193
counterfeit pharmaceuticals 42–3, *44*
counterfeiting and piracy, OECD report on 36–7, 43
cranio-facial reconstruction 190, 202, *207*
CSIA (compound-specific isotope analysis) 78–84
  applications 41, 42, 45, 47–8
  compared with BSIA 81
  and compound identification 79–81
  generic considerations for 116–20
  polar non-volatile organic compounds 83–4
  see also ^{13}C-CSIA; GC/C-IRMS
culture media, stable isotope signatures 215
Curran, Jim xxii
cyclotrimethylenetrinitramine *see* RDX

δ notation 5, 73
  approximation to atomic percent 6
  calculation 6
  listed for various elements (reference materials) *7*
δ^{13}C
  in body pools and tissues 31, *32*
  in reference materials *7*
  scale correction of measured values 88–90
  in testosterone metabolites 226–8
  trophic level shift 30, 31
δ^{13}C/δ^{18}O bivariate plots
  drugs *44*
  paints *97*
δ^2H
  calibration material 5–6
  non-exchangeable hydrogen 108, 110
  reference materials *7*, 87, 88
  relationship with δ^{18}O in meteoric precipitation 18
  scale correction of measured values 87–8
  total versus true values 106, 108–13
  trophic level shift 31–2
δ^2H/δ^{13}C bivariate plots
  matchsticks *188*
  paints *97*
  plastic bags and tapes 217, 219, *220*
  wines *41*
δ^2H/δ^{18}O bivariate plots, paints *97*
δ^{15}N
  factors affecting values 27–30, 199, 203
  in reference materials *7*
  trophic level shift 27–30
δ^{15}N/δ^{13}C bivariate plots
  drugs *157*
  explosives *174*
δ^{15}N/δ^{18}O bivariate plots, ammonium nitrate *171*
δ^{18}O
  in bone/tooth phosphate
    relationship with δ^{18}O in carbonate 24, 239
    relationship with δ^{18}O in source water 22, *23*, 196, 209, 239
  in cooked food 22
  geographical distribution in precipitation *191*, *213*
  in reference materials *7*
  relationship with δ^2H in precipitation 18
δ^{34}S, differences between marine, freshwater and terrestrial sources 34–5

databases 101, 223
  drugs 165, 167
Daubert's principles 147
de Groot, Pier 103
dental enamel *see* tooth enamel
deuterium (^{2}H or D) 4
diagenetic changes
  bio-apatite in bone and teeth 24–5
  structural proteins 32–3
diet
  isotope signatures and 22, 29–32, 34–5, 191–2, 198, 199–200, 203, 211
  marine-based vs freshwater/terrestrially sourced 30, 34
DNA cross-matching 190, 211
Doyle, Sean 169
DR (dynamic range) of δ values 168
drugs
  database 167–8
  differentiation of 149–68
dry samples, analysis of 105–6
dual-inlet IRMS 74
  applications 45
  contrasted with CF-IRMS 118–19
Dublin, δ^2H values for tap water 208
Dublin Royal Canal, murder victim's body 206, 208–11
dynamic scale compression 87

EA-IRMS (elemental analysis isotope ratio mass spectrometry) 75, *76*
  compared with GC-MS *68*
  organic secondary standards *134*

Earth's atmosphere
  components 19, 27
  standard based on 6, 27
Ecstasy
  dataset 167–8
  statistical analysis 92, *93*, 154–5, 168
  see also MDMA
Ehleringer, Jim 20, 150
elephants' eating habits 46
'Emde' synthesis (of methamphetamine from ephedrine) 165, *166*
enrichment factors 12–13, 180
  hexamine/RDX *181*
environmental forensics 43, 45–6
ephedrine
  isotopic signatures 164, *165*, 166
  methamphetamine synthesized from 165–6
  sources 165, *165*
everyday use of stable isotope forensics 26–48
exchangeable hydrogen 108, 110, 183
  equilibration of samples 110–11
  molar exchange fraction 111, 241
expert witnesses, judges' discretion on probing 146–7
explosives 169–83
  bulk isotope analysis 114, 116, 170–8
  laboratories studying 169, 170
  see also ammonium nitrate; PETN; RDX; Semtex

Faraday cup collectors
  advantages 72
  isotope analysis of CO_2 69, *70*, 71, 72
Farmer, Nicola 184, 188, 189, 221
fatty acid esters, IRMS secondary standards *132–3*
FEL (Forensic Explosive Laboratory) 169
femur (bone), turnover rate 23, 193, 210
fertilizers, $\delta^{15}N$ values 28
fire scenes, matching matchsticks from 188–9
FIRMS (Forensic Isotope Ratio Mass Spectrometry) network 101, 169
  inter-laboratory comparisons organized by 100, 103, 108, 112
  members 155, 218
  project on applications of IRMS 100
fish
  determination of source 42
  statistical analysis 93–4
fitness-for-purpose of methods/procedures 145–6, 223

flavours and fragrances, authenticity and provenance 42
folic acid, $\delta^{13}C/\delta^{15}N$ bivariate plots *44*
food chain, trophic level shift in 29–32
food forensics 37–42
forensic IRMS laboratory
  sample reception/storage/preparation 129–30
  security considerations 129
forensic science
  fit-for-purpose methods/procedures used 145–6, 223
  meaning of term 147, **148**
fractionation factors 12, 13, 180
  hexamine/RDX *181*
Fraser, Isla *26*, 208
freshwater animals, $\delta^{13}C$ and $\delta^{15}N$ values 26, 30
fruit juices, authenticity and provenance 40–1
Frye principle 146
FSA (Food Standards Agency – UK) 38, 49
FTIR (Fourier transform infrared) spectroscopy, applications 180, 219, 221
funding of research 100, 101

gas chromatography, peak identification in 230–5
gas supply for IRMS laboratory 126–9
GC/C-IRMS (gas chromatography with combustion/isotope ratio mass spectrometry) 45, 79, *80*
  compared with GC-MS 68
  derivatizing agents for *134*
  interfaced with mass spectrometry 81, *82*, 230
  isotopic calibration during 116–17
  organic secondary standards *134*
  for urinary steroids 230–5
GC-MS (gas chromatography/mass spectrometry)
  combined with IRMS systems 222, 230
  compared with IRMS systems 68
  peak detection in 231–2
geographic location, isotope abundance and 17, 190–1, 198, 203, 205, 211–12
GISP (Greenland Ice Sheet Precipitation) water reference material *7*, 88
GMWL (Global Meteoric Water Line) 18
GNIP (Global Network of Isotopes in Precipitation) data 190–1, 209
grape juice 40
Gröning, Manfred 103
grip-seal plastic bags 217
groundwater contamination 45–6, 94

^{2}H-BSIA
  dry/water-free samples 105–6
  ionization quench effect 113–16
  matchsticks 187–8
  MDMA synthesis products 161, *162*
  particular considerations for 105–16
  total δ^2H vs true/non-exchangeable δ^2H values 106, 108–13
hair *see* human hair
Hatch–Slack cycle 25
heroin 150–2, *153*
hexamine (hexamethylenetetraamine)
  bulk isotope analysis 114, 173–6
  RDX synthesized from 174
hierarchical cluster analysis 91–2
  hexamine samples 174–5, *176*
  MDMA samples *93*, *163*
Hoogewerff, Jurian xxii, *213*
human hair
  ^{13}C isotope analysis *29*, 203, *204*
  ^{2}H isotope analysis 108, 110–13, 201–2, 203, *204*, *205*, 240–1
  δ^{15}N/δ^{13}C bivariate plots *26*, *31*
  ^{15}N isotope analysis *30*, 199–200, 203, *204*
  ^{18}O isotope analysis 200, *204*, *205*
  provenance of person 197–201
  ^{34}S isotope analysis *34*, 203, *204*
  sample preparation procedure 110–11, 240
human provenancing 34, 100, 108, 191, 194, 212, *213*
human remains
  identification of 190–211
  working with 211, *212*
human tissue, stable isotope abundance variation 191–4
hydrogen
  isotopes 4, 6, 16
  sources/precursor pools 18, 19, 192, 193
  stable isotopic distribution 16–19
hydrogen-1 (^{1}H) 4
  natural abundance 6, 16
hydrogen-2 (^{2}H) 4
  calibration and reference material(s) 5–6, 7, 16, 85, 87
  geographical variation 17
  isotope analysis 19, *76*, *77*, *86*, 105–16, 154, *163*, 200–1
  natural abundance 6, 16
  see also δ^2H; ^{2}H-BSIA
hydrogen exchange 108, 110, 183

hydrogen gas, isotope effects on bond strength 10
hydrogen peroxide
  dilution effects on δ^2H and δ^{18}O values 177, *178*
  stable isotope analysis 176–7
hydrological cycle 16
hydroxyapatite 193

IAEA (International Atomic Energy Authority), reference materials 6, *7*
ICP-MS (inductively coupled mass spectrometry), applications 216, 218, 219, 220, 221
IEDs (improvised explosive devices), post-blast debris 217–18
ILCs (International Laboratory Comparisons) 100, 103, 112
infrared spectroscopy, isotopologues 10–11
instrument drift 86
instrumentation 72–84
international reference materials 7
inverse chromatographic isotope effect, in GC/C-IRMS 118–19, 119–20
ionization quench effect 113–16
IRMS (isotope ratio mass spectrometry)
  basis 69–70
  compared with mass spectrometry 67–71
  instrumentation 72–84
  laboratory (CF-IRMS) set-up and running 123–35
  potential pitfalls 104–5, 113–16, 182–3
  reference materials 7
isobaric interference 104–5
isoscapes (isotopic maps) 150, 191
isotope, origin of word 3
isotope effects 10, 11–12
  in GC/C-IRMS during sample injection 117–18
isotope standards 5–6
isotopic calibration, in GC/C-IRMS 116–17
isotopic fractionation
  in acetylation of morphine 151
  causes 10, 11–14, **15**, 16, 20, 26
  causes in GC/C-IRMS 117
  of various light elements 16–17, 20, 113

Jeffreys, Alec J., Prof. Sir xxi

keratin 33, 100
kinetic isotope effects 10, 11–12

Kovats indices 232
Kreuzer-Martin, Helen 214

LA-ICP-MS (laser ablation inductive coupled mass spectrometry), analysis of document paper 216
LA-MC-ICP-MS (laser ablation multicollector inductive coupled mass spectrometry), ^{34}S data from hair 34
lavender oil 41–2
LC-IRMS (liquid chromatography with isotope ratio mass spectrometry) 83–4
lead, sources 197
lead isotope ratio (^{206}Pb/^{207}Pb) analysis 197
likelihood ratios 94
  examples of use 94–9, 173
linear retention indices 232
Lock, Claire 171, *174, 175, 176*, 178
LSVEC (lithium carbonate) reference material *7*, 88, 238

maize oil 38
marijuana 149–50, *153*
marine animals, δ^{13}C and δ^{15}N values 26, 30
marine-based diets 30, 34
mass discrimination 10, **15**
  in GC/C-IRMS sample injection 117–18
mass number 4
mass spectrometry
  compared with isotope ratio mass spectrometry 67–71
  SIM mode 67, *68*, 69, 180, 182
matchsticks 184–9
  ^{13}C-bulk isotope analysis 185–6
  ^{2}H-bulk isotope analysis 187–8
  independent (non-IRMS) analytical techniques 189
  matching fire scene samples 188–9
  ^{18}O-bulk isotope analysis 186–7
  sample preparation 185
MDMA (3,4-methylenedioxy-*N*-methylamphetamine)
  bivariate plots *159*
  impurity profiling 163–4
  isotopic analysis of seized Ecstasy tablets *93*, 154–5, *157*
  precursors 155
  synthesis and isotopic signature 157–64
meat protein-rich diet, δ^{15}N values 208
methamphetamine, synthesis and isotopic signature 164–7

microbial isotope forensics 214–15
molar exchange fraction, of exchangeable hydrogen 111, 241
morphine 150–2, *153*
MSA (Mass Spec Analytical Ltd) 155, 218
MTBE (methyl *tert*-butyl ether) 45–6
multivariate stable isotope analysis
  drugs 155, 167–8, 222
  explosives 172, 173, 183
  foodstuffs 42
  whisky 41
  wildlife 46
mutilated murder victims, identification of 190, 206, 208–11

'Nagai' synthesis (of methamphetamine from ephedrine) 165, *166*
naproxen, δ^{13}C/δ^{18}O bivariate plots *44*
natural and semisynthetic drugs 149–54
Newfoundland (Canada), human remains 201–6, *207*
NFI (Netherlands Forensic Institute) 216, 218
NicDaeid, Niamh 157, 164, 189
nitrogen
  isotopes 6
  sources 30, 192
  stable isotopic distribution 27–33
nitrogen-14 (^{14}N), natural abundance 6
nitrogen-15 (^{15}N)
  calibration and reference material(s) 6, *7*
  dietary influences on abundance in tissues 30, 193
  isotope analysis 27, 30, 46, 77, 149–50
  natural abundance range 6, 27, 171
  see also δ^{15}N
nitrogen cycle, isotopic fractionation in 27, 29
nitrogen-rich compounds, overlap of hydrogen and nitrogen GC peaks 114–16
non-exchangeable δ^{2}H values 108
nutritional stress, effect on δ^{15}N values 199–200

^{18}O-bulk isotope analysis, matchsticks 186–7
ocean water 16
  see also VSMOW standard
OECD (Organization for Economic Cooperation and Development), on counterfeiting and piracy 36–7, 43
oil spills and pollution 45
olive oil, adulteration of 38–9
omnivores, δ^{15}N values 199, 208

On-line Isotopes-in-Precipitation Calculator 191
organic ^{13}C reference materials 7, *89*
organic secondary standards 103, *132–4*, 135
origin of product 8
  example(s) *9*
ovo-lacto 'vegetarian', δ^{15}N values 199
oxygen
  isotopes 6, 20
  sources/precursor pools 20, 186, 192, 193
  stable isotopic distribution 19–25
oxygen-16 (^{16}O), natural abundance 6, 20
oxygen-17 (^{17}O), natural abundance 6, 20
oxygen-18 (^{18}O)
  abundance in source water 20–3, 209–10
  calibration and reference material(s) 5–6, *7*, 20
  geographical variation 17
  isotope analysis *76, 77*, 104, 196, 200, 205, 209, 236–9
  isotopic analysis of CuSO$_4$ 106
  natural abundance 6, 20
  see also δ^{18}O

P2P (1-phenyl-2-propanone) 155
paints
  likelihood ratios used in analysis 94–9
  list of white paints from various sources *95–6*
paper (printer or photocopier) 215–16
  physical characteristics 215
parcel tape 218–21
particulate organic matter (POM), marine compared with freshwater source 26
peak inversions [in gas chromatography] 233
peak overlap [in gas chromatography] 234
per mil scales 5
perchlorates 170
peroxide explosives 176–7
PETN (pentaerythritoltetranitrate) 170, 173
petrochemicals, isotope abundance in 69
pharmaceuticals, counterfeit 42–3, *44*
physicochemical processes, isotopic fractionation in 12, 13–14, **15**
plant material, δ^2H as indicator of source water 18
plastic bags, forensic examination of 216–17
plastics, forensic examination of 216–18
PMK (piperonyl methyl ketone) 155
  MDMA synthesized from 157, *158*
  synthesis from safrole 157, *158*
pollution source differentiation 46, 94

Pong Su heroin samples 151–2
position-specific isotope analysis *see* PSIA
power supply for IRMS laboratory 125–6
precipitation, isotopic composition 18, *19*, 190–1, *213*
precursor pools
  carbon 31
  hydrogen 18–19
  oxygen 20
5β-pregnanediol, as reference compound for testosterone metabolites 228
premium natural products, authenticity and provenance 41–2
pressure-sensitive adhesive (PSA) tapes 218–21
principal component analysis 91
principle of identical treatment 85, 102, 111, 112, 165, 185, 232, 241
proficiency
  FIRMS Network's advice on 101
  importance of 145, 147, 224
provenance (provenancing)
  of drug 150, 152, 155, 156, 167, 168
  of explosives 183
  of foodstuffs 18–19, 38–42
  of human remains and living people 21, 22, 34, 190–213
  of matchsticks 187
pseudoephedrine
  isotopic signatures *165*
  methamphetamine synthesized from *166*
PSIA (position-specific isotope analysis) 81–3
  applications 40, 82–3

quality control, in CF-IRMS 85–6

radioisotopes 3
rain *see* precipitation
ratio-of-ratios approach (in calculation of δ value) 73–4
Rayleigh processes, isotopic fractionation in 13–14, 16–17
RDX (cyclotrimethylenetrinitramine)
  bulk isotope analysis 114, 173, *174*
  synthesis from hexamine 174
reference materials *7*, 74
  organic secondary standards 103, *132–4*, 135
relative retention [gas chromatography] 231
  IUPAC definition 232
relative retention time, use of term in gas chromatography 232

Republic of Ireland
   immigrants' fingerprints records 208
   see also Dublin
retention times [GC], factors affecting 233
rib bones, turnover rate 23, 193

safrole, piperonyl methyl ketone synthesized from 157, *158*
sample preparation 102–3
   human remains 196, 208
   matchsticks 185
sample preparation procedures 236–41
scale correction 86–90
   of measured $\delta^{13}C$ values 88–90
   of measured $\delta^{2}H$ values 87–8
scale definitions 5–6
   for ^{13}C 5, 85
   for ^{2}H 5–6, 85, 87
   for ^{15}N 6
   for ^{18}O 5–6
   for ^{34}S 6
Schimmelman, Arndt 103, 135
'Scissor Sisters' case 206, 208–11
sea bass 93–4
sea spray effect 35
seawater
   dissolution of CO_2 26–7
   evaporation of 17
   see also ocean water
seaweed and kelp 35
Semtex 94, 173
   see also PETN; RDX
silver phosphate, preparation from bio-apatite 196, 209, 236–8
single-seed vegetable oils, authenticity and provenance 38–9
'skull from the sea' 194–7
SLAP (Standard Light Antarctic Precipitation) water reference material 7, 87
snow fall *see* precipitation
Soddy, Frederick 3
soft fruit, $\delta^{2}H$ as indicator of source water 18–19
source of product 8
   example(s) 9
source water
   and bone bio-apatite 20–2
   and bone ^{18}O 20–3, 209–10
   and $\delta^{2}H$ for soft fruit 18–19
   and $\delta^{2}H$ for wood/matchsticks 187
   geographical variation of ^{2}H and ^{18}O abundance 17, 190–1

stable isotope dilution assay 120
stable isotopes
   meaning of term 3–4
   natural abundance variation 5–7
standard operating procedures 100, 102
statistical analysis 91–9
steroids, detection of exogenous 47–8
strontium isotope ratio ($^{87}Sr/^{86}Sr$) analysis 197, 203
structural proteins
   $\delta^{13}C$ and $\delta^{15}N$ values affected by diet 30–1
   diagenetic changes 32–3
   see also bone collagen; hair keratin
sugar
   beet- or cane-sourced 7, 9, 25, 69
   isotope variation 6–7, *9*
sulfur
   isotopes 6
   stable isotopic distribution 33–5
sulfur-32 (^{32}S), natural abundance 6
sulfur-33 (^{33}S), natural abundance 6
sulfur-34 (^{34}S)
   calibration material 6
   natural abundance 6, 34
   see also $\delta^{34}S$
sulfur isotope analysis, techniques used 33–4
synthetic drugs 154–68
SYSTAT software 92, 162

TC/EA-IRMS (thermal conversion/elemental analysis isotope ratio mass spectrometry) 77, *78*
   organic secondary standards *134*
   sample preparation 197
temperature control, IRMS laboratory 124, 125
testosterone
   detection of exogenous 47–8, 224–35
   metabolic pathway *227*
testosterone metabolism, and ^{13}C isotopic composition 226–8
testosterone metabolites
   $\delta^{13}C$ values 226
   $\delta^{2}H$ values 48
thermodynamic isotope effects 10, 12
tooth enamel
   diagenetic changes 24–5, 203
   ^{18}O abundance 193
   strontium isotope ratios 203, 205
triacetone triperoxide 176
trinitrotoluene, ^{2}H-BSIA 116
trivariate plots

advantages 222
Ecstasy tablets *92*
RDX precursor (hexamine) *175*
synthetic MDMA samples *162*
trophic ecology, isotopic fractionation and 26–7, 27–8
trophic level shift
   isotopic fractionation associated with 28, 29–32
   practical example 30

Urey, Harold C. 4

VCDT (Vienna Canyon Diablo Troilite) standard 6
   $^{34}S/^{32}S$ isotope ratio 6, 34
vegan, $\delta^{15}N$ values 30, 199
vegetable oils, authenticity and provenance 38–9
VPDB (Vienna Pee Dee Belemnite) standard 5, 85, 239
   $^{13}C/^{12}C$ isotope ratio 6
   lower-end anchor for 7, 88
VSMOW (Vienna-Standard Mean Ocean Water) standard 6, 74, 85, 239
   $^{2}H/^{1}H$ isotope ratio 6, 16
   lower-end anchor for 7, 87
   $^{18}O/^{16}O$ isotope ratio 6

WADA (World Anti-Doping Agency)
   accredited laboratories 225
   certified method for testosterone metabolites 224
water
   calibration material for ^{2}H and ^{18}O 6, 74, 85, 87
   $\delta^{2}H$ as indicator of source 17, 19
   isotopic fractionation between liquid and vapour 16–17, 113
   reference materials for ^{2}H and ^{18}O 7, 74, 87
   volume on Earth 16
   see also source water
water-free samples, analysis of 105–6
whisky, authenticity and provenance 41
wildlife forensics 46–7
wines, authenticity and provenance 40, *41*
wood
   bulk isotope analysis 185–8
   see also matchsticks
World Health Organization (WHO), on counterfeit pharmaceuticals 42–3

XRF (X-ray fluorescence spectrometry), applications 216, 219, 220, 221

Zero-Blank autosampler 106, *107*

Lightning Source UK Ltd.
Milton Keynes UK
UKOW07n1107130116

266269UK00001B/2/P